Series on Directions in Condensed Matter Physics — Volume 20

LUTTINGER MODEL
The First 50 Years and Some New Directions

SERIES ON DIRECTIONS IN CONDENSED MATTER PHYSICS

ISSN: 1793-1444

*For the complete list of the published titles in this series, please visit:
www.worldscientific.com/series/SDCMP

Series on Directions in Condensed Matter Physics — Volume 20

LUTTINGER MODEL

The First 50 Years and Some New Directions

edited by

Vieri Mastropietro
University of Milan, Italy

Daniel Charles Mattis
University of Utah, USA

W World Scientific

NEW JERSEY · LONDON · SINGAPORE · BEIJING · SHANGHAI · HONG KONG · TAIPEI · CHENNAI

Published by

World Scientific Publishing Co. Pte. Ltd.

5 Toh Tuck Link, Singapore 596224

USA office: 27 Warren Street, Suite 401-402, Hackensack, NJ 07601

UK office: 57 Shelton Street, Covent Garden, London WC2H 9HE

British Library Cataloguing-in-Publication Data
A catalogue record for this book is available from the British Library.

Series on Directions in Condensed Matter Physics — Vol. 20
LUTTINGER MODEL
The First 50 Years and Some New Directions

Copyright © 2014 by World Scientific Publishing Co. Pte. Ltd.

ISBN 978-981-4520-71-3

Printed in Singapore

Other World Scientific Titles by Prof. D. C. Mattis

The Many-Body Problem
An Encyclopedia of Exactly Solved Models in One Dimension
(3rd Printing with Revisions and Corrections)

Statistical Mechanics Made Simple
A Guide for Students and Researchers

The Theory of Magnetism Made Simple
An Introduction to Physical Concepts and to Some Useful
Mathematical Methods

Statistical Mechanics Made Simple
2nd Edition

Other World Scientific Title by Prof. Vieri Mastropietro

Non-Perturbative Renormalization

CONTENTS

PREFACE

The Luttinger Model is the only model of many-fermion physics with legitimate claims to be both exactly and completely solvable. In several respects it plays the same role in many-body theory as does the 2D Ising model in statistical physics. Interest in the Luttinger model has increased steadily ever since its introduction half a century ago.

The present volume celebrates the five decades of the impact of the Luttinger model presenting a number of papers constituting a comprehensive and updated introduction to the subject, from the main historical achievements to the more modern research lines.

In *Chapter 1* are reprinted two milestones, namely the paper of *J. M. Luttinger* in which the model was introduced, and the one by *D. C. Mattis and E. H. Lieb* in which the model was completely solved and all its correlation functions were explicitly computed. The Luttinger model owes its solubility to a number of peculiar features, like the absence of a lattice and the presence of linear dispersion relation, which allow an exact solution by bosonization. The many body interaction produces a drastic change in the physical properties, manifesting for instance in the presence of anomalous exponents in the large distance decay of correlations.

Chapter 2 is devoted to the natural question of how much of the properties of the Luttinger model are generic and describe the typical behavior of many body systems. Several contributions in this chapter analyze such problem under different perspectives, focusing on the physics of one dimensional systems. According to the Luttinger liquid conjecture, formulated by D. M. Haldane (Phys. Rev. Lett. 45, 1358 (1980)), a number of features of the Luttinger model remain valid in a class of systems. As explained in the review by *J. Sirker*, the Luttinger liquid conjecture can be partially tested in integrable models like the XXZ chain which can be solved by Bethe ansatz, and information on transport can be obtained. Even in the absence of solution, a large part of the theory of Luttinger liquids can be proven to be true with full mathematical rigor, using the Renormalization Group methods developed in Constructive Quantum Field Theory; in particular, as explained in the contribution by *V. Mastropietro*, the scaling relations between critical exponents and other thermodynamical quantities are proven to be true in a large class of non solvable models. On the other hand, the Luttinger liquid theory has limitations and the curvature in the dispersion relations produce important new effects in the dynamical properties, as explained in the contribution by *L. Pereira*. The presence of extra bosonic modes in presence of a finite curvature of the dispersion relation, and the possibility of an exact solution also in such a case are discussed in a reprint by *D. C. Mattis*.

In *Chapter 3* are reported several contributions in which different aspects of the physics of the Luttinger model are experimentally observed. The review by *T. Giamarchi* reports the observed evidence of the Luttinger model power laws in organic compounds, suggesting new experiments for even more quantitative evidence; in particular it is would be extremely important to experimentally verify the exact relations between exponents predicted by the Luttinger model. In the contribution of *F. Li* applications to mesoscopic systems in a non equilibrium situation are discussed. Luttinger liquid behavior emerges also in experiments on a wide range of different physical systems, from quantum Hall edge states, as explained in the contribution of *O. Ciftja*, to Helium-filled nanopores, as reported in the contribution by *A. Del Maestro*.

Chapter 4 is devoted to the possibility of observing Luttinger liquid behavior in dimensions greater than one, a question which is considered of extreme importance in the context of high-T_c superconductivity, as explained in the review of *C. Di Castro and S. Caprara*. In the case of fermionic system with flat Fermi surfaces there is an exactly solvable two dimensional analogue of the Luttinger model, called the Mattis model, which is reviewed and analyzed by bosonization in the contribution by *J. De Woul and E. Langmann*. Finally, a comprehensive account of a formal extension of the Luttinger model solution to an interacting fermionic system with circular or spherical Fermi surface is given in the paper by *D. C. Mattis*.

The papers collected in this volume provide insights into aspects of the Luttinger model that have evolved over the years. These range from the most advanced theoretical topics to its most esoteric experimental realizations. While celebrating achievements of the past the papers in this volume should also help point out issues for future research.

Vieri Mastropietro

INTRODUCTION

It was the early 1960s. Elliott H. Lieb and the present author were colleagues at the IBM Research Laboratories in Yorktown Heights, occupying adjoining cubicles. We collaborated in the study, creation and solution of mathematical models of many-body phenomena. In the course of our work we ultimately came to the realization that solvable models — whether they dealt with interacting bosons, Heisenberg spins, "X-Y" spins, or whatever — typically shared an intrinsic property: all were restricted to one spatial dimension with no obvious generalizations to higher dimensions — a tantalizing challenge.

It seemed natural to exploit this commonality. An initial idea of founding a *Journal of One-Dimensional Physics* was quickly dismissed once we became convinced there was insufficient material. But a *book* entitled *Mathematical Physics in One Dimension*? *That* should nail down the topic of "one-dimensional models," allow papers to be indexed under that rubric and provide a platform for future research. That is when we decided upon, and that is where the "Luttinger Model" first came in. The topics and chapters of that book[1] were then boiled down to seven:

(1) Classical statistical mechanics,
(2) Disordered chains of harmonic oscillators,
(3) Electron energy bands in ordered and disordered crystals,
(4) The many-Fermion gas,
(5) The Bose gas,
(6) Magnetism,
(7) Time-dependent phenomena.

Each chapter opened with a presentation of difficulties that had been encountered in the named topic when confronting what is ordinarily an insoluble N-body problem ($N \to \infty$). This was followed by a description of the specific methods (Bethe's ansatz, the transfer matrix, fermionization in the X-Y model, etc.) that had been devised to overcome these issues, and ended with reprints of what we considered to be the seminal papers solving the problem. We stumbled, however, when it came to Chapter 4 dealing with the notoriously fractious "many-Fermion problem." At the time the best-known contribution to this topic

---

[1]E. H. Lieb and D. C. Mattis, *Mathematical Physics in One Dimension*, Academic Press, New York, 1966.

came from an approximately solved model[2] by Sin-itiro Tomonaga dating back to 1950. But then, the present author recalled having seen Joaquin "Quin" Luttinger's contribution in the *Journal of Mathematical Physics*.[3] Its title "Exact Solution ..." suggested that it fit the requirements although, upon further examination, this "exact" solution did seem flawed in several respects — notably in its failure to satisfy the so-called "concavity theorem," its lack of a stable ground state, etc.

Clearly, Luttinger, unaware at the time either of Tomonaga's work or of Jordan's discoveries of a generation earlier,[4] had been attempting to transfer the Thirring model[5] to condensed matter physics without extensive re-examination. It was purely in the interest of completing Chapter 4 that we set about to *understand* this model — if that were at all possible — before finalizing the book. In the process we serendipitously (re)discovered[4] "bosonization."[6] Once this was figured this out, it was a bonus that our new solution seemingly remained valid for arbitrary two-body interactions — not just delta function interactions — while still subject to the very same physical requirements, e.g. the filling of the sea of negative-energy particles, that originally had been finessed. The main difference between the physical and all other, unphysical,[7] solutions was addressed briefly in footnote 5 of our paper,[6] and concerned the correct choice between two inequivalent representations of fermionic creation/annihilation operators.

This is not a mere theoretical "nicety." Many seemingly novel features flow out of a mathematically rigorous, physically acceptable, solution of Luttinger's model that might not have seemed believable had they been derived from approximate solutions. Among these: correlation functions that satisfy power laws with exponents that depend on the strength of the two-body interactions, the decoupling of the particles' spin and charge into modes that propagate at distinct speeds, and other anomalies. Properties that once might have been dismissed as unusual offshoots of an unusual model have now been identified in *all* physically sensible systems of interacting fermions in 1D and thus have the aura of universality that helps distinguish one-dimensional systems of fermions (now dubbed "Luttinger Liquids") from the "Fermi Liquids" in higher dimensions.

Subsequently, even as other, ever more complex, models of interacting fermions were analyzed, including E. H. Lieb and F. Y. Wu's remarkable solution of the one-dimensional

---

[2]S. Tomonaga, *Progr. Theoret. Phys.* (Kyoto) **5**, 544 (1950).

[3]J. M. Luttinger, *Math. Phys.* **4**, 1154 (1963).

[4]In the prehistory of high-energy physics, P. Jordan proposed a theory, since discredited, of *light* (photons are bosons) starting from a one-dimensional field of *neutrinos* (fermions). See P. Jordan, *Z. Phys.* **93**, 464 (1935), **98**, 759 (1936), **99**, 109 (1936), **102**, 243 (1936), **105**, 114 (1937), **105**, 229 (1937), *inter alia*.

[5]W. Thirring, *Ann. Phys.* (NY) **3**, 91 (1958).

[6]D. C. Mattis and E. H. Lieb, "Exact Solution of a Many-Fermion System and its Associated Boson Field," *J. Math. Phys.* **6**, 304 (1965).

[7]It was fun to find that in addition to the original exact (but unphysical) solution in Ref. 3, Luttinger's model[3] *also* admits *other* exact (but equally unphysical) solutions such as the "sausage" in D. C. Mattis and B. Sutherland, "Strange Solutions ...," *J. Math. Phys.* **22**, 1692 (1981). Such are the vagaries of infinite sets of coupled linear differential equations.

Hubbard model,[8] Luttinger's remains the gold standard by virtue of its simplicity and general applicability. The present volume celebrating its Jubilee contains numerous applications and improvements, not only in 1D but also in higher dimensions. It is intended to be a fitting tribute to the versatility of the concept and as a springboard to its Centennial.

Daniel C. Mattis

---

[8]E. H. Lieb and F. Y. Wu, *Phys. Rev. Lett.* **20**, 1445 (1968). For additional references and reprints of studies in the 1D many-fermion problem, see Chapter 4 in the successor volume to Ref. 1: D. C. Mattis, *The Many-Body Problem, An Encyclopedia of Exactly Solved Models in One Dimension*, World Scientific Publ. Co., Singapore, 2009. It is notable that the number of reprints included with that chapter swelled from three in the original volume (with just the one[6] on the Luttinger model) to 19 in the successor volume (including seven *specifically* on that topic.)

Chapter I

The Luttinger Model and Its Solution

JOURNAL OF MATHEMATICAL PHYSICS   VOLUME 4, NUMBER 9   SEPTEMBER 1963

An Exactly Soluble Model of a Many-Fermion System*

J. M. LUTTINGER

Department of Physics, Columbia University, New York, New York
(Received 2 April 1963)

An exactly soluble model of a one-dimensional many-fermion system is discussed. The model has a fairly realistic interaction between pairs of fermions. An exact calculation of the momentum distribution in the ground state is given. It is shown that there is no discontinuity in the momentum distribution in this model at the Fermi surface, but that the momentum distribution has infinite slope there. Comparison with the results of perturbation theory for the same model is also presented, and it is shown that, for this case at least, the perturbation and exact answers behave qualitatively alike. Finally, the response of the system to external fields is also discussed.

I. INTRODUCTION

WE shall be concerned in this paper with a model of a many-fermion system which is exactly soluble. The model is quite unrealistic for two reasons: it is one-dimensional and the fermions are massless. On the other hand, it has the realistic feature that there is a true pair interaction between the particles. It is very closely related to the well-known Thirring Model[1] in field theory, though slightly more general. Our main interest in the model is in connection with the question of whether or not a sharp Fermi Surface (F.S.) exists in the exact ground state.

This question has only been investigated previously[2] by a special sort of many-body perturbation theory, when it has been shown for the usual realistic three-dimensional many-fermion system that each term of the series does give rise to a sharp F.S. This, of course, proves nothing about the entire series unless one can also prove something about its convergence, which has not been possible so far. The main point of this investigation therefore is to see if in this soluble model the exact solution and the perturbation solution (via propagators) behave in an essentially different fashion.

We now consider the exact formulation of the model. Consider first the case of no interaction between the particles. These are taken to be spinless, massless, fermions moving in a one-dimensional space. The analogue of the relativistic Dirac Hamiltonian is $v_0\sigma_3 p$ (σ_3 is the usual Pauli spin matrix; units such that $\hbar = 1$ are chosen). v_0 is the velocity

of the particles, which would be c in the relativistic case. Then the Hamiltonian is

$$H_0 = v_0 \int_0^L \psi^+(x)\sigma_3 p\psi(x)\, dx. \quad (1)$$

Here ψ is the two component spinor

$$\psi = \begin{pmatrix} \psi_1 \\ \psi_2 \end{pmatrix}, \quad (2)$$

and we are assuming that the particles are confined to a length L along the x axis. The quantity p is of course the ordinary momentum operator $1/i\, \partial/\partial x$.

Written out, (1) becomes

$$H_0 = v_0 \int_0^L [\psi_1^+ p\psi_1 - \psi_2^+ p\psi_2]\, dx. \quad (3)$$

If we go into momentum space via

$$\psi_j(x) = \sum_k a_{jk}e^{ikx}/L^{\frac{1}{2}} \quad (4)$$

(where the allowed values of k are

$$k = (2\pi/L)n, \qquad n = 0, \pm1, \pm2, \cdots, \pm\infty \quad (5)$$

since we shall impose periodic boundary conditions on our sample), we obtain

$$H_0 = v_0 \sum_k (a_{1k}^+ a_{1k} - a_{2k}^+ a_{2k})k. \quad (6)$$

The creation and destruction operators a, a^+ satisfy the commutation relationship

$$a_{jk}^+ a_{j'k'} + a_{jk'} a_{jk}^+ = \delta_{jj'}\delta_{kk'}. \quad (7)$$

Since the allowed values of $a_{jk}^+ a_{jk}$ are zero and unity, the lowest state of H_0 is $-\infty$ since we can choose all the $j = 1$, $k < 0$ and the $j = 2$, $k > 0$ states occupied. This is the usual problem occurring in Dirac theory and requires a redefinition of the creation and destruction operators so that we deal only with "particles" and "holes". Define

---

* Work supported in part by the Office of Naval Research.
[1] W. Thirring, Ann. Phys. **3**, 91 (1958). See also V. Glaser, Nuovo Cimento **9**, 990 (1958); T. Pradhan, Nucl. Phys. **9**, 124 (1961); K. Johnson, Nuovo Cimento **21**, 773 (1961).
[2] J. M. Luttinger and J. C. Ward, Phys. Rev. **118**, 1417 (1960); J. M. Luttinger, *ibid.* **119**, 1153 (1960); **121**, 942 (1961).

$$a_{1k} = b_k \quad k > 0$$
$$= c_k^+ \quad k < 0,$$
$$a_{2k} = b_k \quad k < 0 \tag{8}$$
$$= c_k^+ \quad k > 0.$$

We may also write this as

$$b_k = \theta_k^+ a_{1k} + \theta_k^- a_{2k}, \qquad c_k = \theta^- a_{1k}^+ + \theta_k^+ a_{2k}^+, \tag{9}$$

where

$$\theta_k^+ = \begin{array}{ll} 1 & k > 0 \\ 0 & k < 0, \end{array}$$
$$\theta_k^- = \begin{array}{ll} 1 & k < 0 \\ 0 & k > 0. \end{array} \tag{10}$$

From (9) we see at once that b_k, c_k also have the commutation rules of fermions, i.e.,

$$b_k^+ b_{k'} + b_{k'} \cdot b_k^+ = \delta_{kk'}, \quad c_k^+ c_{k'} + c_{k'} \cdot c_k^+ = \delta_{kk'}, \tag{11}$$

and all the rest anticommute.

Inserting (8) in (6) we obtain

$$H_0 = v_0 \sum_k (b_k^+ b_k + c_k^+ c_k) |k|$$
$$+ v_0 (\sum_{k<0} k - \sum_{k>0} k). \tag{12}$$

The last term is infinite, but a constant, and as usual we simply redefine H_0 without it, i.e., we take

$$H_0 = v_0 \sum_k (b_k^+ b_k + c_k^+ c_k) |k|. \tag{13}$$

We shall call the operators b_k and c_k the destruction operators for particles and holes respectively. The vacuum state ϕ_0 is clearly defined by

$$b_k \phi_0 = 0, \qquad c_k \phi_0 = 0. \tag{14}$$

The interaction Hamiltonian H' is taken to be (this special choice is what makes the model soluble)

$$H' = 2\lambda v_0 \iint_0^L \psi_1^+(x) \psi_1(x) V(x - y)$$
$$\times \psi_2^+(y) \psi_2(y) \, dx \, dy. \tag{15}$$

$V(x - y)$ is an arbitrary two-body potential at this point. If we write this in momentum space [assuming also that $V(x - y)$ satisfies periodic boundary conditions], we obtain

$$H' = \frac{2\lambda v_0}{L} \sum_{k_1 \cdots k_4} \delta_{k_1 - k_2 + k_3 - k_4, 0}$$
$$\times v(k_3 - k_4) a_{1k_1}^+ a_{1k_2} a_{2k_3}^+ a_{2k_4}, \tag{16}$$

where

$$V(x) = \frac{1}{L} \sum_k v(k) e^{-ikx},$$
$$v(k) = \int_0^L dx e^{ikx} V(x). \tag{17}$$

The term in (16) corresponding to $k_3 = k_4$, say H'', is given by

$$H'' = \frac{2\lambda}{L} v_0 v(0) (\sum_k a_{1k}^+ a_{1k})(\sum_k a_{2k}^+ a_{2k}). \tag{18}$$

This term clearly gives rise to divergent effects, since for the unperturbed vacuum the number of "1" and "2" particles are infinite. To avoid this difficulty, we shoose $v(0)$ to be zero, which is the same as taking the average value of the potential ($\bar{V}$) equal to zero. We also express this by saying that in (15) we replace $V(x - y)$ by $V(x - y) - \bar{V}$.

The total Hamiltonian of the problem is now given by

$$H = H_0 + H'. \tag{19}$$

II. EXACT SOLUTION OF THE MODEL

We shall show that (19) can be diagonalized by a very simple canonical transformation. Consider

$$\tilde{H} = e^{i\lambda S} H e^{-i\lambda S}, \tag{20}$$

where

$$S \equiv \iint_0^L dx \, dy \, \psi_1^+(x) \psi_1(x) E(x - y) \psi_2^+(y) \psi_2(y). \tag{21}$$

Here $E(x)$ is defined by

$$dE(x)/dx = V(x) - \bar{V}. \tag{22}$$

Writing

$$V(x) - \bar{V} = \frac{1}{L} \sum_k' v(k) e^{-ikx}, \tag{23}$$

we obtain

$$E(x) = \frac{i}{L} \sum_k' \frac{v(k)}{k} e^{-ikx}. \tag{24}$$

Let us define

$$N_i(x) = \psi_i^+(x) \psi_i(x). \tag{25}$$

Then from the commutation rules it follows at once that

$$(N_i(x), N_{i'}(x')) = 0, \tag{26}$$

so that

$$e^{i\lambda S} H' e^{-i\lambda S} = H'. \tag{27}$$

(In the non-second quantized version of the theory H' and S are just functions of position.)

Therefore

$$\tilde{H} = H + i\lambda(S, H_0)$$
$$+ [(i\lambda)^2/2'](S, (S, H_0)) + \cdots . \quad (28)$$

Using the commutation rules for the ψ_j, we obtain at once

$$(S, H_0) = v_0 \iint\limits_0^L dx\, dy \left\{ \frac{\partial N_1(x)}{\partial x} E(x-y) N_2(y) \right.$$
$$\left. - N_1(x) E(x-y) \frac{\partial N_2(y)}{\partial y} \right\}. \quad (29)$$

Integrating by parts and using the periodic boundary conditions to drop the surface terms, we obtain

$$(S, H_0) = -\frac{2v_0}{i} \iint\limits_0^L dx\, dy N_1(x) E'(x-y) N_2(y)$$
$$= -\frac{2v_0}{i} \iint\limits_0^L dx\, dy N_1(x)(V(x-y) - \bar{V}) N_2(y)$$
$$= -\frac{1}{i\lambda} H'. \quad (30)$$

Since this commutes with S, there are no higher terms in the series (28), and we obtain

$$\tilde{H} = \tilde{H} - H' = H_0. \quad (31)$$

(Again these results are seen very easily by going over to the non-second quantized representation.)

Now $\tilde{H}$ is trivial to diagonalize, just being the noninteracting Hamiltonian. Therefore, all the energy levels of H are the same as those of H_0. (This is very unrealistic indeed.) On the other hand, the wavefunctions of H are very different from the free-particle ones. If ψ_n^0 is a wavefunction of H_0 corresponding to energy E_n^0, then the corresponding wavefunction for H (say, ψ_n) is

$$\psi_n = e^{-i\lambda S} \psi_n^0. \quad (32)$$

Therefore, although the energy levels do not change as a result of the interaction, other properties depending on more details of the wavefunction may be profoundly altered.

We next want to formulate the many-body problem for our system. We at once have the following problem: since particle-hole pairs can be produced by the interaction, the number of particles in an eigenstate of H is not fixed. However, we clearly must have that the number of particles minus the number of holes (call this n) is fixed in an eigenstate. Writing

$$n = \sum_k (b_k^+ b_k - c_k^+ c_k), \quad (33)$$

we can easily verify by direct calculation that n is a constant of the motion.

The noninteracting case for the N-particle problem is clearly the case of n having the eigenvalue N. Similarly, we define the N-particle problem for the interacting case as the system for which n has the value N. There will always be a certain number of holse present, but the smaller the interaction, the smaller this number will be.

The exact ground state of the N-particle system may be obtained as follows. Certainly the lowest state (ψ_N^0) of $\tilde{H}$ for which $n = N$ is obtained by having no holes present. Then the first N particle states will be occupied. That is

$$b_k^+ \psi_N^0 = 0, \qquad |k| < k_F,$$
$$b_k \psi_N^0 = 0, \qquad |k| > k_F, \quad (34)$$
$$c_k \psi_N^0 = 0,$$

where the Fermi momentum k_F is determined by

$$N = \sum_{|k| < k_F} 1 = \frac{L}{2\pi} \int_{-k_F}^{k_F} dk = \frac{L}{\pi} k_F. \quad (35)$$

We may also write

$$\psi_N^0 = b_{k_N}^+ \cdots b_{k_1}^+ \phi_0, \quad (36)$$

where ϕ_0 is the unperturbed vacuum and $k_1 \cdots k_N$ are the N allowed momenta between $-k_F$ and k_F.

Therefore the exact ground-state wavefunction (ψ_N) is given by

$$\psi_N = e^{-i\lambda S} \psi_N^0. \quad (37)$$

In order to study the sharpness of the F.S., we must investigate[2] the mean number of particles with momentum k, say $\bar{n}_k$. We have, of course,

$$n_k = b_k^+ b_k,$$

so

$$\bar{n}_k = (\psi_N, b_k^+ b_k \psi_N) = (\psi_N^0, e^{i\lambda S} b_k^+ b_k e^{-i\lambda S} \psi_N^0). \quad (38)$$

If we wanted to know the average number of holes $\bar{N}_h$ present we may use

$$\bar{N}_h = \sum_k \bar{n}_k - N. \quad (39)$$

Clearly $\bar{n}_k$ is an even function of k, so we shall restrict ourselves to $k > 0$. Then, by (9),

$$b_k^+ b_k = a_{1k}^+ a_{1k} = \frac{1}{L} \iint_0^L d\xi \, d\eta e^{ik(\xi-\eta)} \psi_1^+(\xi) \psi_1(\eta), \quad (40)$$

as one sees, by direct integration,

$$\bar{n}_k = \frac{1}{L} \iint_0^L d\xi \, d\eta e^{ik(\xi-\eta)}$$
$$\times (\psi_N^0 |e^{i\lambda S} \psi_1^+(\xi) \psi_1(\eta) e^{-i\lambda S}| \psi_N^0). \quad (41)$$

Now we have the following operator identity

$$\exp\left[i\lambda \int_0^L g(x) N_1(x) \, dx\right] \psi_1(\eta)$$
$$\times \exp\left[-i\lambda \int_0^L g(x) N_1(x) \, dx\right] = e^{-i\lambda g(\eta)} \psi_1(\eta), \quad (42)$$

if $g(x)$ commutes with $\psi_1(\eta)$. This is most easily proved by differentiating with respect to λ and making use of the fact that

$$(\psi_1(\eta), N_1(x)) = \delta(x - \eta) \psi_1(\eta). \quad (43)$$

Using (42), (41) becomes

$$\bar{n}_k = \frac{1}{L} \iint_0^L d\xi \, d\eta e^{ik(\xi-\eta)}$$
$$\times \left(\psi_N^0 \left| \psi_1^+(\xi) \psi_1(\eta) \exp\left\{i\lambda \int_0^L dy N_2(y)\right.\right.\right.$$
$$\times [E(\xi - y) - E(\eta - y)]\Big\} \Big| \psi_N^0\Big). \quad (44)$$

Expressed in terms of a_{ik}, (34) becomes

$$a_{1k}^+ \psi_N^0 = 0, \quad k < k_F, \quad (45)$$
$$a_{1k} \psi_N^0 = 0, \quad k > k_F,$$

and

$$a_{1k}^+ \psi_N^0 = 0, \quad k > -k_F, \quad (46)$$
$$a_{2k} \psi_N^0 = 0, \quad k < -k_F,$$

Writing $\psi_N^0 = \Psi_1 \Psi_2$ where Ψ_1, depends on the variables of the field "1" and is given by (45), and Ψ_2 depends on the variables of the field "2" and is given by (46), we have

$$\bar{n}_k = \frac{1}{L} \iint_0^L d\xi \, d\eta e^{ik(\xi-\eta)} (\Psi_1 |\psi_1^+(\xi) \psi_1(\eta)| \Psi_1)$$
$$\times \left(\Psi_2 \left| \exp\left\{i\lambda \int_0^L dy N_2(y)\right.\right.\right.$$
$$\times [E(\xi - y) - E(\eta - y)]\Big\} \Big| \Psi_2\Big). \quad (47)$$

From (45) we have at once

$$(\Psi_1 |\psi_1^+(\xi) \psi_1(\eta)| \Psi_1) = \frac{1}{L} \sum_{k'<k_F} e^{-ik'(\xi-\eta)}. \quad (48)$$

The second factor in (47) is also not difficult to reduce to simpler form. We have, in fact,

$$\left(\Psi_2 \left| \exp\left\{i\lambda \int_0^L dy N_2(y)\right.\right.\right.$$
$$\times [E(\xi - y) - E(\eta - y)]\Big\} \Big| \Psi_2\Big) = \text{Det}(g). \quad (49)$$

Det (g) is the determinant of the matrix $g_{\alpha\alpha'}$, where

$$g_{\alpha\alpha'} = \frac{1}{L} \int_0^L dy e^{-i(k_\alpha - k_{\alpha'})y}$$
$$\times \exp\{i\lambda[E(\xi - y) - E(\eta - y)]\}, \quad (50)$$

the k_α being the occupied states of the "2" particles in (46), i.e., the k_α are the set of discrete allowed k values greater than $-k_F$. The proof of (50) is given in the Appendix. The remarkable thing is that this (infinite) determinant can in fact be evaluated and the answer reduced to quadratures.

Writing Det $(g) = G(\xi, \eta)$, (47) becomes

$$\bar{n}_k = \frac{1}{L^2} \iint_0^L d\xi \, d\eta \sum_{k'<k_F} e^{+i(k-k')(\xi-\eta)} G(\xi, \eta)$$
$$= \frac{2\pi}{L} \sum_{k'<k_F} F(k - k'), \quad (51)$$

where

$$F(\kappa) = \frac{1}{2\pi L} \iint_0^L d\xi \, d\eta e^{i\kappa(\xi-\eta)} G(\xi, \eta). \quad (52)$$

III. EXPLICIT EVALUATION OF MOMENTUM DISTRIBUTION

We now must consider the determinant $G(\xi, \eta)$ in more detail. Since $k_\alpha = (2\pi/L)n$,

$$k_\alpha - k_{\alpha'} = (2\pi/L)(n - n');$$
$$n, n' = -n_F, -n_F + 1, \cdots, \infty, \quad (53)$$

we may write

$$G = \begin{vmatrix} g_0 & g_{-1} & g_{-2} & \cdots \\ g_1 & g_0 & g_{-1} & \cdots \\ g_2 & g_1 & g_0 & \cdots \\ g_3 & g_2 & g_1 & \cdots \\ \cdots & \cdots & \cdots & \cdots \end{vmatrix}, \quad (54)$$

where

$$g_m = \frac{1}{L} \int_0^L dy e^{-2\pi i m y/L}$$

$$\times \exp\{i\lambda[E(\xi - y) - E(\eta - y)]\}. \quad (55)$$

[(54) incidentally, is independent of k_F.]

This type of determinant has been studied extensively, and is known as a *Toeplitz* determinant.[3] For very large order, an asymptotic formula can be given for them, which in our case (infinite-determinant) becomes exact. The result is the following: for a finite Toeplitz determinant

$$\mathbf{D}_M = \begin{vmatrix} g_0 & g_{-1} & \cdots & g_{-M} \\ g_1 & g_0 & \cdots & g_{-M+1} \\ \cdot & \cdot & & \cdot \\ \cdot & \cdot & & \cdot \\ g_M & g_{M-1} & \cdots & g_0 \end{vmatrix},$$

we have[4]

$$\lim_{M \to \infty} \frac{\mathbf{D}_M}{\mathbf{D}^{M+1}} = \exp\left(\sum_{l=1}^{\infty} K_l K_{-l} l\right), \quad (56)$$

where

$$\mathbf{D} = \exp\left[\frac{1}{2\pi} \int_0^{2\pi} d\theta \log f(\theta)\right], \quad (57)$$

$$K_l = \frac{1}{2\pi} \int_0^{2\pi} d\theta e^{-il\theta} \log f(\theta), \quad (58)$$

$$f(\theta) = \sum_{m=-\infty}^{\infty} g_m e^{im\theta}. \quad (59)$$

In the proof, $\log f(\theta)$ is defined by

$$\log f(\theta) = \log \{1 - [1 - f(\theta)]\}$$

$$= -\sum_{n=1}^{\infty} \frac{[1 - f(\theta)]^n}{n}, \quad (60)$$

and it is assumed that this series converges.

In our case, this leads to particularly simple results. Changing variables in (55) from y to θ where

$$\theta = 2\pi y/L, \qquad 0 \le \theta \le 2\pi,$$

we obtain

$$g_m = \frac{1}{2\pi} \int_0^{2\pi} d\theta e^{-im\theta}$$

$$\times \exp\left\{i\lambda\left[E\left(\xi - \frac{L\theta}{2\pi}\right) - E\left(\eta - \frac{L\theta}{2\pi}\right)\right]\right\}. \quad (61)$$

[3] See, for example, V. Grenander and G. Szegö, *Toeplitz Forms and their Applications*, (University of California Press, Berkeley and Los Angeles, 1958), especially P. 176 ff. See also M. Kac, *Probability and Related Topics in Physical Sciences*, (Interscience Publishers, London and New York, 1959), p. 60 ff.
[4] The formula given in Grenander and Szegö, (reference 3) contains K_L^* instead of K_{-L} as given in (56). I am indebted to Professor M. Kac for pointing out to me that if $f(\theta)$ is complex, rather than real as Grenander and Szegö assume, this simple change is all that is necessary.

Therefore from (59) we see at once that

$$f(\theta) = \exp\left\{i\lambda\left[E\left(\xi - \frac{L\theta}{2\pi}\right) - E\left(\eta - \frac{L\theta}{2\pi}\right)\right]\right\}. \quad (62)$$

Thus for sufficiently small λ, (60) is clearly satisfied since as one easily sees from (23) or (24), $E(x)$ is a bounded function of x. We shall for simplicity assume that λ is sufficiently small, and therefore we may write

$$\log f(\theta) = i\lambda\left[E\left(\xi - \frac{L\theta}{2\pi}\right) - E\left(\eta - \frac{L\theta}{2\pi}\right)\right]. \quad (63)$$

Now

$$\frac{1}{2\pi} \int_0^{2\pi} d\theta \log f(\theta)$$

$$= \frac{i\lambda}{L} \int_0^L dy[E(\xi - y) - E(\eta - y)] = 0, \quad (64)$$

since, by (24), the average of $E(x)$ is zero. Therefore

$$\mathbf{D} = 1. \quad (65)$$

Further,

$$K_l = \frac{1}{2\pi} \int_0^{2\pi} d\theta e^{-il\theta} \log f(\theta)$$

$$= \frac{i\lambda}{L} \int_0^L dy[E(\xi - y) - E(\eta - y)]e^{-2\pi i l y/L}$$

$$= \frac{\lambda}{L}\left[(e^{-ik\eta} - e^{-ik\xi})\frac{v(k)}{k}\right]_{k=2\pi l/L}. \quad (66)$$

Then

$$\sum_{l=1}^{\infty} K_l K_{-l} l = -\frac{\lambda^2}{2\pi}\frac{1}{L}\sum_{k>0}^{\infty} |v(k)|^2 \frac{|e^{-i\eta k} - e^{-i\xi k}|^2}{k}$$

$$= -\frac{\lambda^2}{\pi}\frac{1}{L}\sum_{k>0}^{\infty} |v(k)|^2 \frac{1 - \cos k(\xi - \eta)}{k}$$

$$\equiv -Q(\xi - \eta). \quad (67)$$

So finally we have

$$G(\xi, \eta) = e^{-Q(\xi-\eta)}. \quad (68)$$

Using the periodicity of Q in ξ and η, we see that (52) may be written

$$F(\kappa) = \frac{1}{2\pi} \int_{-\frac{1}{2}L}^{\frac{1}{2}L} d\xi e^{+L\kappa\xi} e^{-Q(\xi)}$$

$$= \frac{1}{2\pi} \int_{-\infty}^{\infty} d\xi e^{i\kappa\xi} e^{-Q(\xi)}. \quad (69)$$

Finally, replacing the sum by an integral in (67) we obtain

$$Q(\xi) = \frac{\lambda^2}{2\pi^2} \int_0^{\infty} dk \frac{1 - \cos k\xi}{k} |v(k)|^2. \quad (70)$$

We cannot go further in the evaluation of $Q(\xi)$ without some further information on the potential. However, the nature of the discontinuity at the F.S. can be investigated.

We may write

$$\bar{n}_k = \frac{2\pi}{L} \sum_{k' < k_F} F(k - k') = \int_{k-k_F}^{\infty} d\kappa F(\kappa) \qquad (71)$$

$$= \int_0^{\infty} d\kappa F(\kappa) + \int_{k-k_F}^0 d\kappa F(\kappa).$$

The first term of (71) is a constant. To study the behavior of $\bar{n}_k$ near the F.S. ($k \cong k_F$) we therefore need $F(\kappa)$ only for very small κ. This in turn, from (69), requires the behavior of $Q(\xi)$ for large ξ. Since $Q(\xi)$ is an even function of ξ, we consider it for large positive ξ. We have

$$\frac{\partial Q(\xi)}{\partial \xi} = \frac{\lambda^2}{2\pi^2} \int_0^{\infty} dk \sin k\xi \, |v(k)|^2$$

$$= \frac{\lambda^2}{2\pi^2} \left[|v(0)|^2 \frac{1}{\xi} + O\left(\frac{1}{\xi^2}\right) \right], \qquad (72)$$

by successive integrations by parts. Integrating, we get

$$Q(\xi) = (\lambda^2/2\pi^2)[|v(0)|^2 \log \xi + C + O(1/\xi^2)], \qquad (73)$$

where C is a constant which is in principle calculable from the potential. This may be written in the following way:

$$Q(\xi) = \alpha \log (\xi/a)^2 + O(1/\xi^2), \qquad (74)$$

where

$$\alpha \equiv (\lambda^2/4\pi^2) \, |v(0)|^2, \qquad (75)$$

and a is a constant with the dimensions of a length, which depends only on the shape of the potential (it is a measure of its range). Inserting (74) into (69) we obtain, for $|\kappa a| \ll 1$,

$$F(\kappa) = \frac{1}{\pi} \int_0^{\infty} d\xi \, \frac{\cos \kappa \xi}{(\xi/a)^{2\alpha}}$$

$$= \frac{\Gamma(1 - 2\alpha) \sin \pi\alpha}{\pi} \frac{a^{2\alpha}}{|\kappa|^{1-2\alpha}}. \qquad (76)$$

Thus we obtain, for $|k - k_F| \, a \ll 1$,

$$\int_{k-k_F}^0 d\kappa F(\kappa) = -\frac{\Gamma(1 - 2\alpha) \sin \pi\alpha}{2\pi\alpha}$$

$$\times |(k - k_F)a|^{2\alpha} \, \sigma(k - k_F), \qquad (77)$$

where

$$\sigma(x) = 1, \quad x > 0$$
$$= -1, \quad x < 0.$$

Therefore we see that there is, for $\alpha \neq 0$, no discontinuity at the F.S. (because the factor $|(k - k_F)a|^{2\alpha}$ vanishes there) though the slope is infinite at this point. On the other hand, if $\alpha = 0$, (77) behaves like $-\frac{1}{2}\sigma(k - k_F)$, which just gives the usual discontinuity at the F.S. Thus, in this model, the smallest amount of interaction *always* destroys the discontinuity of $\bar{n}_k$ at the F.S.

The behavior of $\bar{n}_k$ for large k[i.e., $(k - k_F)a \gg 1$] is also not difficult to obtain. From (71) we need $F(\kappa)$ for large κ, which is the same as knowing $Q(\xi)$ for small ξ. From (70) this may be obtained by expanding

$$Q(\xi) = \frac{\lambda^2}{2\pi^2} \frac{1}{2} \int_0^{\infty} dk k |v(k)|^2 \xi^2 + \cdots, \qquad (78)$$

as long as the integral converges, which we shall assume. Writing this as

$$Q(\xi) \cong \frac{1}{2}\xi^2/b^2, \qquad (79)$$
$$b^2 \equiv \frac{\lambda^2}{2\pi^2} \int_0^{\infty} dk \cdot k \cdot |v(k)|^2,$$

we obtain

$$F(\kappa) = b/(2\pi)^{\frac{1}{2}} e^{-b^2\kappa^2/2}. \qquad (80)$$

Therefore, for large k, we have

$$\bar{n}_k \cong \frac{b}{(2\pi)^{\frac{1}{2}}} \int_k^{\infty} d\kappa e^{-\kappa^2 b^2/2} \cong \frac{1}{(2\pi)^{\frac{1}{2}}} \frac{1}{kb} e^{-k^2 b^2/2}. \qquad (81)$$

Therefore the momementum distribution decreases exponentially for large k.

For k close to the origin we may write

$$\bar{n}_k = \int_{k-k_F}^{\infty} F(\kappa) \, d\kappa$$

$$= \int_{-\infty}^{\infty} F(\kappa) \, d\kappa + \int_{-(k-k_F)}^{-\infty} F(\kappa) \, d\kappa. \qquad (82)$$

From (69),

$$\int_{-\infty}^{\infty} d\kappa F(\kappa) = \int_{-\infty}^{\infty} d\xi \, \delta(\xi) e^{-Q(\xi)} = e^{-Q(0)} = 1. \qquad (83)$$

Further, $F(\kappa)$ is an even function of κ. Thus

$$\bar{n}_k = 1 - \int_{k_F-k}^{\infty} F(\kappa) \, d\kappa = 1 - \bar{n}_{2k_F-k}, \qquad (84)$$

$$\bar{n}_0 = 1 - \bar{n}_{2k_F}. \qquad (85)$$

Therefore $\bar{n}_0 < 1$. If the interaction is such that $k = 2k_F$ is already in the asymptotic region for large k, then $\bar{n}_{2k_F}$ is exponentially small, and $\bar{n}_0$ is very close to unity.

Finally, we should like to conclude this section

with a remark about the case where $V(x) = \delta(x)$, the Dirac δ function. In this case $v(k)$ is a constant, so that (70) diverges logarithmically.

If one regards the δ function as the limit of a smooth function [a very convenient choice, with which one can calculate explicitly, is $v(k) = e^{-|k|a/2}$, letting a approach zero in the final answer], it is easy to see that the result is simply $\bar{n}_k = \frac{1}{2}$. The anomalous behavior of the δ-function case is not surprising as it looks at first. Since the particle mass is zero and λ (as may easily be verified) is dimensionless, the only length which can come into the problem is the mean distance between particles or, equivalently k_F^{-1}. However, from (54), k_F does not enter into $F(\kappa)$, so that $\bar{n}_k$ is a function of $k - k_F$ alone, which must be dimensionless. One such example is the unperturbed distribution, which depends only on whether $|k| > k_F$ or not. Another is a constant, which is what we actually obtain for the δ-function potential. The physical origin of this distribution which extends to infinite k, is that the high fourier components of the δ function produce infinitely many pairs, so that infinitely many particles are present.

IV. COMPARISON WITH PERTURBATION THEORY

According to the general formulas[2] the momentum distribution in the ground state is given by

$$\bar{n}_k = \frac{1}{2\pi i} \int_{\mu-i\infty}^{\mu+i\infty} d\zeta \, \frac{e^{\zeta 0^+}}{\zeta - \epsilon_k - G_k(\zeta)}, \quad (86)$$

where $G_k(\zeta)$ is the proper self-energy part of the particle propagator. In this formalism one should calculate the correct propagator at finite temperature (including "anomalous" diagrams) and also use the correct chemical potential μ. It was found there that if the F.S. does not distort (spherical case) this is the same as using ordinary Goldstone perturbation theory (no anamolous diagrams) and the the unperturbed chemical potential. We shall assume that this is also the case here, there being nothing comparable to F.S. distortion in one dimension. Then we replace μ by $v_0 k_F$ and take for $G_k(\zeta)$ the lowest nonvanishing contribution. This is second order. A straightforward calculation yields, for $k > 0$,

$$G(\zeta) = \frac{\lambda}{2} \left(\frac{\lambda}{\pi}\right)^2 v_0 \left(\int_{|k'|}^{\infty} + \int_{-\infty}^{-|k'|}\right) d\kappa$$
$$\times \frac{|v[\frac{1}{2}(\kappa + |k'|)]|^2 \, |\frac{1}{2}(\kappa + |k'|)|}{z + \kappa}, \quad (87)$$

where

$$k' = k - k_F, \qquad z = (\zeta - k_F v_0)/v_0.$$

This function is analytic in the cut z plane, the cuts extending from $-\infty$ to $-|k'|$ and from $|k'|$ to ∞.

If this is inserted in (86) (with μ replaced by $v_0 k_F$), the resulting integral is quite complicated to discuss, even in the neighborhood of $k = k_F$, for an arbitrary potential, and we shall limit ourselves to a special case.

Writing $z = x - i0^+$, we have

$$G_k(\zeta) = v_0[K_{k'}(x) + iJ_{k'}(x)]. \quad (88)$$

It is easy to see that by suitably deforming the contour in (86) we may write

$$\bar{n}_k = \frac{1}{2\pi i} \int_{-\infty}^{0} dx$$
$$\times \left[\frac{1}{x - k' - K_{k'}(x) - iJ_{k'}(x)} - \text{c.c.}\right] k' > 0 \quad (89)$$

$$= 1 - \frac{1}{2\pi i} \int_{0}^{\infty} dx$$
$$\times \left\{\frac{1}{x - k' - K_{k'}(x) - iJ_{k'}(x)} - \text{c.c.}\right\} k' < 0. \quad (90)$$

Now choosing

$$|v(\kappa)|^2 = 1 \quad |\kappa| < \frac{1}{2}q$$
$$= 0 \quad |\kappa| > \frac{1}{2}q, \quad (91)$$

one easily sees

$$K_{k'}(x) = \alpha |k'| \left[-2 + \left(1 - \frac{x}{|k'|}\right)\right.$$
$$\left.\times \log\left|\frac{q^2 - (x - |k'|)^2}{x^2 - k'^2}\right|\right], \quad (92)$$

$$J_{k'}(x) = 0 \quad \text{unless} \quad -q + |k'| < x < -|k'|,$$
$$\text{or} \quad |k'| < x < q + |k'|$$
$$= \pi\alpha \, |x - |k'|| \quad \text{otherwise.}$$

We want to investigate $\bar{n}_k$ for small k'. It is not difficult, using (92), to show that, for small α and $|k'|$, $\bar{n}_k$ takes the form

$$\bar{n}_k = \frac{1}{2}\{1 - \sigma(k')/(1 - 2\alpha \log |k'a|)\}, \quad (93)$$

where $a = 1/q$.

This expression is, just as the exact expression, continuous at $k = k_F$, and has infinite slope there. In fact if we write

$$|k'a|^{2\alpha} = (e^{-2\alpha \log |k'a|})^{-1}$$
$$= (1 - 2\alpha \log |k'a| + \cdots)^{-1},$$

forcing an expansion of the exact result (77) for

small α, we see that in this sense (93) agrees exactly with the exact answer to the order involved.

Thus, unlike the realistic three-dimensional case, perturbation theory predicts no discontinuity at the F.S. Since the exact answer behaves in the same way, perturbation theory (for the proper self-energy part) in this problem at least is a reliable guide to the behavior of $\bar{n}_k$.

V. RESPONSE TO EXTERNAL FIELDS

If one considers particles to have a charge e, we can induce currents to flow by applying an external field. It follows at once from the commutation relationships that

$$\dot{\rho} + \partial j/\partial x = 0, \tag{94}$$

where

$$\rho(x) = e\psi^+(x)\psi(x) = e[N_1(x) + N_2(x)], \tag{95}$$

$$j(x) = ev_0\psi^+(x)\sigma_3 \quad \psi(x) = ev_0[N_1(x) - N_2(x)], \tag{96}$$

$$\dot{\rho} \equiv i[H, \rho]. \tag{97}$$

This is clearly the equation of continuity of charge, and we can identify ρ and j with the charge and current densities, respectively.[5]

Suppose we couple to our system an external electric field described by a potential $\varphi(x, t)$. The interaction is described by a Hamiltonian H_{ext} given by

$$H_{ext} = \int_0^L \rho(x)\varphi(x, t) \, dx, \tag{98}$$

$$H_T = H + H_{ext}. \tag{99}$$

If we again make the canonical transformation (20),

$$\tilde{H}_T = e^{i\lambda S}H_T e^{-i\lambda S}, \tag{100}$$

we find, since S commutes with H_{ext},

$$\tilde{H}_T = H_0 + H_{ext}. \tag{101}$$

Therefore, for a static field, all the energy levels are identical with the noninteracting case. In particular, this means that the Kohn effect[6] (which predicts a logarithmic singularity in $|q - 2k_F|$ for the change in energy of the system in the presence of an external field of wavenumber q) is completely unaltered by the interaction, this, in spite of the fact that the behavior of $\bar{n}_k$ in the neighborhood of $k = k_F$ is profoundly altered.

If we calculate the linear response, (i.e., the current

that flows to terms linear in the external field) by means of (say) the Kubo formula,[7] then one sees immediately that the result is the same as in the unperturbed case. Again this result is due to the fact that both the charge and current densities depend only on N_1 and N_2, which commute with S. This is also true if the external field couples to the current or when there are impurities present which act on the individual particles.

Finally, we may consider "positron annihilation" in this model.[8] Usually this is thought of as an effect which gives a direct experimental measurement of $\bar{n}_k$. In the one-dimensional case one cannot measure an angular correlation between the photons which come out. However, one can ask questions about the probability of one of them having a momentum between q and $q + dq$. We do not want to enter into a long discussion of the various possibilities here. We mention, however, that if one couples massless "photons" described by a scalar field ϕ having velocities $u_0(<v_0)$, via an effective interaction for pair annihilation,

$$H''' = g \int_0^L dx \rho(x)\phi^2(x); \tag{102}$$

then again only the unperturbed momentum distribution plays a role. However, if one takes more complicated couplings (depending for example on other bilinear expressions than ρ or j) one can get a large effect from the interaction.

Thus we see that although the momentum distribution is very much altered by the interaction in this model, it is by no means true that effects due to "particles at the Fermi Surface" are correspondingly altered. In other words, the naive association of the existence of a discontinuity in the momentum distribution, and the quasiparticle-like behavior of a weakly excited system of interacting fermions is shown to be unjustified for this model.

APPENDIX

We want to evaluate expressions of the following type:

$$I = (\Psi, \Lambda\Psi),$$
$$\Lambda \equiv \exp\left[i \int_0^L Q(y)\psi^+(y)\psi(y) \, dy\right], \tag{A1}$$

where $Q(y)$ is an ordinary function, and where Ψ

[5] In reality these definitions should be modified by the subtraction of infinite constants corresponding to the redefinition of the vacuum state as that with no holes and no electrons. We imagine this done in what follows.
[6] W. Kohn, Phys. Rev. Letters, 2, 393 (1959).

[7] R. Kubo, Can. J. Phys. 34, 1274 (1956).
[8] See, for example, R. Ferrell, Rev. Mod. Phys. 28, 308 (1956).

represents a wavefunction in which the single-particle states $n = 1, 2, \cdots, M$ are occupied. If we write

$$\psi(y) = \sum_n a_n \varphi_n(y), \qquad (A2)$$

where the $\varphi_n(y)$ are a complete orthonormal set of single-particle states, then clearly

$$\Psi = a_1^+ \cdots a_M^+ \Psi_0. \qquad (A3)$$

Ψ_0 is the unperturbed vacuum.

We may write (A1) as

$$I = (\Psi_0, a_M \cdots a_1 \Lambda a_1^+ \cdots a_M^+ \Psi_0). \qquad (A4)$$

Writing

$$a_1 = \int_0^\infty dz_1 \varphi_1^*(z_1) \psi(z_1), \qquad (A5)$$

we get, making use of (42),

$$a_1 \Lambda = \int dz_1 \varphi_1^*(z_1) \psi(z_1) \Lambda$$
$$= \Lambda \int dz_1 \varphi_1^*(z_1) e^{iQ(z_1)} \psi(z_1). \qquad (A6)$$

Therefore, (A4) becomes

$$I = \left(\Psi_0, \Lambda \int d^M z \left(\prod_{n=1}^{M} \varphi_n^*(z_n) e^{iQ(z_n)} \right) \right.$$
$$\left. \times \psi(z_M) \cdots \psi(z_1) a_1^+ \cdots a_M^+ \middle| \Psi_0 \right). \qquad (A7)$$

Since Ψ_0 is the unperturbed vacuum,

$$\Lambda \Psi_0 = \Psi_0, \qquad (A8)$$

so that (A7) becomes

$$I = \int d^M z \left(\prod_{n=1}^{M} \varphi_n^*(z_n) e^{iQ(z_n)} \right)$$
$$\times (\Psi_0, \psi(z_M) \cdots \psi(z_1) a_1^+ \cdots a_M^+ \Psi_0) \qquad (A9)$$
$$= \int d^M z \, d^M z' \left(\prod_{n=1}^{M} \varphi_n^*(z_n) \varphi_n(z_n') e^{iQ(z_{n'})} \right)$$
$$\times (\Psi_0 | \psi(z_M) \cdots \psi(z_1) \psi^+(z_1') \cdots \psi^+(z_M') | \Psi_0). \qquad (A10)$$

The expectation value in (A10) is a familiar one in the many-body problem. It can be obtained by taking the sum of the products of the corresponding expectation value for all possible ψ, ψ^+ pairs. The sign of each term is given by a plus if the permutation necessary to bring them to the required position is even, a minus if it is odd. Clearly then

$$I = \int d^M z \, d^M z' \sum_P (-)^P P \left(\prod_1^M \varphi_n^*(z_n) \varphi_n(z_n') e^{iQ(z_n)} \right)$$
$$\times (\Psi_0 | \psi(z_1) \psi^+(z_1') | \Psi_0) \cdots (\Psi_0 | \psi(z_M) \psi^+(z_M') | \Psi_0). \qquad (A11)$$

The sum on P is over all possible permutations of the variables. Now

$$(\Psi_0 | \psi(z_1) \psi^+(z_1') | \Psi_0)$$
$$= (\Psi_0 | \psi(z_1) \psi^+(z_1') + \psi^+(z_1') \psi(z_1) | \Psi_0)$$
$$= \delta(z_1 - z_1')(\Psi_0, \Psi_0) = \delta(z_1 - z_1'), \qquad (A12)$$

$$I = \int d^M z \sum_P (-)^P \varphi_1^*(z_1) e^{iQ(z_1)} \cdots \varphi_M^*(z_M) e^{iQ(z_M)}$$
$$\times \varphi_{i_1}(z_1) \cdots \varphi_{i_M}(z_M), \qquad (A13)$$

where

$$P(1, 2, \cdots, M) = (i_1, i_2, \cdots, i_M),$$

or

$$I = \sum_P (-)^P (\varphi_1, e^{iQ(z)} \varphi_{i_1}) \cdots (\varphi_M, e^{iQ(z)} \varphi_{i_M}). \qquad (A14)$$

This, however, is just the definition of the determinant of the matrix $g_{nn'}$, where

$$g_{nn'} = (\varphi_n, e^{iQ(z)} \varphi_{n'}). \qquad (A15)$$

Therefore,

$$I = \text{Det} (g).$$

If we take for the φ_n plane wave states, we get just the result used in the text.

Incidently, if one does this in configuration space and uses determinental wavefunctions, this becomes a well-known theorem about the integral over products of determinants.

ACKNOWLEDGMENTS

I should like to express my appreciation to the Brookhaven National Laboratory for its hospitality during the summer of 1962, when this work was initiated. I should also like to express my thanks to Professor W. Kohn for many valuable conversations relating to the general question of the Fermi surface, out of which this work originated.

JOURNAL OF MATHEMATICAL PHYSICS   VOLUME 6, NUMBER 2   FEBRUARY 1965

Exact Solution of a Many-Fermion System and Its Associated Boson Field

Daniel C. Mattis

International Business Machines Corp., Thomas J. Watson Research Center,
Yorktown Heights, New York

AND

Elliott H. Lieb*

Belfer Graduate School of Science, Yeshiva University, New York, New York
(Received 22 September 1964)

Luttinger's exactly soluble model of a one-dimensional many-fermion system is discussed. We show that he did not solve his model properly because of the paradoxical fact that the density operator commutators $[\rho(p), \rho(-p')]$, which always vanish for any finite number of particles, no longer vanish in the field-theoretic limit of a filled Dirac sea. In fact the operators $\rho(p)$ *define* a boson field which is *ipso facto* associated with the Fermi–Dirac field. We then use this observation to solve the model, and obtain the exact (and now nontrivial) spectrum, free energy, and dielectric constant. This we also extend to more realistic interactions in an Appendix. We calculate the Fermi surface parameter $\bar{n}_k$, and find: $\partial \bar{n}_k / \partial k|_{k_F} = \infty$ (i.e., there exists a sharp Fermi surface) only in the case of a sufficiently weak interaction.

I. INTRODUCTION

THE search for a soluble but realistic model in the many-electron problem has been just about as unfruitful as the historic quest for the philosopher's stone, but has equally resulted in valuable byproducts. For example, 15 years ago Tomonaga[1] published a theory of interacting fermions which was soluble only in one dimension with the provision that certain truncations and approximations were introduced into his operators. Nevertheless he had success in showing approximate boson-like behavior of certain collective excitations, which he identified as "phonons." (Today we would denote these as "plasmons," following the work of Bohm and Pines.[2]) Lately, Luttinger[3] has revived interest in the subject by publishing a variant model of spinless and massless one-dimensional interacting fermions, which demonstrated a singularity at the Fermi surface, compatible with the results of the modern many-body perturbation theory.[4]

Unfortunately, in calculating the energies and wavefunctions of his model Hamiltonian, Luttinger fell prey to a subtle paradox inherent in quantum field theory[5] and *therefore did not achieve a correct* solution of the problem he himself had posed. In the present paper we shall give the solution to his interesting problem and calculate the free energy. We shall show the existence of collective plasmon modes, and shall calculate the singularity at the Fermi surface (which may in fact disappear if the interaction is strong enough), the energy of the plasmons, and the (nontrivial) dielectric constant of the system. In an Appendix we shall show how the model may be generalized in such a manner as to remove certain restrictions on the interactions which Luttinger had found necessary to impose.

It is fortunate that solid-state and many-body theorists have so far been spared the plagues of quantum field theory. Second quantization has been often just a convenient bookkeeping arrangement to save us from writing out large determinantal wavefunctions. However there is a difference between very large determinants and *infinitely* large ones; and we shall show that one of the important differences *is the failure of certain commutators to vanish* in the field-theoretic limit when common sense and experience based on finite N tells us they *should* vanish! (Here N refers to the number of particles in the field.)

* Research supported by the U. S. Air Force Office of Scientific Research.

[1] S. Tomonaga, Progr. Theoret. Phys. (Kyoto) 5, 544 (1950).

[2] D. Bohm and D. Pines, Phys. Rev. 92, 609 (1953).

[3] J. M. Luttinger, J. Math. Phys. 4, 1154 (1963). Note that we set his $v_0 = 1$, thereby fixing the unit of energy. References to this paper will be frequent, and will be denoted by L (72), for example, signifying his Eq. (72).

[4] J. M. Luttinger and J. C. Ward, Phys. Rev. 118, 1417 (1960).

[5] Luttinger made a transformation, L (8), which was canonical in appearance only. But in the language of G. Barton [*Introduction to Advanced Field Theory*, (Interscience

Publishers, Inc., New York, 1963), pp. 126 *et seq.*] this transformation connected two "unitarily inequivalent" Hilbert spaces, which has as a consequence that commutators, among other operators, must be reworked so as to be well-ordered in fermion field operators. It was first observed by Julian Schwinger [Phys. Rev. Letters 3, 296 (1959)] that the very fact that one postulates the existence of a ground state (i.e., the filled Fermi sea) *forces* certain commutators to be nonvanishing even though in first quantization they automatically vanish. The "paradoxical contradictions" of which Schwinger speaks seem to anticipate the difficulties in the Luttinger model.

We shall show that these nonvanishing commutators *define* boson fields which must *ipso facto* always be associated with a Fermi–Dirac field, and we shall use the ensuing commutation relations to solve Luttinger's model exactly. Because this model is soluble both in the Hilbert space of finite N and also in the Hilbert space $N = \infty$, with different physical behavior in each, we believe it has applications to the *theory of fields* which go beyond the study of the many-electron problem. The model can be extended to the case of electrons with spin. This has interesting consequences in the band *theory of ferromagnetism*, as will be discussed in some detail in an article under preparation.[5a]

II. MODEL HAMILTONIAN

We recall Luttinger's Hamiltonian[3] and recapitulate some of his results:

$$H = H_0 + H', \tag{2.1}$$

where the "unperturbed" part is

$$H_0 = \int_0^L dx \; \psi^+(x)\sigma_3 p\psi(x) \tag{2.2a}$$

$$= \sum_k (a_{1k}^* a_{1k} - a_{2k}^* a_{2k})k, \tag{2.2b}$$

and the interaction is

$$H' = 2\lambda \iint_0^L dx \, dy \; \psi_1^+(x)\psi_1(x)$$

$$\times V(x - y)\psi_2^+(y)\psi_2(y) \tag{2.3a}$$

$$= \frac{2\lambda}{L} \sum \delta_{k_1+k_2,k_3+k_4} v(k_3 - k_4)$$

$$\times a_{1k_1}^* a_{1k_2} a_{2k_3}^* a_{2k_4}. \tag{2.3b}$$

Here ψ is a two-component field and the form (b) of the operator is obtained from (a) by setting

$$\psi = \frac{1}{\sqrt{L}} \sum_k e^{ikx}\binom{a_{1k}}{a_{2k}}$$

and

$$\psi^+ = \frac{1}{\sqrt{L}} \sum_k e^{-ikx}(a_{1k}^*, a_{2k}^*), \tag{2.4}$$

with a_{jk}'s defined to be anticommuting fermion operators which obey the usual relations

$$a_{jk}a_{j'k'} + a_{j'k'}a_{jk} \equiv \{a_{jk}, a_{j'k'}\} = 0$$
$$\{a_{j,k}^*, a_{j'k'}^*\} = 0, \text{ and } \{a_{jk}, a_{j'k'}^*\} = \delta_{jj'}\delta_{kk'}. \tag{2.5}$$

Luttinger noted that for an appropriate operator

5a D. Mattis, Physics 1, 184 (1964).

S_0, the canonical transformation

$$\tilde{H} = e^{i\lambda S_0}He^{-i\lambda S_0} \tag{2.6}$$

gave the result that

$$\tilde{H} = H_0, \tag{2.7}$$

and consequently that *the spectrum of $H = H_0 + H'$ was the same as that of H_0, independent of the interaction $V(x - y)$.* This can be explicitly verified for his choice of

$$S_0 = \iint_0^L dx \, dy \; \psi_1^+(x)\psi_1(x)E(x - y)\psi_2^+(y)\psi_2(y), \tag{2.8}$$

where $E(x)$, not to be confused with the energy E, is defined by:

$$\partial E(x - y)/\partial x \equiv V(x - y), \tag{2.9}$$

assuming that

$$\bar{V} \equiv \frac{1}{L}\int_0^L V(x) \, dx = 0. \tag{2.10}$$

In the Appendix we shall show among other things how to generalize to $\bar{V} \neq 0$. It is also simple and instructive to verify Eqs. (2.6) and (2.7) somewhat differently by using the *first* quantization,

$$H_0 = -i\sum_{n=1}^N \frac{\partial}{\partial x_n} + i\sum_{m=1}^M \frac{\partial}{\partial y_m} \tag{2.11}$$

and

$$H' = 2\lambda \sum_{n=1}^N \sum_{m=1}^M V(x_n - y_m), \tag{2.12}$$

where N and M are, respectively, the total number of "1" particles and "2" particles, with coordinates x_n and y_m, respectively. The properly antisymmetrized wavefunctions are given by

$$\Psi = \det |e^{ik_ix_j}| \det |e^{iq_iy_j}|$$

$$\times \exp\left\{\sum_{n=1}^N \sum_{m=1}^M i\, E(x_n - y_m)\right\}. \tag{2.13}$$

Using Eqs. (2.9) and (2.10), Ψ is readily seen to obey Schrödinger's equation

$$H\Psi = E\Psi \tag{2.14}$$

with just the unperturbed eigenvalue

$$E = \sum_{n=1}^N k_n - \sum_{m=1}^M q_m. \tag{2.15}$$

The wavenumbers are of the form

$$k_i \text{ or } q_j = 2\pi \text{ integer}/L, \tag{2.16}$$

as required for periodic boundary conditions. This is in exact agreement with the results of Ref. 3, and can also be checked in perturbation theory; first-order perturbation theory also gives vanishing results, and indeed, it is easy to verify that to every order in λ the cancellation is complete, in accordance with the exact result given above.

Up to this point, Luttinger's analysis (which we have briefly summarized) is perfectly correct. It is the next step that leads to difficulty. The Hamiltonian discussed so far has no ground-state energy; in order to remove this obstacle, and thereby establish contact with a real electron gas, Luttinger proposed modifying the model by "filling the infinite sea" of negative energy levels (i.e., all states with $k_1 <$ and $q_2 > 0$). Following L(8) we define b's and c's obeying the usual anticommutators, such that

and
$$a_{1k} = \begin{cases} b_k & k \geq 0 \\ c^*_k & k < 0, \end{cases} \qquad (2.17)$$
$$a_{2k} = \begin{cases} b_k & k < 0 \\ c^*_k & k \geq 0. \end{cases}$$

Using this notation the total particle-number operator becomes

$$\mathfrak{N} = \sum_{\text{all } k} b^+_k b_k - c^+_k c_k \qquad (2.17a)$$

(i.e., the number of particles minus the number of holes).

Since the Hamiltonian commutes with $\mathfrak{N}$ we can demand that $\mathfrak{N}$ have eigenvalue N_0. In the noninteracting ground state there are no holes and the b particles are filled from $-k_F$ to k_F where $k_F = \pi(N_0/L) = \pi\rho$. The noninteracting ground-state energy is $N_0\pi\rho +$ energy of the filled sea (W).

The kinetic energy assumes the form

$$H_0 = \sum_{\text{all } k} (b^*_k b_k + c^*_k c_k) \, |k| + W, \qquad (2.18)$$
where
$$W = (\sum_{k<0} k - \sum_{k>0} k) \qquad (2.18a)$$

is the infinite energy of the filled sea, an uninteresting c number which we drop henceforth in accordance with Luttinger's prescription. The interaction [H', Eq. (2.3) and the operator S_0, Eq. (2.8)] can also be expressed in the new language by means of the substitution (2.17). The reader will no doubt be surprised, as indeed we were, to find that now with the new operators, *Eq. (2.7), with $\tilde{H}$ defined in (2.6), is no longer obeyed.*

Upon further reflection one sees that this must be so, on the basis of very general arguments. In the new Hilbert space defined by the transformation to the particle–hole language (2.17), H is no longer unbounded from below and now has a ground state.

A general and inescapable *concavity theorem* states that if $E_0(\lambda)$ is the ground-state energy in the presence of interactions, (2.3), then

$$\partial^2 E_0(\lambda)/\partial\lambda^2 < 0. \qquad (2.19)$$

This inequality is incompatible with the previous result, viz. all $E =$ independent of λ, which was possible only in the strange case of a system without a ground state.

The same thing can be seen more trivially using second-order perturbation theory (first-order perturbation theory vanishes). It is easily seen that

$$E_0^{(2)} = -\left(\frac{2\lambda}{L}\right)^2 \sum_k \frac{|v(k)|^2}{2k} \, n_1(k)n_2(-k), \qquad (2.20)$$

where $n_1(k)$ and $n_2(k)$ are the number of ways of shifting a particle of type "1" and type "2" respectively by an amount k to an unoccupied state. A simple geometric exercise will convince the reader of the following facts: (1) if we start with a state having a finite number of particles, then n_1 and n_2 are *always* even functions of k (i.e., there are just as many ways to increase the momentum by k as to decrease it by the same amount.) (2) If we start with a filled infinite sea then there is no way to decrease the momentum of the "1" particles nor to increase the momentum of "2" particles. Hence for this second case $n_1(k)n_2(-k)$ is nonzero only for $k > 0$. Thus $E_0^{(2)}$ vanishes for a state with a finite number of particles, but it is negative for a filled sea.

If the reader is unconvinced by perturbation theory, then he can easily prove that E_0 is lowered by doing a variational calculation.

What has gone wrong? We turn to some algebra to resolve this paradox, and following this, present a solution of the field-theoretic problem defined by $H_0 + H'$ in the representation of b's and c's.

III. CASE OF THE FILLED DIRAC SEA

The various relevant operators are given below; the form (a) of each equation will *not* be used in the bulk of the paper, and is just given here for completeness. In the following equations, $p > 0$.

$$\rho_1(+p) \equiv \sum_k a^*_{1\,k+p} a_{1\,k} \qquad (3.1a)$$

$$= \sum_{k<-p} c_{k+p} c^*_k + \sum_{-p \leq k < 0} b^*_{k+p} c^*_k + \sum_{k \geq 0} b^*_{k+p} b_k, \qquad (3.1b)$$

$$\rho_1(-p) \equiv \sum_k a^*_{1\,k} a_{1\,k+p} \qquad (3.2a)$$

$$= \sum_{k<-p} c_k c^*_{k+p} + \sum_{-p \leq k < 0} c_k b_{k+p} + \sum_{k \geq 0} b^*_k b_{k+p}. \qquad (3.2b)$$

$$\rho_2(+p) \equiv \sum_k a^*_{2\,k+p} a_{2\,k} \qquad (3.3a)$$

$$= \sum_{k<-p} b^*_{k+p} b_k + \sum_{-p\le k<0} c_{k+p} b_k + \sum_{k\ge0} c_{k+p} c^*_k, \qquad (3.3b)$$

$$\rho_2(-p) \equiv \sum_k a^*_{2\,k} a_{2\,k+p} \qquad (3.4a)$$

$$= \sum_{k<-p} b^*_k b_{k+p} + \sum_{-p\le k<0} b^*_k c^*_{k+p} + \sum_{k>0} c_k c^*_{k+p}. \qquad (3.4b)$$

Equations (3.1a)–(3.4a) give the density operators in the original representation, so let us calculate in this language a commutator such as (assume $p \ge p' \ge 0$ for definiteness)

$$[\rho_1(-p),\, \rho_1(p')] = \sum_{k,k'} [a^*_{1\,k} a_{1\,k+p},\, a^*_{1\,k'+p'} a_{1\,k'}]$$

$$= \sum_{k=-\infty}^{+\infty} a^*_{1\,k} a_{1\,k+p-p'} - \sum_{k=-\infty}^{+\infty} a^*_{1\,k+p'} a_{1\,k+p} = 0. \qquad (3.5)$$

The zero result could have been expected by writing the operators in first quantization:

$$\rho_1(-p) = \sum_n e^{-ipx_n} \quad \text{and} \quad \rho_2(p) = \sum_m e^{ipy_m}, \qquad (3.6)$$

whence they evidently commute. Nevertheless, the zero result is achieved in (3.5) only through the almost "accidental" cancellation of two operators, each of which may diverge in the field-theory limit when $N = \infty$. We now show that in that limit the operators in fact no longer cancel, by evaluating the commutator using form (b) for the density operators. It is a matter of only some minor manipulation to obtain the important new result:

$$[\rho_1(-p),\, \rho_1(p')] = [\rho_2(p),\, \rho_2(-p')]$$

$$= \delta_{p,p'} \sum_{-p<k<0} 1 = \frac{pL}{2\pi} \delta_{p,p'}, \quad (p' > 0). \qquad (3.7a)$$

In addition,

$$[\rho_1(p),\, \rho_2(p')] = 0. \qquad (3.7b)$$

A quick check is provided by evaluating the vacuum expectation value

$$\langle 0| \,[\rho_1(-p),\, \rho_1(p)]\, |0\rangle$$

$$= \sum_{-p<k,k'<0} \langle 0| \, c_k b_{k+p} b^*_{k'+p} c^*_{k'} \,|0\rangle = pL/2\pi, \qquad (3.8)$$

which is exactly what is expected on the basis of the previous equation. Evidently the form (b) of the operators $(2\pi/pL)^{+\frac12}\rho_1(+p)$ and $(2\pi/pL)^{+\frac12}\rho_2(-p)$ have properties of boson raising operators [call them $A^*(p)$ and $B^*(-p)$] and $(2\pi/pL)^{+\frac12}\rho_1(-p)$ and $(2\pi/pL)^{+\frac12}\rho_2(+p)$ have properties of boson lowering operators [$A(p)$ and $B(-p)$], i.e.,

$$[A,\, B] = [A^*,\, B] = 0, \qquad (3.9)$$

$$[A(p),\, A^*(p')] = [B(-p),\, B^*(-p)] = \delta_{p,p'}.$$

The B field is the continuation of the A field to negative p; therefore together they form a *single* boson field defined for all p.

The relationship of the $\rho(p)$'s to Luttinger's $N(x)$'s, L(25), is obtained by using (2.4):

$$N_1(x) = \psi^*_1(x)\psi_1(x) = \frac1L \sum_p \rho_1(p)e^{-ipx},$$

$$N_2(x) = \psi^*_2(x)\psi_2(x) = \frac1L \sum_p \rho_2(p)e^{-ipx}. \qquad (3.10)$$

IV. SOLUTIONS OF THE MODEL HAMILTONIAN

Before making use of the results of the previous section, we remark that $\rho_1(+p)$ and $\rho_2(-p)$ are exact raising operators of H_0, and $\rho_1(-p)$ and $\rho_2(p)$ are exact lowering operators of H_0 corresponding to excitation energies p. That is,

$$[H_0,\, \rho_1(\pm p)] = \pm p\rho_1(\pm p),$$

$$[H_0,\, \rho_2(\pm p)] = \mp p\rho_2(\pm p). \qquad (4.1)$$

The identification of the ρ's with boson operators made in the previous section suggested to us the possibility of constructing a new operator T which obeys the same equations (4.1), as H_0. This is indeed possible, if we define T as follows:

$$T \equiv \frac{2\pi}{L} \sum_{p>0} \{\rho_1(p)\rho_1(-p) + \rho_2(-p)\rho_2(p)\} \qquad (4.2)$$

[the ρ's being defined here and in the remainder of the paper by Eqs. (3.1b)–(3.4b), i.e., always in the hole-particle representation]. It follows that

$$[T,\, \rho_1(\pm p)] = \pm p\rho_1(\pm p) \qquad (4.3)$$

as required, and similarly for $\rho_2(\mp p)$. Therefore, let us decompose H into two parts

$$H = H_1 + H_2 \qquad (4.4)$$

with

$$H_1 = H_0 - T = \left\{ \sum_k |k|\, (b^*_k b_k + c^*_k c_k) - \frac{2\pi}{L} \sum_{p>0} \{\rho_1(p)\rho_1(-p) + \rho_2(-p)\rho_2(p)\} \right\}, \qquad (4.5)$$

and

$$H_2 = H' + T$$

$$= \frac1L \left[2\lambda \sum_{p>0} \{v(p)\rho_1(-p)\rho_2(p) + v(-p)\rho_1(p)\rho_2(-p)\} + 2\pi \sum_{p>0} \{\rho_1(p)\rho_1(-p) + \rho_2(-p)\rho_2(p)\} \right] \qquad (4.6)$$

with $v(p) =$ real, even function of p. By actual construction, all the ρ operators which appear in H_2

commute with H_1. This will be an important feature in constructing an exact solution of the model. We define an Hermitian operator S,

$$S = \frac{2\pi i}{L} \sum_{\text{all } p} \frac{\varphi(p)}{p} \rho_1(p)\rho_2(-p), \qquad (4.7)$$

where $\varphi(p)$ is also a real, even, function of p to be determined subsequently by imposing a condition that the unitary transformation e^{iS} diagonalize H_2. First we evaluate the effect of such a transformation on various operators. It commutes with H_1,

$$e^{iS}H_1e^{-iS} = H_1 = H_0 - T, \qquad (4.8)$$

because both ρ_1 and ρ_2 appearing in S commute with H_1, as noted above. In the following, p can have either sign:

$$e^{iS}\rho_1(p)e^{-iS} = \rho_1(p)\cosh\varphi(p) + \rho_2(p)\sinh\varphi(p), \qquad (4.9)$$

$$e^{iS}\rho_2(p)e^{-iS} = \rho_2(p)\cosh\varphi(p) + \rho_1(p)\sinh\varphi(p). \qquad (4.10)$$

We have verified that this transformation is a proper unitary transformation and preserves commutation relations (3.7) as well as anticommutation relations (2.5), and the reader may easily check this point. H_2 is brought into canonical form by requiring that in $(\exp iS)\ H_2\ (\exp -iS)$ there be no cross terms such as $\rho_1(p)\rho_2(-p)$. This leads to the equation

$$\tanh 2\varphi = -\lambda v(p)/\pi, \qquad (4.11)$$

which cannot be obeyed unless

$$|\lambda v(p)| < \pi \quad \text{for all} \quad p. \qquad (4.12)$$

Equation (4.12) serves to limit the magnitude of potentials capable of having well-behaved solutions (e.g., a real ground-state energy). For the more realistic potentials discussed in the Appendix, there is also a more realistic bound on $v(p)$: there, $v(p)$ may not be *too* attractive, but it can have any magnitude when it is repulsive, i.e., positive.

With the choice of φ in (4.11), the evaluation of H_2 becomes

$$e^{iS}H_2e^{-iS} = \frac{2\pi}{L} \sum_{p>0} \text{sech } 2\varphi(p)\{\rho_1(p)\rho_1(-p)$$
$$+ \rho_2(-p)\rho_2(p)\} - \sum_{p>0} p(1 - \text{sech } 2\varphi). \qquad (4.13a)$$

The second term is the vacuum renormalization energy

$$W_1 = -\sum_{p>0} p(1 - \text{sech } 2\varphi)$$
$$= \frac{L}{2\pi} \int_0^\infty dp\ p\left\{\left(1 - \frac{\lambda^2 v^2(p)}{\pi^2}\right)^{\frac{1}{2}} - 1\right\}. \qquad (4.13b)$$

It may be expanded in powers of λ to effect a comparison with Goldstone's many-body perturbation theory[4]; we have checked that they agree to third order.

The problem is now formally solved, for we can find all the eigenfunctions and eigenvalues by studying Eqs. (4.4), (4.8), and (4.13). First notice that the operator T does not depend upon the interaction and that if there is *no interaction* we could write the Hamiltonian either as

$$H = H_0, \qquad (4.14a)$$

or as

$$H = (H_0 - T) + T = H_1 + H_2. \qquad (4.14b)$$

Since H_1 and H_2 commute, every eigenstate, Ψ, of H may be assumed to be an eigenfunction of H_1 and H_2 separately. Moreover, Ψ may also be assumed to be an eigenfunction of each $\alpha_p = A_p^+A_p$ and $\beta_p = B_{-p}^+B_{-p}$ for all $p > 0$, since these operators commute with H and $\mathfrak{R}$.

Evidently (4.14a) and (4.14b) provide two different ways of viewing the noninteracting spectrum. H_0 is quite degenerate: the raising operators of H_0 are the b^+'s and c^+'s. By requiring that Ψ also be an eigenstate of α_p, β_p and H, we are merely attaching quantum numbers to the degenerate levels of H_0. If $\alpha_p\Psi = n_p\Psi$ and $\beta_p\Psi = m_p\Psi$ (where n_p and m_p are of course integers), we say that we have n_p plasmons of momentum p and m_p plasmons of momentum $-p$. With no interaction the energy of a plasmon is

$$\epsilon(p) = |p|. \qquad (4.15)$$

We may speak of H_1 as the quasiparticle part of the Hamiltonian; in H_1 the operator T plays the role of subtracting the plasmon part of the energy from H_0.

When we turn on the interaction, the above description of the energy levels is still valid, except that now we are *forced* to use the form (4.14b) because H_2 is no longer T. The degeneracy of H is partially removed by the interaction, because now the energy of a plasmon is

$$\epsilon'(p) = |p|\ \text{sech } 2\varphi(p). \qquad (4.16)$$

Notice that the plasmon energy is always *lowered* [and therefore the plasmons cannot propagate faster than the speed of light $c = 1$, i.e., $d\epsilon'/dp \leq 1$. In the more realistic case discussed in the Appendix, the plasmon energy *can* be increased by the interaction although $d\epsilon'/dp \leq 1$ is always obeyed.] by the interaction; if (4.12) is violated the plasmon energy is no longer real and the system becomes unstable. Note, there are no plasmons in the ground state, so that W_1 (4.13), is the shift in the ground-state energy of the system.

There is one important point, however, that requires some elucidation. We would like to be able to say that in view of the fact that H_1, $\alpha(p)$, and $\beta(p)$

conserve particle number, the most general energy level of H (fixed N_0) is the sum of *any* energy of H_1 (same N_0, and no plasmons) plus *any* (plasmon) energy of H_2 (note: the plasmon spectrum is independent of N_0). Were we dealing with a finite-dimensional vector space, such a statement would not be true, for even though H_1 and H_2 commute they could not possibly be independent. Thus, if H_2 had n eigenvalues $e_1, \cdots, e_n$, and if H_1 had an equal number $E_1, \cdots, E_n$ the general total eigenvalue would not be *any* combination of $e_i + E_i$ for this would give too many values (viz. n^2 instead of n.) But we are dealing with an infinite-dimensional Hilbert space and the additivity hypothesis is in fact true for the present model.

To prove this assertion we consider any eigenstate Ψ which is necessarily parameterized by the integers n_p and m_p. Consider the state $\Phi = \{\prod_p (A_p)^{n_p}(B_p)^{m_p}\}\Psi$. The state Φ is nonvanishing and has quantum numbers $n_p = 0 = m_p$. It is also an eigenstate of H_1 with energy $E_1(\Psi)$. In addition (and this is the important point) the state Ψ may be recovered from Φ by the equation

$$\Psi = \text{const} \times \{\prod_p (A_p^+)^{n_p}(B_p^+)^{m_p}\}\Phi.$$

To every state Ψ, therefore, there corresponds a *unique* state Φ from which it may be obtained using raising operators. Conversely, to any eigenstate of H_1 (for fixed N_0) we may apply raising operators as often as we please and obtain a new (nonvanishing) eigenstate. Thus the general energy is an arbitrary sum of quasiparticle and plasmon energies.

It may be wondered where we used the fact that the Hilbert space is infinite-dimensional in the above proof. The answer lies in the boson commutation relations of the A's and B's. It is impossible to have such relations in a finite-dimensional vector space.

The eigenvalues corresponding to these states Φ will be labeled in some order, E_i ($i = 1, 2, \cdots$), so that the total canonical partition function $Z(\lambda)$ and the free energy $F(\lambda)$ are given by

$$Z(\lambda) = e^{-F(\lambda)/kT}$$
$$= (\sum_i e^{-E_i/kT})(e^{-W_1/kT}) \prod_{\substack{\text{all } p \\ \neq 0}} \left(\sum_{n=0}^{\infty} e^{-n\epsilon'(p)/kT}\right).$$
$$(4.17)$$

The first factor is difficult to evaluate directly. However it can be obtained circuitously by noting that the energies E_i are independent of λ and therefore

$$Z(0) = e^{-F(0)/kT}$$
$$= (\sum_i e^{-E_i/kT}) \prod_{\substack{\text{all } p \\ \neq 0}} \left(\sum_{n=0}^{\infty} e^{-n\epsilon(p)/kT}\right). \quad (4.18)$$

But the second factor can be trivially evaluated, as can $F(0) =$ free energy of noninteracting fermions. Therefore we use (4.18) to eliminate the trace involving the E_i's in (4.17), with the final result:

$$F(\lambda) = F(0) + W_1$$
$$+ 2kT \sum_{p>0} \ln \{(1 - e^{-\epsilon'(p)/kT})/(1 - e^{-\epsilon(p)/kT})\}, \quad (4.19)$$

where ϵ and ϵ' are given in (4.15) and (4.16). It is noteworthy that the ground state and free energy both diverge in the case of a δ-function potential.

V. EVALUATION OF THE MOMENTUM DISTRIBUTION

In this section we calculate the mean number of particles with momentum k. This quantity is $\bar{n}_k$ and is the expectation value of

$$n_k = b_k^+ b_k \quad (5.1)$$

in the ground state. Since $\bar{n}_k$ is an even function of k we need only consider $k > 0$, and it is further convenient to introduce a Fourier transform so that [using (2.4)]

$$\bar{n}_k = \frac{1}{L} \iint_0^L ds\, dt\, e^{ik(s-t)} I(s, t). \quad (5.2)$$

Here

$$I(s, t) = \langle \Psi | \psi_1^+(s)\psi_1(t) | \Psi \rangle$$
$$= \langle \Psi_0 | e^{iS}\psi_1^+(s)e^{-iS}e^{iS}\psi_1(t)e^{-iS} | \Psi_0 \rangle, \quad (5.3)$$

where S is given by (4.7), Ψ is the new ground state, and Ψ_0 is the noninteracting ground state which is filled with b particles between $-k_F$ and k_F and has no holes (or c particles). This assignment depends on there having been no level crossing, which can be readily verified using (4.7)-(4.13).

In order to calculate the quantity $e^{iS}\psi_1(t)e^{-iS}$ we introduce the auxiliary operator

$$f_\sigma(t) = e^{i\sigma S}\psi_1(t)e^{-i\sigma S}, \quad (5.4)$$

where σ is a c number. We observe that $f_1(t)$ is the desired quantity while

$$f_0(t) = \psi_1(t). \quad (5.5)$$

In addition,

$$\partial f/\partial \sigma = e^{i\sigma S}i[S, \psi_1(t)]e^{-i\sigma S}$$
$$= e^{i\sigma S}[2\pi/L \sum_p \rho_2(-p)\varphi(p)p^{-1}e^{ipt}]e^{-i\sigma S}f_\sigma(t), \quad (5.6)$$

where we have used the commutation relations (3.7) as well as the fact that ψ_1 commutes with ρ_2. Equation (5.6) is a differential equation for $f_e(t)$ and (5.5) is the boundary condition. The solution is

$$f_e(t) = W_e(t)R_e(t)\psi_1(t), \tag{5.7}$$

where

$$W_e(t) = \exp \{2\pi/L \sum_{p>0} [\rho_1(-p)e^{ipt}$$
$$- \rho_1(p)e^{-ipt}]p^{-1}[\cosh \sigma\varphi(p) - 1]\} \tag{5.8}$$

and

$$R_e(t) = \exp \{2\pi/L \sum_{p>0} [\rho_2(-p)e^{ipt}$$
$$- \rho_2(p)e^{-ipt}]p^{-1} \sinh \sigma\varphi(p)\} \tag{5.9}$$

The reader may verify that (5.7) satisfies (5.5) and (5.6) by using the commutation relations (3.7). We recall the well-known rule that

$$\exp (A + B) = \exp (A) \exp (B) \exp (-1/2[A, B]) \tag{5.10}$$

when $[A, B]$ commutes with A and B. From here on we shall set $\sigma = 1$ and drop it as a subscript. We note that since $\rho_1(p)^+ = \rho_1(-p)$ and $\rho_2(p)^+ = \rho_2(-p)$,

$$R^+(t) = R^{-1}(t) \quad \text{and} \quad W^+(t) = W^{-1}(t). \tag{5.11}$$

We also note that R and W commute with each other. Thus, (5.3) becomes

$$I(s, t) = \langle \Psi_0| \psi_1^+(s)R^{-1}(s)W^{-1}(s)W(t)R(t)\psi_1(t) |\Psi_0\rangle$$
$$= I_1(s, t)I_2(s, t), \tag{5.12}$$

where

$$I_1(s, t) = \langle \Psi_1| \psi_1^+(s)W^{-1}(s)W(t)\psi_1(t) |\Psi_1\rangle, \tag{5.13}$$
$$I_2(s, t) = \langle \Psi_2| R^{-1}(s)R(t) |\Psi_2\rangle.$$

We have used the fact that the ground state is a product state: $\Psi_0 = \Psi_1 * \Psi_2$ where Ψ_1 is a state of the "1" field and Ψ_2 is a state of the "2" field. Ψ_1 is filled with b particles up to $+k_F$ and has no c particles; Ψ_2 is filled with b particles down to $-k_F$ and has no c particles.

Now, using the definition (5.8) and the rule (5.10) we easily find that

$$W^{-1}(s)W(t) = W_-(s, t)W_+(s, t)Z_1(s, t), \tag{5.14}$$

with

$$W_+(s, t) = \exp \{2\pi/L \sum_{p>0} \rho_1(-p)[\cosh \varphi(p) - 1]$$
$$\times p^{-1}(e^{ipt} - e^{ips})\},$$
$$W_-(s, t) = \exp \{2\pi/L \sum_{p>0} \rho_1(p)[\cosh \varphi(p) - 1]$$
$$\times p^{-1}(e^{-ips} - e^{-ipt})\},$$

$$Z_1(s, t) = \exp \{2\pi/L \sum_{p>0} [\cosh \varphi(p) - 1]^2$$
$$\times p^{-1}(e^{ip(s-t)} - 1)\}. \tag{5.15}$$

Likewise,

$$R^{-1}(s)R(t) = R_-(s, t)R_+(s, t)Z_2(s, t), \tag{5.16}$$

with

$$R_+(s, t) = \exp \{2\pi/L \sum_{p>0} \rho_2(p)[\sinh \varphi(p)]$$
$$\times p^{-1}(e^{-ips} - e^{-ipt})\},$$
$$R_-(s, t) = \exp \{2\pi/L \sum_{p>0} \rho_2(-p)[\sinh \varphi(p)]$$
$$\times p^{-1}(e^{ipt} - e^{ips})\},$$
$$Z_2(s, t) = \exp \{2\pi/L \sum_{p>0} [\sinh \varphi(p)]^2$$
$$\times p^{-1}(e^{ip(t-s)} - 1)\}. \tag{5.17}$$

We see at once from the definition (3.1b), (3.2b), of $\rho_1(p)$ that, for $p > 0$, $\rho_1(-p) |\Psi_1\rangle = 0$. Similarly $\langle \Psi_1| \rho(p) = 0$, $\rho_2(p) |\Psi_2\rangle = 0$, and $\langle \Psi_2| \rho_2(-p) = 0$. Hence,

$$I_2(s, t) = Z_2(s, t)$$

and

$$I_1(s, t) = Z_1(s, t)\langle \Psi_1| W_-^{-1}\psi_1^+(s)W_-W_+\psi_1(t)W_+^{-1} |\Psi_1\rangle. \tag{5.18}$$

If we now define

$$h_+(y) = 2\pi/L \sum_{p>0} [\cosh \varphi(p) - 1]$$
$$\times p^{-1}(e^{ipt} - e^{ips})e^{-ipy},$$
$$h_-(y) = 2\pi/L \sum_{p>0} [\cosh \varphi(p) - 1]$$
$$\times p^{-1}(e^{-ipt} - e^{-ips})e^{ipy}, \tag{5.19}$$

combining (3.10) and (5.15) we have that

$$W_+(s, t) = \exp \int_0^L N_1(y)h_+(y) \, dy,$$
$$W_-(s, t) = \exp -\int_0^L N_1(y)h_-(y) \, dy. \tag{5.20}$$

Since

$$[\psi_1(x), N_1(y)] = \delta(x - y)\psi_1(x),$$
$$[\psi_1^+(x), N_1(y)] = -\delta(x - y)\psi_1^+(x), \tag{5.21}$$

it follows that

$$W_+(s, t)\psi_1(t)W_+^{-1}(s, t) = \psi_1(t) \exp [-h_+(t)]$$
$$W_-^{-1}(s, t)\psi_1^+(s)W_-(s, t) = \psi_1^+(s) \exp [+h_-(s)]. \tag{5.22}$$

Finally,

$$\langle \Psi_1 | \ \psi_1^+(s) \psi_1(t) \ |\Psi_1\rangle = 1/L \sum_{p \leq k_F} e^{ip(t-s)}$$

$$\equiv Z_3(s, t). \tag{5.23}$$

Combining all these results, we conclude that

$$I(s, t) = Z_0(s, t) Z_1(s, t) Z_2(s, t) Z_3(s, t), \tag{5.24}$$

where

$$Z_0(s, t) = \exp (h_-(s) - h_+(t))$$

$$= \exp \{-4\pi/L \sum_{p>0} [\cosh \varphi(p) - 1]$$

$$\times (1 - e^{ip(s-t)})\}. \tag{5.25}$$

In order to make a comparison with Luttinger's calculation of $\bar{n}_k$, we first observe that the functions $Z_i(s, t)$ are really functions of $r = s - t$ and that they are periodic in s and t in $(0, L)$. We then define the functions $G(r)$ and $Q(r)$ as follows:

$$\exp [-Q(r)] \equiv G(r) \equiv Z_0(r) Z_1(r) Z_2(r). \tag{5.26}$$

Substituting (5.26), (5.24), and (5.23) into (5.2) we obtain

$$\bar{n}_k = 2\pi/L \sum_{p \leq k_F} F(k - p), \tag{5.27}$$

where

$$F(k) = 1/2\pi \int_{-\frac{1}{2}L}^{\frac{1}{2}L} dr \, e^{ikr} e^{-Q(r)} \tag{5.28}$$

$$\cong 1/2\pi \int_{-\infty}^{\infty} dr \, e^{ikr} e^{-Q(r)}. \tag{5.29}$$

In (5.29) we have passed to the bulk limit N, $L \to \infty$, not an approximation.

At this point our expression for $\bar{n}_k$ is formally the same as Luttinger's [cf. L (52), L (69)]. The difference is that our Q is different from his. He obtains Q by evaluating an infinite Toeplitz determinant with the result that [L (70)]

$$Q(r) = \lambda^2/2\pi^2 \int_0^{\infty} dp \frac{1 - \cos pr}{p} |v(p)|^2. \quad \text{(Luttinger)} \tag{5.30}$$

Our Q, which is the correct one to use, is obtained by combining (5.15), (5.17), and (5.25), replacing sums by integrals in the usual way, and using the definition (4.11) of $\varphi(p)$. The result is

$$Q(r) = \lambda^2/2\pi^2 \int_0^{\infty} dp \frac{1 - \cos pr}{p} |u(p)|^2, \tag{5.31}$$

where

$$|u(p)|^2 = (2\pi^2/\lambda^2)\{(1 - (\lambda v(p)/\pi)^2)^{-\frac{1}{2}} - 1\}. \tag{5.32}$$

It is worth noting that (5.30) agrees with (5.31) to leading order in λ^2.

Since we have not yet specified $v(p)$, we may now follow Luttinger's discussion from this point on with the proviso that we use the correct (λ dependent) $u(p)$ instead of $v(p)$. The reader is referred to pages 1159 and 1160 of Luttinger's paper.

There are two main conclusions one can draw. The first is that if we start with a δ-function interaction [so that $v(p)$ and hence $u(p)$] are constants, it can be shown that $\bar{n}_k = \frac{1}{2}$ for all k. Such a result is quite unphysical, but it is not unreasonable because the ground-state energy W (4.13a) diverges when $v(p) = $ constant at large p. Also, the result would be the same if we started with the more physical interaction

$$H' = 1/L \sum_p \{\rho_1(p) + \rho_2(p)\}\{\rho_1(-p) + \rho_2(-p)\}v(p)$$

discussed in the Appendix. This is indeed unfortunate, because relativistic field theories usually begin with local (δ-function) interactions.

The second conclusion is that if one makes a reasonable assumption about $v(p)$, and hence about $u(p)$ and $Q(r)$, one finds that for k in the vicinity of k_F, $\bar{n}_k$ behaves like

$$\bar{n}_k \sim d - e \, |k - k_F|^{2\alpha} \, \sigma(k - k_F), \tag{5.33}$$

where

$$\sigma(k) = 1, \qquad k > 0$$
$$= -1, \qquad k < 0 \tag{5.34}$$

and d, e, and α are certain positive constants. Now in Luttinger's calculation

$$\alpha = \lambda^2/4\pi^2 v(0)^2, \quad \text{(Luttinger)} \tag{5.35}$$

$$[\text{cf. } L(75)], \quad \text{where} \quad v(0) \equiv \lim_{p \to 0} v(p).$$

If $2\alpha < 1$, then the conclusion to be drawn is that although the interaction removes the discontinuity in $\bar{n}_k$ at the Fermi surface, we are left with a function that has an infinite slope there. There is, so to speak, a residual Fermi surface. In Sec. IV of his paper, Luttinger shows that at least for one example of $v(p)$ perturbation theory gives the same qualitative result as (5.33) with the same value of α, (5.35).

If, on the other hand, $2\alpha > 1$ then there is no infinite derivative at the Fermi surface. $\bar{n}_k$ is perfectly smooth there (although, technically speaking, it is nonanalytic unless $2\alpha =$ odd integer.) In this case virtually all trace of the Fermi surface has been eliminated. But notice that the correct α to use is obtained by replacing $v(0)$ by $u(0) \equiv \lim_{p \to 0} u(p)$ in (5.35), i.e.,

D. C. MATTIS AND E. H. LIEB

$$2\alpha = \{1 - [\lambda v(0)/\pi]^2\}^{-\frac{1}{2}} - 1. \qquad (5.36)$$

Thus, even subject to the requirement that $|\lambda v(0)|$ be less than π, 2α can become as large as one pleases. Yet perturbation theory predicts (5.35) which yields 2α always less than $\frac{1}{2}$.

We may conclude that a strong enough interaction can eliminate the Fermi surface, while perturbation theory predicts that is always there.

VI. DIELECTRIC CONSTANT

Because the response to external fields of wave vector q only depends on an interaction expression linear in the density operators, we can immediately obtain for the generalized static susceptibility function or *dielectric constant* (response ÷ driving force), for any temperature, T

$$\chi_\lambda(q, T) = \chi_0(q, T)\{\sinh \varphi(q) + \cosh \varphi(q)\}^2 \cosh 2\varphi_q$$

$$= \chi_0(q, T) \frac{1}{1 + \lambda v(q)/\pi} \qquad (6.1)$$

in terms of the "unperturbed" susceptibility $\chi_0(q, T)$. It is also a simple exercise to calculate exactly the time dependent susceptibility in terms of the "unperturbed" quantity.

It is interesting to note that the susceptibility can diverge (which is symptomatic of a phase transformation) only for

$$\lambda v(q) \rightarrow -\pi, \qquad (6.2)$$

i.e. only for sufficiently *attractive* interactions and not for repulsive $[v(q) > 0]$ interactions.

Recently Ferrell[6] advanced plausible arguments why a one-dimensional metal cannot become superconducting. We can prove this rigorously in the present model. The electron–phonon interaction is

$$H_{el-ph} = \sum_p g(p)[\rho_1(p) + \rho_2(p)] \cdot [\xi_p + \xi_{-p}^+], \qquad (6.3)$$

where ξ and ξ^+ are the phonon field operators. In the "filled-sea" limit this coupling is bilinear in harmonic-oscillator operators, and therefore the Hamiltonian continues to be exactly diagonalizable. The new normal modes can be calculated and there is found to be no phase transition at any finite temperature.

APPENDIX

We shall be interested in extending Luttinger's model in two ways. Firstly, we note that the restriction $\bar{V} = 0$ is really not necessary. Turning back to Eqs. (2.13) *et seq.* we impose periodic boundary conditions $\Psi(\cdots, x_i + L, \cdots) = \Psi(\cdots, x_i, \cdots)$, and find that

[6] R. A. Ferrell, Phys. Rev. Letters 13, 330 (1964).

$(q + N\lambda \bar{V})$ and $(k + M\lambda \bar{V}) = 2\pi/L \times$ integer (A1) replace the usual condition (2.16), where $N =$ number of "1" particles and $M =$ number of "2" particles. However, when $N, M \rightarrow \infty$ in the field-theoretic limit the problem evidently becomes ill-defined unless $\bar{V} \equiv 0$.

A less trivial observation concerns the form of the interaction potential. There is no reason to restrict it to the form $\propto \rho_1 \rho_2$, and in fact the more realistic two-body interaction

$$H' = \frac{\lambda}{L} \sum_p v(p)\{\rho_1(-p) + \rho_2(-p)\}\{\rho_1(p) + \rho_2(p)\} \qquad (A2)$$

is fully as soluble as the one assumed in the text, for any strength positive $v(p)$, and provided only

$$\lambda v(p) > -\tfrac{1}{2}\pi, \qquad (A3)$$

i.e. provided no Fourier component is *too* attractive. The shift in the ground-state energy is now given by

$$W_2 = \sum_{p>0} p\left\{\left(1 + \frac{2\lambda v(p)}{\pi}\right)^{\frac{1}{2}} - 1\right\}. \qquad (A4)$$

The plasmon energy is now

$$\epsilon''(p) \equiv |p| (1 + 2\lambda v(p)/\pi)^{\frac{1}{2}} \qquad (A5)$$

and *for the important case of the Coulomb repulsion,* $v(p) = p^{-2}$, the plasmons describe a relativistic boson field with mass

$$m^* \equiv (2\lambda/\pi)^{\frac{1}{2}} \qquad (A6)$$

and dispersion

$$\epsilon''(p) = (p^2 + m^{*2})^{\frac{1}{2}}. \qquad (A7)$$

Here, too, $d\epsilon''/dp < 1$.

ACKNOWLEDGMENTS

We thank D. Jepsen, L. Landovitz, and T. Schultz for interesting discussions during various stages of this work. We especially thank J. Luttinger for performing a detailed review of his own work and ours, and thus corroborating the results of the present article. While reading our manuscript, Professor Luttinger called our attention to Schwinger's relevant, indeed prophetic remarks (cf. Ref. 5), and we thank him for this reference. We wish to thank Dr. A. S. Wightman, for pointing out to us that P. Jordan first discovered the bosons associated with fermion fields (in his attempts to construct a neutrino theory of light), and for a list of references.[7]

[7] P. Jordan, Z. Physik 93, 464 (1935); 98, 759 (1936); 99, 109 (1936); 102, 243 (1936); 105, 114 (1937); 105, 229 (1937). M. Born and N. Nagendra-Nath, Proc. Ind. Acad. Sci. 3, 318 (1936). A. Sokolow, Phys. Z. der Sowj. 12, 148 (1937).

Chapter II
Lattice, Dynamical and Nonlinear Effects

LUTTINGER MODEL AND LUTTINGER LIQUIDS*

VIERI MASTROPIETRO

University of Rome Tor Vergata, Mathematics Department,
Viale della Ricerca Scientifica 00133, Rome, Italy

The Luttinger model owes its solvability to a number of peculiar features, like its linear relativistic dispersion relation, which are absent in more realistic fermionic systems. Nevertheless according to the Luttinger liquid conjecture a number of relations between exponents and other physical quantities, which are valid in the Luttinger model, are believed to be true in a wide class of systems, including tight binding or jellium one-dimensional fermionic systems. Recently a rigorous proof of several Luttinger liquid relations in nonsolvable models has been achieved; it is based on exact Renormalization Group methods coming from Constructive Quantum Field Theory and its main steps will be reviewed below.

Keywords: Luttinger model; Luttinger liquids; renormalization group; Ward Identities.

1. Introduction

The crucial observation that the low energy excitations of one-dimensional metals are well described in terms of $1+1$ massless Dirac fermions dates back to Tomonaga[1] and it was one of the earliest examples of *emerging* Quantum Field Theory (QFT) description in condensed matter physics; an idea which, later on, found several applications including the recent one to graphene (which is described by Dirac fermions in $d = 2+1$). Subsequently Luttinger,[2] unaware of the Tomonaga results, proposed a model, nowadays called *Luttinger model*, based on the same ideas of Tomonaga: the quasi-particle excitations are described in terms of two kinds of spinless fermions with linear dispersion relation, and a Dirac sea is introduced filling the infinite states with negative energy. Luttinger was inspired by the Thirring model,[3] proposed a little earlier, as it was noted by Luttinger himself; both models describe massless Dirac fermions in $d = 1 + 1$ with an interaction quartic in the fermionic fields. There are however important differences: while in the Thirring model the quartic current–current interaction is *local*, the Luttinger interaction is short ranged; moreover the ultraviolet cut-off (to be removed) is only on the momenta and not on the energies. As a consequence, the Thirring model is plagued by *ultraviolet divergences* and is Lorentz invariant, while the Luttinger model has no ultraviolet divergences and Lorentz invariance is lost.

Luttinger correctly observed that his model should be solvable, but unfortunately the solution he proposed was not correct. Curiously, the fate of the Luttinger model

---

*This article first appeared in International Journal of Modern Physics B, Vol. 26, No. 22 (2012).

was then somewhat similar to the one of the Thirring model, whose exact solution was anticipated by several uncorrect attempts, including the one of Thirring himself; the solution of the Thirring model was finally found by Johnson[4] assuming the validity of anomalous commutators as functions of two parameters which are fixed by a self-consistence argument.

The solution of the Luttinger model was found by Mattis and Lieb[5] showing that the density operators verify *anomalous* commutation relations, a crucial point which was missed by Luttinger, and showing that the fermionic Luttinger Hamiltonian can be mapped in a quadratic boson Hamiltonian; this allows the determination of all the correlations functions and of several physical quantities. It was found that the interaction radically alters the physical behavior. The occupation number, which has a discontinuity in the absence of interaction, becomes continuous and the wavefunction renormalization diverges at the Fermi points with a critical exponent nontrivial function of the coupling. By the solution in Ref. 5 one can compute the n-point functions and the response functions, whose large distance behavior have also anomalous exponents.

The exact determination of all its correlations makes the Luttinger model rather unique with respect to other solvable fermionic models, like the Hubbard model, in which only the ground state energy or the energy of some excited states can be computed. On the other hand, the Luttinger model has been proposed as a caricature of a true many body fermionic Hamiltonian, and owes its exact solvability to certain peculiar features (for instance the linear dispersion relation and the introduction of a Dirac sea) which are surely unrealistic. It is then natural to ask which features of the Luttinger model are still true in a realistic Hamiltonian.

It is reasonable to expect, according to Ref. 1, that anomalous critical exponents are generically present (at least in the absence of the spin) in 1D metals. The exponents will depend on the model details and of course differ from the Luttinger model ones; moreover for generic models there is in general no exact solution, or even if it is (as by Bethe ansatz in the *XXZ* chain) only certain eigenvalues of the energy can be computed and not the exponents. It is then clear the importance of the *Luttinger liquid conjecture* proposed by Haldane[6,7] (see also Ref. 8), according to which a number of exact simple relations between exponents and other physical quantities, which are valid in the Luttinger model, are still true in a wide class of more realistic one-dimensional models. In particular, according to such conjecture, one can determine all the exponents in terms of a single one, or in terms of (say) the Fermi velocity and the susceptibility.

The validity of the Luttinger liquid conjecture is a nontrivial statement, connected with the extended scaling relations proposed by Kadanoff[9] for two-dimensional classic statistical mechanics models like the *Eight vertex* or the *Ashkin–Teller* model. A justification of the Luttinger liquid conjecture is sometimes considered the fact that, according to a Renormalization Group (RG) analysis, interacting 1D fermionic systems with nonlinear dispersion relation differ from the Luttinger model through *irrelevant terms* in the RG sense. However this simple argument is not correct; the irrelevant terms change the exponents and break a number of symmetries which are instead true in the Luttinger model and which are essential for its solvability. Therefore, one has to prove that the symmetry-breaking effects due to the lattice or nonlinear bands, while changing the exponents, do not change the relations between exponents and other physical parameters. Usual approximate RG methods neglect the irrelevant terms so that they are not suitable to face this problem. In recent years the methods developed in constructive QFT have been applied to many body theory. Such methods are *exact* (that is the irrelevant terms are taken into account) and *rigorous*, as the physical observables can be written in terms of *renormalized* expansion, whose convergence can be proved at least if the coupling is not too large. Using this approach several Luttinger liquid relations in a wide class of models in generic fermionic one-dimensional models has been recently proven, and here we will review the main ideas behind such proof. In the second section we recall the Luttinger model and its solution. In the third section we introduce a generic model for 1D fermions on a lattice and we recall the Luttinger liquid conjecture. In the fourth section we review the exact RG analysis of a generic (nonsolvable) 1D lattice fermionic system, leading to a *convergent* expansion of the physical obervables in terms of running coupling constants. In the fifth section we will show that such running coupling constants do not increase under RG iterations as consequence of Ward Identities. In order to do that, one has to face and solve a well known problem, which is present in any Wilsonian RG approach; the momentum cut-off breaks the local symmetries and produces additional terms in the Ward identities which must be taken into account. In the sixth section we will derive *emerging* Ward Identities for the lattice model related to the validity of asymptotic symmetries (broken by irrelevant terms) which will be used in the seventh section for proving the Luttinger liquid relations for nonsolvable lattice models.

2. The Luttinger Model Exact Solution

The Luttinger model Hamiltonian is:

$$H = H_0 + V = \int_0^L dx \sum_{\varepsilon = \pm} : \tilde{\psi}_{x,\varepsilon}^+ \varepsilon \partial \tilde{\psi}_{x,\varepsilon}^- :$$

$$+ \lambda \int dx dy v(x-y) \left[\sum_{\varepsilon = \pm} q_{1,\varepsilon} : \tilde{\psi}_{\mathbf{x},\varepsilon}^+ \tilde{\psi}_{\mathbf{x},\varepsilon}^- : \right] \left[\sum_{\varepsilon = \pm} q_{2,\varepsilon} : \tilde{\psi}_{\mathbf{y},\varepsilon}^+ \tilde{\psi}_{\mathbf{y},\varepsilon}^- : \right] \quad (1)$$

where if $\varepsilon = \pm$

$$\psi_{x,\varepsilon}^\pm = e^{\pm i \varepsilon p_F x} \tilde{\psi}_{x,\varepsilon}^\pm = \frac{e^{\pm i \varepsilon p_F x}}{\sqrt{L}} \sum_k e^{\pm i k x} \hat{a}_{k,\varepsilon}^\pm \quad (2)$$

and $\tilde{\psi}_{x,\varepsilon}^{\pm}$ is a fermionic field with periodic boundary conditions, p_F is the Fermi momentum, $v(x - y)$ is a short range potential $|\hat{v}(p)| \leq Ce^{-\zeta|p|}$ with $\zeta > 0$ a constant, and the Wick ordering is defined rearranging the order so that $a_{-k,+}^+$, $a_{k,+}^-$, $a_{k,-}^+$, $a_{-k,-}^-$, $k \geq 0$ are always to the right of the other operators, and the new product is multiplied by the parity sign necessary to produce it. Finally $q_{i,\varepsilon}$ are charges, and for definiteness we will follow the original choice of Ref. 2

$$q_{1,+} = q_{2,-} = 1 \tag{3}$$

and zero otherwise; this choice is done only for definiteness and the model is solvable for a wide class of charges $q_{i,\varepsilon}$. With this choice of the charges the Hamiltonian can be rewritten as:

$$
\begin{aligned}
H_0 &= \sum_{\varepsilon, k>0} k(\hat{a}_{\varepsilon k,\varepsilon}^+ \hat{a}_{\varepsilon k,\varepsilon}^- + \hat{a}_{-\varepsilon k,\varepsilon}^- \hat{a}_{-\varepsilon k,\varepsilon}^+) \\
V &= \frac{\lambda}{L} \sum_{p>0} \hat{v}(p)[\rho_+(p)\rho_-(-p) + \rho_+(-p)\rho_-(p)] + \frac{\lambda\hat{v}(0)}{L} N_1 N_{-1} ,
\end{aligned}
\tag{4}
$$

where

$$\rho_\varepsilon(p) = \sum_k \hat{a}_{k+p,\varepsilon}^+ \hat{a}_{k,\varepsilon}^- \quad N_\varepsilon = \sum_{k>0} (\hat{a}_{\varepsilon k,\varepsilon}^+ \hat{a}_{\varepsilon k,\varepsilon}^- - \hat{a}_{-\varepsilon k,\varepsilon}^- \hat{a}_{-\varepsilon k,\varepsilon}^+) . \tag{5}$$

The regularization which is implicit in the above Hamiltonian is the suppression of the modes $|k| \geq \Lambda$ for the fermion with momentum k, where Λ is a momentum cut-off to be removed $\Lambda \to \infty$; that is $\rho_\varepsilon(p)$ has been thought as: $\sum_k \chi_\Lambda(k + p)\chi_\Lambda(k)\hat{a}_{k+p,\varepsilon}^+ \hat{a}_{k,\varepsilon}^-$ where $\chi_\Lambda(k) = 1$ for $|k| \leq \Lambda$ and zero otherwise.

The Hamiltonian (1) can be regarded as an operator defined on the Hilbert space $\mathcal{H}$ constructed as follows. If $|0\rangle$ is a state (the "Dirac sea" state) such that $\hat{a}_{k,-}^+|0\rangle = 0$ and $\hat{a}_{-k,+}^+|0\rangle = 0$ for $k \geq 0$, $\mathcal{H}_0$ is the linear span of all the states obtained applying finitely many creation or annihilation operators on $|0\rangle$ and $\mathcal{H}$ is the completion of $\mathcal{H}_0$. Note that H is a bounded operators over $\mathcal{H}$ provided that $\hat{v}(p)$ verifies the stability condition $|\lambda\hat{v}(p)| \leq 2\pi$. The crucial observation of Mattis and Lieb was that the *density operators* verify the following *anomalous commutation relations*

$$[\rho_\varepsilon(-p), \rho_\varepsilon(p')] = \varepsilon\frac{pL}{2\pi}\delta_{p,p'} \quad p > 0 \tag{6}$$

This follows noting that the commutator when $p = p'$ is equal to

$$-\sum_{k=-\Lambda+p}^{\Lambda} \hat{a}_{k,\varepsilon}^+ \hat{a}_{k,\varepsilon}^- + \sum_{k=-\Lambda}^{\Lambda-p} \hat{a}_{k,\varepsilon}^+ \hat{a}_{k,\varepsilon}^- = \sum_{k=-\Lambda}^{-\Lambda+p} \hat{a}_{k,\varepsilon}^+ \hat{a}_{k,\varepsilon}^- - \sum_{k=\Lambda-p}^{\Lambda} \hat{a}_{k,\varepsilon}^+ \hat{a}_{k,\varepsilon}^- \tag{7}$$

which on any state in $\mathcal{H}$ is, in the limit $\Lambda \to \infty$, equal to $\varepsilon(pL/2\pi)$. Other important relations are, for $p > 0$

$$[H_0, \rho_\varepsilon(p)] = -\varepsilon p \varepsilon_\varepsilon(p) \quad \left[\rho_\varepsilon(p), \sum_{\varepsilon, p>0} \rho_\varepsilon(\varepsilon p)\rho_\varepsilon(-\varepsilon p)\right] = -\varepsilon p \frac{L}{2\pi}\rho_\varepsilon(p) \qquad (8)$$

and $\rho_\varepsilon(-\varepsilon p)|0\rangle = 0$. We can write:

$$H = H_1 + H_2 , \qquad (9)$$

where, if $T = (2\pi/L)\sum_{\varepsilon, p>0}\rho_\varepsilon(\varepsilon p)\rho_\varepsilon(-\varepsilon p)$

$$H_1 = H_0 - T + \frac{\lambda \hat{v}(0)}{L}N_+ N_-$$

$$H_2 = T + \frac{\lambda}{L}\sum_{p>0}\hat{v}(p)[\rho_+(p)\rho_+(-p) + \rho_-(-p)\rho_-(p)] \qquad (10)$$

and $[H_1, \rho_\varepsilon(\pm p)] = 0$ for $p > 0$, while H_2 can be easily diagonalized; if $S = (2\pi/L)\sum_{p\neq 0}p^{-1}\rho_+(p)\rho_-(-p)$

$$e^{iS}H_2 e^{-iS} = \frac{2\pi}{L}\sum_{p>0}\varepsilon(p)[\rho_+(p)\rho_+(-p) + \rho_-(p)\rho_-(-p)] + E_0 , \qquad (11)$$

where E_0 is a constant and

$$\varepsilon(p) = \operatorname{sech}(2\phi(p)) , \quad \tanh 2\phi = -\frac{\lambda \hat{v}(p)}{2\pi} . \qquad (12)$$

The above formula (12) refers to the specific choice (3) of the $q_{i,\varepsilon}$, but a general formula can be easily obtained. The set of states $|j, n_1, n_2\rangle$ obtained applying operators $\rho_+(p)$, $\rho_-(-p)$ an arbitrary number of times on the state in which all the levels are filled up to n_1 with fermions of type $+$ and down to level n_2 is *complete*[6,10] and H_1 is a constant in the subspace with fixed n_1, n_2, from the fact that $[H_1, \rho_\varepsilon(\pm p)] = 0$. Therefore $e^{-iS}|0\rangle$ is the ground state of H as H_1 is constant on the subspace with $n_1 = n_2 = 0$ while

$$H_2 e^{-iS}|0\rangle = e^{-iS}(e^{iS}H_2 e^{-iS})|0\rangle = E_0 e^{-iS}|0\rangle \qquad (13)$$

Defining $\tilde{\psi}_{\mathbf{x},\varepsilon} = e^{Ht}\tilde{\psi}_{x,\varepsilon}e^{-Ht}$, $\mathbf{x} = (t, x)$ the n-point function is given by

$$\langle \tilde{\psi}^{\sigma_1}_{\mathbf{x}_1,\varepsilon_1} \cdots \tilde{\psi}^{\sigma_n}_{\mathbf{x}_n,\varepsilon_n}\rangle \equiv \langle 0|e^{iS}\mathcal{T}(\tilde{\psi}^{\sigma_1}_{\mathbf{x}_1,\varepsilon_1} \cdots \tilde{\psi}^{\sigma_n}_{\mathbf{x}_n,\varepsilon_n})e^{-iS}|0\rangle \qquad (14)$$

where $\mathcal{T}$ is the time ordering. In Ref. 5 only the case $n = 2$, $t_1 = t_2 = 0$ was studied, but it is only a matter of algebra to deduce from the exact solution the explicit form of the time-dependent n-point correlation; this was done in Ref. 11 for the 2-point and in Refs. 12 and 13 for the n-point function. It is found

$$\langle \tilde{\psi}^{\sigma_1}_{\mathbf{x}_1,\varepsilon_1} \cdots \tilde{\psi}^{\sigma_n}_{\mathbf{x}_n,\varepsilon_n} \rangle = \langle \tilde{\psi}^{\sigma_1}_{\mathbf{x}_1,\varepsilon_1} \cdots \tilde{\psi}^{\sigma_n}_{\mathbf{x}_n,\varepsilon_n} \rangle_0 e^{-Q_n} , \tag{15}$$

where $\langle \tilde{\psi}^{\sigma_1}_{\mathbf{x}_1,\varepsilon_1} \cdots \tilde{\psi}^{\sigma_n}_{\mathbf{x}_n,\varepsilon_n} \rangle_0$ is the noninteracting $\lambda = 0$ (imaginary time) n-point function (expressed by the Wick rule in terms of the 2-point free function) and

$$
\begin{aligned}
Q_n = \frac{2\pi}{L} \sum_{p>0} \sum_{\varepsilon=\pm} &\left\{ s^2(p) \left[\frac{n}{2} + \sum_{i,j \in I_\varepsilon, i<j} \sigma_i \sigma_j e^{-p|t_j-t_j|\varepsilon(p)} \cos p(x_i - x_j) \right] \right. \\
&- c(p)s(p) \sum_{i \in I_\varepsilon, j \in I_{-\varepsilon}} \sigma_i \sigma_j e^{-p|t_j-t_j|\varepsilon(p)} \cos p(x_i - x_j) \\
&- \sum_{i,j \in I_\varepsilon, i<j} [e^{-|t_i-t_j|} - e^{-p|t_i-t_j|\varepsilon(p)}] \cos p(x_i - x_j) \\
&\left. -i\varepsilon \sum_{i,j \in I_\varepsilon, i<j} \sigma_i \sigma_j \frac{t_j - t_j}{|t_i - t_j|} [e^{-|t_i-t_j|} - e^{-p|t_i-t_j|\varepsilon(p)}] \sin p(x_i - x_j) \right\}
\end{aligned}
\tag{16}
$$

where

$$s(p) = \sinh \phi(p), \quad c(p) = \cosh \phi(p) \tag{17}$$

In particular the 2-point function is given by

$$\langle \tilde{\psi}^-_{\mathbf{x},\varepsilon} \tilde{\psi}^+_{0,\varepsilon} \rangle = \frac{1}{2\pi} \frac{1}{i\varepsilon x + t} e^{-Q_2(\mathbf{x})} \tag{18}$$

$$
\begin{aligned}
Q_2(\mathbf{x}) = \int_0^\infty \frac{dp}{p} &\left[2s^2(p)(1 - e^{-p\varepsilon(p)|t|} \cos px) \right. \\
&\left. + \left(\cos px - i\varepsilon \frac{t}{|t|} \sin px \right) (e^{-p|t|} - e^{-p\varepsilon(p)|t|}) \right].
\end{aligned}
\tag{19}
$$

From the above expression it is easy to see that the nonlocality of the two body potential is essential to have a finite 2-point function; if we consider a local potential then $\hat{v}(p) = 1$ and the above expression is diverging. Note also that (19) has *the same* ultraviolet divergence at $\mathbf{x} = \mathbf{0}$ than the *free* 2-point function as $Q_2(\mathbf{0}) = 0$. On the contrary the large distance behavior is different with respect to the noninteracting case and given by:

$$\langle \tilde{\psi}^-_{\mathbf{x},\varepsilon} \tilde{\psi}^+_{0,\varepsilon} \rangle \sim_{|\mathbf{x}| \to \infty} \frac{1}{i\varepsilon x + v_F t} \frac{A(\lambda)}{|x^2 + v_F^2 t^2|^{\eta/2}} \tag{20}$$

with $A(\lambda)$ a suitable constant and

$$v_F = \varepsilon(0) = \sqrt{1 - (\lambda \hat{v}(0)/2\pi)^2}$$

$$\eta = 2\sinh^2 \phi(0) = \frac{1}{2}[K + K^{-1} - 2],$$

$$K = e^{2\phi} = \sqrt{\frac{1 + \lambda \hat{v}(0)/2\pi}{1 - \lambda \hat{v}(0)/2\pi}}.$$

(21)

From (21) we see that the interaction not only modifies the Fermi velocity but also changes qualitatively the asymptotic infrared behavior of the 2-point function, producing an anomalous dimension with exponent $1 + \eta$, with $\eta > 0$ and nontrivial function of the coupling. The presence of anomalous exponents is one of the most interesting feature of the Luttinger model, and implies a rather different physical behavior with respect to the noninteracting system; for instance the occupation number is not discontinuous as in the $\lambda = 0$ case, but it becomes continuous and such that $n_{k+p_F} - n_{p_F} \sim |k|^\eta$ for small k.

It is also important to compare the the 2-point function of the Luttinger model with the 2-point function $S_{\text{th}}(\mathbf{x})$ of the Thirring model, as computed in Ref. 4, which is given by

$$S_{\text{th}}(\mathbf{x}) = \frac{1}{2\pi} \frac{1}{i\varepsilon x + t} \frac{1}{|x^2 + t^2|^\eta}.$$

(22)

Despite the similar long distance behavior, there are important differences; the *ultraviolet* short distance behavior is different with respect to the free one, and the light velocity is not changed due to Lorentz invariance.

From (15) we can write the response function; if we defined the total density $\rho_{\mathbf{x}}^{\text{CDW}} = \lim_{\alpha \to 0} \sum_\varepsilon e^{2i\varepsilon p_F x} \tilde{\psi}_{x,t+\alpha\varepsilon}^+ \tilde{\psi}_{\mathbf{x},-\varepsilon}^-$ then, using a point-splitting procedure from (15), for $|\mathbf{x}| \to \infty$, if T denotes truncation

$$\langle \rho_{\mathbf{x}}^{\text{CDW}} \rho_0^{\text{CDW}} \rangle_T \sim (1 + A(\lambda)) \frac{\cos(2p_F x)}{2\pi^2 (v_F^2 t^2 + x^2)^{X_+}}$$

(23)

with X_+ is a critical exponent given by

$$X_+ = K.$$

(24)

Similarly we can introduce the Cooper pair density

$$\rho_{\mathbf{x}}^c = \lim_{\alpha \to 0} \sum_{\varepsilon,\sigma} \psi_{x,t+\alpha,\varepsilon}^\sigma \psi_{x,x_0,-\varepsilon}^\sigma$$

(25)

and one gets

$$\langle \rho_{\mathbf{x}}^c \rho_0^c \rangle_T \sim (1 + A(\lambda)) \frac{1}{2\pi^2 (v_F^2 t^2 + x^2)^{X_-}}$$

(26)

with

$$X_- = K^{-1}.\tag{27}$$

Note that (21), (24) and (27) relates the exponents by simple relations; while the exponents depend from the details of the model (1) (for instance from the charges $q_{i,\varepsilon}$) such relations are true for any choice of the charges for which the solvability holds. For instance if $q_{i,\varepsilon} = 1/2$ then (21), (24) and (27) are still true even if K has a different expression

$$K = \sqrt{1 + \frac{\lambda \hat{v}(0)}{2\pi}}.\tag{28}$$

As we will see in the following section such relations between exponents have been proposed to be true in a wide class of models, even when an exact solution is lacking.

Defining the *density* and *current* operators as

$$\rho_{\mathbf{x}} = \sum_{\varepsilon=\pm} \rho_{\mathbf{x},\varepsilon}, \quad j_{\mathbf{x}} = \sum_{\varepsilon=\pm} \varepsilon \rho_{\mathbf{x},\varepsilon}\tag{29}$$

we get, as an immediate consequence of the commutation rules (1)

$$\frac{\partial \hat{\rho}_{t,p}}{dt} = e^{Ht}[H, \hat{\rho}_p]e^{-Ht} = p v_N(p)\hat{j}_{t,p},$$

$$\frac{\partial \hat{j}_{t,p}}{dt} = e^{Ht}[H, \hat{j}_p]e^{-Ht} = p v_J(p)\hat{\rho}_{t,p}$$

with

$$v_J(p) = \left(1 - \frac{\lambda \hat{v}(p)}{2\pi}\right), \quad v_N(p) = \left(1 + \frac{\lambda \hat{v}(p)}{2\pi}\right)\tag{30}$$

and we have used that, from the commutation rules (1), $[V, \rho_p] = \lambda \hat{v}(p)/2\pi j_p$, $[V, j_p] = -\lambda \hat{v}(p)/2\pi \rho_p$. As a consequence, the following Ward Identities[14] can be derived

$$-i\omega \langle \hat{\rho}_{\mathbf{p}} \hat{\psi}^-_{\mathbf{k},\varepsilon} \hat{\psi}^+_{\mathbf{k+p},\varepsilon} \rangle_T + \varepsilon v_J \langle \hat{j}_{\mathbf{p}} \hat{\psi}^-_{\mathbf{k},\varepsilon} \hat{\psi}^+_{\mathbf{k+p},\varepsilon} \rangle_T$$
$$= \langle \hat{\psi}^-_{\mathbf{k},\varepsilon} \hat{\psi}^+_{\mathbf{k},\varepsilon} \rangle - \langle \hat{\psi}^-_{\mathbf{k+p},\varepsilon} \hat{\psi}^+_{\mathbf{k+p},\varepsilon} \rangle\tag{31}$$

and

$$-i\omega \langle \hat{j}_{\mathbf{p}} \hat{\psi}^-_{\mathbf{k},\varepsilon} \hat{\psi}^+_{\mathbf{k+p},\varepsilon} \rangle_T + \varepsilon v_N p \langle \hat{\rho}_{\mathbf{p}} \hat{\psi}^-_{\mathbf{k},\varepsilon} \hat{\psi}^+_{\mathbf{k+p},\varepsilon} \rangle_T$$
$$= \langle \hat{\psi}^-_{\mathbf{k},\varepsilon} \hat{\psi}^+_{\mathbf{k},\varepsilon} \rangle - \langle \hat{\psi}^-_{\mathbf{k+p},\varepsilon} \hat{\psi}^+_{\mathbf{k+p},\varepsilon} \rangle.\tag{32}$$

The quantities $v_N(p)$ and $v_J(p)$ are velocities associated with the charge and current excitations; they appear also related to the Drude weight D and the susceptibility

κ^7 (see next section) by the relations $D = v_J(0)/\pi$ and $\kappa = \pi/v_N(0)$ so that:

$$\frac{D}{\kappa} = v_F^2 \,. \tag{33}$$

3. Nonsolvable Lattice Models and the Luttinger Liquid Conjecture

The exact solvability of the Luttinger model is due to certain peculiar features, like its linear "relativistic" dispersion relation, which are lost in more realistic models of one-dimensional metals. It is then natural to ask which features of the Luttinger model are still true in more realistic lattice Hamiltonian like the following tight-binding model

$$H = H_0 + V$$

$$H_0 = -\frac{1}{2} \sum_{x=1}^{L-1} [a_x^+ a_{x+1}^- + a_{x+1}^+ a_x^-] + \mu \sum_{x=1}^{L} \left(a_x^+ a_x^- - \frac{1}{2} \right) \tag{34}$$

$$V = \lambda \sum_{1 \le x,y \le L} v(x-y) \left(a_x^+ a_x^- - \frac{1}{2} \right) \left(a_y^+ a_y^- - \frac{1}{2} \right),$$

where $a_x^\pm$, $x = 1, 2, \ldots, L$, are fermionic creation or destruction operators, $\rho_x = a_x^+ a_x^-$, μ is the chemical potential, and $|v(x-y)| \le e^{-\kappa|x-y|}$. Contrary to what happens in the Luttinger model, the Hamiltonian of the chain model (34) cannot be rewritten as a quadratic boson Hamiltonian. An exact solution by Bethe ansatz[18] exists in the special case (XXZ model)

$$v(x-y) = \delta_{|x-y|,1}/2 \,, \quad \mu = 0 \,. \tag{35}$$

The solution provides the excitation spectrum and several thermodynamic properties but not an explicit form for the correlations.

We define $O_{\mathbf{x},\varepsilon} = e^{Ht} O_x e^{-Ht}$, $\mathbf{x} = (t,x)$ and

$$\langle O_{\mathbf{x}_1} \cdots O_{\mathbf{x}_n} \rangle_\beta = \frac{\mathrm{Tr} e^{-\beta H} \mathcal{T} O_{\mathbf{x}_1} \cdots O_{\mathbf{x}_n}}{\mathrm{Tr} e^{-\beta H}} \Big|_T \,, \tag{36}$$

where $\mathcal{T}$ is the time ordering operator and T denotes truncation (the label T is understood in the l.h.s. of (36)); we call $\lim_{\beta \to \infty} \langle O_{\mathbf{x}_1} \cdots O_{\mathbf{x}_n} \rangle_\beta = \langle O_{\mathbf{x}_1} \cdots O_{\mathbf{x}_n} \rangle$. In addition to the density ρ_x it is convenient to introduce the current

$$J_x = \frac{1}{2i} [a_{x+1}^+ a_x^- - a_x^+ a_{x+1}^-] \,. \tag{37}$$

From the continuity equation

$$\frac{\partial \rho_{\mathbf{x}}}{\partial t} = e^{Ht}[H, \rho_x] e^{-Ht} = -i(J_{x,t} - J_{x-1,t}) \,, \tag{38}$$

where we have used that $[V, \rho_x] = 0$, we can derive the following *Ward Identity*

$$-i\omega\langle\hat{\rho}_{\mathbf{p}}\hat{a}_{\mathbf{k}}^-\hat{a}_{\mathbf{k+p}}^+\rangle - i(1 - e^{ip})\langle J_{\mathbf{p}}\hat{a}_{\mathbf{k}}^-\hat{a}_{\mathbf{k+p}}^+\rangle$$

$$= \langle\hat{a}_{\mathbf{k}}^-\hat{a}_{\mathbf{k}}^+\rangle - \langle\hat{a}_{\mathbf{k+p}}^-\hat{a}_{\mathbf{k+p}}^+\rangle \tag{39}$$

The l.h.s. of the above Ward Identity coincides with the one of the free $\lambda = 0$ case, as consequence of $[V, \rho_x] = 0$, while in its analogue for the Luttinger model (31) there is an explicit dependence from the coupling. The fact that the Luttinger model has an extra Ward Identity (32) is related to chiral gauge invariance which is absent in the model (34).

Other Ward Identities for the model (34) are:

$$-i\omega\langle\hat{\rho}_{\mathbf{p}}\hat{\rho}_{-\mathbf{p}}\rangle - i(1 - e^{-ip})\langle\hat{\rho}_{\mathbf{p}}\hat{J}_{-\mathbf{p}}\rangle = 0$$

$$-i\omega\langle\hat{\rho}_{\mathbf{p}}\hat{J}_{-\mathbf{p}}\rangle - i(1 - e^{-ip})\langle\hat{J}_{\mathbf{p}}\hat{J}_{-\mathbf{p}}\rangle = i(1 - e^{-ip})\Delta, \tag{40}$$

where $\Delta = \langle\Delta_x\rangle$ and $\Delta_x = -(1/2)[a_x^+ a_{x+1}^- + a_{x+1}^+ a_x^-]$. Note the last term in the second of (40), coming from the fact that $[\rho_x, J_y] = -i\delta_{x,y}\Delta_x + i\delta_{x-1,y}\Delta_y$; it is called *Schwinger term* and is related to the diamagnetic part of the current Δ_x, whose mean value $\langle\Delta_x\rangle$ is indeed independent of x.

The susceptibility is defined as, if $\mathbf{p} = (\omega, p)$

$$\kappa = \lim_{p\to 0}\lim_{\omega\to 0}\langle\hat{\rho}_{\mathbf{p}}\hat{\rho}_{-\mathbf{p}}\rangle \tag{41}$$

while the conductivity is:

$$\sigma(\omega) = \frac{D(i\omega, p)}{i\omega + 0^+}, \quad D(\mathbf{p}) = -\Delta - \langle\hat{J}_{\mathbf{p}}\hat{J}_{-\mathbf{p}}\rangle \tag{42}$$

while the *Drude weight* is $\lim_{\omega\to 0}D(\mathbf{p}) = D$; a finite Drude weight implies an infinite dc conductivity.

The WI (40) and the fact that $\langle\rho_{\mathbf{p}}J_{-\mathbf{p}}\rangle = \langle J_{\mathbf{p}}\rho_{-\mathbf{p}}\rangle$ implies that

$$\langle\rho_{\mathbf{p}}\rho_{-\mathbf{p}}\rangle|_{\omega,0} = 0, \quad D(0, p) = 0. \tag{43}$$

The fact that κ and D are not vanishing is related to the fact that $\langle\rho_{\mathbf{p}}\rho_{-\mathbf{p}}\rangle$ and $D(\mathbf{p})$ are not continuous, a fact which can be easily checked in the noninteracting case. Indeed if $\lambda = 0$ the 2-point function $\langle a_{\mathbf{x}}^- a_{\mathbf{y}}^+\rangle_0 \equiv g(\mathbf{x} - \mathbf{y})$ is given by:

$$g(\mathbf{x}, \mathbf{y}) = \int d\mathbf{k}\frac{e^{i\mathbf{k}\mathbf{x}}}{-ik_0 + \cos k - \cos p_F} = \int d\mathbf{k}e^{i\mathbf{k}\mathbf{x}}\hat{g}(\mathbf{k}) \tag{44}$$

with $\cos p_F = \mu$; $\hat{g}(\mathbf{k})$ is singular at $k_0 = 0$ and $k = \pm p_F$ and close to each one of the two Fermi points $\pm p_F$ the 2-point function is identical, up to small correction, to the Luttinger model noninteracting two point function

$$\hat{g}(\mathbf{k}' \pm \mathbf{p}_F) = \frac{1}{-ik_0 \pm \sin p_F k'}(1 + O(\mathbf{k}')) \qquad (45)$$

for small $\mathbf{k}'$, where $\mathbf{p}_F = (0, p_F)$. The density–density correlation in the noninteracting case $\lambda = 0$ is:

$$\langle \rho_{\mathbf{p}} \rho_{-\mathbf{p}} \rangle_0 = \int \frac{dk}{2\pi} \frac{1}{-ip_0 + \cos k - \cos(k + p)} [n_F(k) - n_F(k + p)], \qquad (46)$$

where $n_F(k)$ is the Fermi distribution; therefore for small p

$$\langle \rho_{\mathbf{p}} \rho_{-\mathbf{p}} \rangle_0 = \frac{1}{\pi} \frac{\sin p_F p^2}{\omega^2 + \sin^2 p_F p^2} \left(1 + O\left(\frac{p^3}{|\mathbf{p}|^2}\right)\right). \qquad (47)$$

In the presence of interaction it is reasonable to expect, following Ref. 1, the presence of critical exponents but there is no reason to expect that they coincide with the Luttinger model ones. On the other hand, the exponents in the Luttinger model verify (21), (24), (27) and the content of the *Luttinger liquid conjecture*[6-8] is that the same relations are true in a wide class of models, including models with nonlinear bands like the model (34). In particular according to such conjecture there exists a (model dependent) function K such that:

$$\eta = \frac{1}{2}[K + K^{-1} - 2], \quad X_+ = K, \quad X_- = K^{-1} \qquad (48)$$

and K can be obtained from thermodynamic functions by the relations

$$\kappa = \frac{K}{\pi v_F}, \quad D = \frac{v_F K}{\pi}, \qquad (49)$$

where v_F is the interacting Fermi velocity.

As the validity of (48), (49) in the Luttinger model relies on its exact mapping in a system of free bosons (a property which is true only for its linear dispersion relation), their validity in the model (1), with a sinusoidal dispersion relation, is far from to be obvious. A very important test came from the exact solution of the XXZ; indeed by the Bethe ansatz solution it is found,[7,18,47] if $\cos \bar{\mu} = \lambda$,

$$v_F = \frac{\pi}{\bar{\mu}} \sin \bar{\mu}, \quad \kappa = \frac{1}{2\pi(\pi/\bar{\mu} - 1) \sin \bar{\mu}} \qquad (50)$$

and

$$D = \frac{\pi}{\bar{\mu}} \frac{\sin \bar{\mu}}{2(1 - \bar{\mu}/\pi)}. \qquad (51)$$

Even if such expressions are different (as functions of the interaction) from the Luttinger ones, the relation (33) follows. Moreover in Ref. 17 the correlation length exponent ν in the XYZ model has been computed finding $\nu = \pi/2\bar{\mu}$; therefore using the scaling relation $\nu = (2 - X_+^{-1})$ proposed in Ref. 8 we can compute the other exponents, for instance

$$X_+ = K = \left(2 - \frac{2\bar{\mu}}{\pi} \right)^{-1} = 1 + \frac{2\lambda}{\pi} + O(\lambda^2). \tag{52}$$

The same values is found by the value of v_F and D using (49), and this provides a check of the Luttinger liquid conjecture, as stressed in Ref. 7.

In the model (34) with a generic short range interaction there is no exact solution. To give support to the validity of the Luttinger liquid relations (48), (49) in a wide class of models, including nonsolvable ones, in Ref. 6 a perturbation of the Luttinger model was considered in which the linear dispersion relation is replaced by:

$$\varepsilon_\pm(k) = \pm k + \frac{1}{2m}k^2 + \frac{\lambda}{12m^2}k^3 \tag{53}$$

with $\lambda \geq (3/4)$. While such model reduces to the Luttinger model as $m \to \infty$, for finite m *nonquadratic* terms appear in the boson Hamiltonian; however remarkably the Luttinger liquid relations (48), (49) still hold. This provides evidence in support of the conjecture, but on the other hand the Luttinger liquid relations could be checked up to $O(m^{-2})$ and the model with dispersion relation (53) is still only an approximate effective description for the model (34).

Instead of using boson variables, one could try to prove the Luttinger liquid relations (48), (49) using an expansion in Feynman graphs, see e.g., Ref. 20; its use is however problematic for 1D interacting fermions as the expansion is plagued by *infrared* divergences. Dzyaloshinski and Larkin[21] considered 1D fermions with bandwidth cut-off; contrary to the Luttinger model this model, called Tomonaga model, is not solvable. They performed a resummations of the perturbative expansions finding the anomalous decay of correlations, but at the price of several (uncontrolled) approximations. The use of (multiplicative) RG methods for the Tomonaga model dates back to Solyom[22]; irrelevant terms are simply discarded and the presence of anomalous exponents results from a property, called vanishing of beta function, which was proved only at lowest orders in perturbation theory. Metzner and Di Castro[14] pointed out that the vanishing of the Beta function in the multiplicative RG follows from the Ward Identities (31), (32), which are however true only in the Luttinger and not in the Tomonaga model.

With respect to multiplicative RG, exact Wilsonian[23] RG method, used by Polchinski[24] and Gallavotti[25] for the renormalization of ϕ^4 models and therefore applied by to nonrelativistic interacting fermions in $d = 1, 3$ by Benfatto and Gallavotti,[26] has two main advantages. They are : (i) *exact*, in the sense that it does not need any relativistic approximation or continuum limit and it allow us to keep the full lattice structure of the problem; (ii) *nonperturbative*, in the sense that it involves expansions whose convergence can be mathematically proved. The fact that such methods are exact and can take into account the irrelevant terms is a crucial feature: the Luttinger or the lattice model (34) differ for irrelevant terms and the proof of the Luttinger liquid relations (48), (49) consists essentially in showing that such irrelevant terms, while contributing to the exponents or other physical quantities, leave unchanged the Luttinger liquid relations.

Such exact RG methods have been used in *Constructive QFT* to prove the existence of the continuum limit of Quantum Theory model at $d = 1 + 1$ dimensions, like the massless Gross–Neveu model with $N > 1$ colors[27,28] or the massive Yukawa model.[29] The main extra difficulty in analyzing a model like (1), with respect to QFT model like the Gross–Neveu model analyzed in Ref. 27, is that the theory is *not* asymptotically free; it belongs to a class of models with *vanishing* beta function. Such models can be constructed only exploiting nontrivial cancellations at *all orders* in the renormalized expansion; this is a crucial difference with respect to the asymptotically free models in which a second order computation is enough for establishing the nature of the flow of the effective coupling. The first example of rigorous construction of a model with vanishing beta function was in Ref. 30 and it regards the Jellium model in 1D, describing interacting nonrelativistic fermions in the continuum. The crucial property of vanishing of Beta function was proved using an indirect argument based on comparison with the *exact solution* of the Luttinger model.[12] The construction of models with vanishing Beta function *without* any use of exact solutions was finally realized in Refs. 31–33 solving the well known technical problem related to the basic conflict between Wilsonian RG and Ward Identities; the decomposition in momentum space necessary in Wilson scheme breaks the local gauge invariance necessary to establish Ward Identities.

In the following sections we will resume the main achievements of the application of such exact RG methods to systems like (34) and the main steps of the proof of the Luttinger liquid relations (48), (49) for nonsolvable models, finally achieved in Refs. 34–38.

4. Exact Renormalization Group Analysis

The starting point of the exact RG analysis is the expression of the correlations of the lattice model (34) in terms of *Grassmann integrals*. We introduce a set of *Grassmann variables* $\psi_{\mathbf{x}}^{\pm}$ such that $\{\psi^{\pm}, \psi^{\pm}\} = \{\psi^{\pm}, \psi^{\mp}\} = 0$ and

$$\int d\psi_{\mathbf{x}}^{\pm} = 0, \quad \int d\psi_{\mathbf{x}}^{\pm} \psi_{\mathbf{x}}^{\pm} = 0. \tag{54}$$

The integral of any analytic functions can be obtained by (54) by linearity. The correlations of a model like (34) can be written by suitable derivatives of *generating functional* expressed by the following Grassmann integral

$$e^{\mathcal{W}(A,\phi)} = \int P(d\psi) e^{-\mathcal{V}(\psi) + B(A,\psi) + \int d\mathbf{x}[\phi_{\mathbf{x}}^+ \psi_{\mathbf{x}}^- + \psi_{\mathbf{x}}^- \psi_{\mathbf{x}}^+]}, \tag{55}$$

where $\phi_{\mathbf{x}}^{\pm}$ are Grassmann variables, $\int d\mathbf{x}$ is a shortcut for $\sum_x \int_{-\beta/2}^{\beta/2} dx_0$, $P(d\psi)$ is a Grassmann Gaussian measure in the field variables $\psi_{\mathbf{x}}^{\pm}$ with covariance (the free propagator) given by:

$$g(\mathbf{x} - \mathbf{y}) = \frac{1}{\beta L} \sum_{\mathbf{k}} \frac{e^{i\mathbf{k}(\mathbf{x}-\mathbf{y})}}{-ik_0 + (\cos p_F - \cos k)}, \qquad (56)$$

with

$$\cos p_F = -\lambda - \mu - \nu \qquad (57)$$

and the interaction is

$$\mathcal{V}(\psi) = \lambda \int d\mathbf{x} d\mathbf{y}\, v(\mathbf{x} - \mathbf{y}) \psi_{\mathbf{x}}^+ \psi_{\mathbf{y}}^+ \psi_{\mathbf{y}}^- \psi_{\mathbf{x}}^- + \nu \int d\mathbf{x} \psi_{\mathbf{x}}^+ \psi_{\mathbf{x}}^- \qquad (58)$$

with $\mathbf{e_1} = (0,1)$, $v(\mathbf{x} - \mathbf{y}) = \delta(t - s)v(x - y)$ while the source is:

$$B(A, \psi) = \int d\mathbf{x} \left\{ \psi_{\mathbf{x}}^+ \psi_{\mathbf{x}}^- A_0(\mathbf{x}) + \frac{1}{2i}[\psi_{\mathbf{x}+\mathbf{e_1}}^+ \psi_{\mathbf{x}}^- - \psi_{\mathbf{x}}^+ \psi_{\mathbf{x}+\mathbf{e_1}}^-]A_1(\mathbf{x}) \right\}. \qquad (59)$$

Contrary to what happens in the Luttinger model, the value of the Fermi momentum is modified by the interaction; it is convenient to analyze (55) considering p_F as a parameter, choosing ν as a function of λ and p_F so that the singularity of the Fourier transform of the interacting two-point function is fixed at $\mathbf{k} = (\pm p_F, 0)$. Finally from (57) we get the Fermi momentum p_F as function of λ and μ.

We briefly describe the exact RG analysis of (55), as developed in Ref. 31; for a tutorial introduction to the properties of Grassman integrals, we refer to Ref. 39. The starting point is to use the *addition property* for a Gaussian Grassmann measure saying that:

$$\int P(d\psi)F(\psi) = \int P(d\psi^{(1)}) \int P(d\psi^{(2)})F(\psi^{(1)} + \psi^{(2)}), \qquad (60)$$

where $P(d\psi)$, $P(d\psi^{(1)})$, $P(d\psi^{(2)})$ are Grassmann Gaussian measures respectively with propagator g, $g^{(1)}$, $g^{(2)}$ and $g = g^{(1)} + g^{(2)}$. We can introduce the following smooth decomposition of the unity

$$1 = \sum_{\varepsilon = \pm} \chi(k - \varepsilon p_F, k_0) + f_1(\mathbf{k}), \qquad (61)$$

where $\chi(k', k_0)$ is a smooth compact-support function such that $\chi(k', k_0) = 1$ for $|\mathbf{k}'|_{T^1} \leq t_0$ and $\chi(k', k_0) = 0$ for $|\mathbf{k}'|_{T^1} \geq \gamma t_0$, if $t_0, \gamma > 1$ are suitable constants; in other words, $\chi(k - \varepsilon p_F, k_0)$ is a compact support function nonvanishing around the two Fermi points $(0, \pm p_F)$. We can decompose the propagator as:

$$\hat{g}(\mathbf{k}) = \sum_{\varepsilon=\pm} \chi(k - \varepsilon p_F, k_0) \hat{g}(\mathbf{k}) + f^{(1)}(\mathbf{k}) \hat{g}(\mathbf{k}) \equiv \sum_{\varepsilon=\pm} \hat{g}_\varepsilon^{(\leq 0)}(\mathbf{k}) + \hat{g}^{(1)}(\mathbf{k}) \qquad (62)$$

leading, according to (60), to a decomposition of the Grassmann field of the form:

$$\psi_{\mathbf{x}} = \sum_{\varepsilon=\pm} e^{i\varepsilon p_F x} \psi_{\mathbf{x},\varepsilon}^{(\leq 0)} + \psi_{\mathbf{x}}^{(1)}, \qquad (63)$$

where $\psi_{\mathbf{x},\varepsilon}^{(\leq 0)}$, $\psi_{\mathbf{x}}^{(1)}$ are independent Grassmann fields with propagator $e^{i\varepsilon p_F(x-y)}$. $\hat{g}_\varepsilon^{(\leq 0)}(\mathbf{x} - \mathbf{y})$ and $\hat{g}^{(1)}(\mathbf{x} - \mathbf{y})$ respectively. Note that, while the propagator of the fields $\psi_{\mathbf{x},\varepsilon}^{(\leq 0)}$ has a slow $O(|\mathbf{x} - \mathbf{y}|^{-1})$ decay for large distances, the propagator of $\psi_{\mathbf{x}}^{(1)}$ decays faster than any power; that is for any N

$$|g^{(1)}(\mathbf{x} - \mathbf{y})| \leq \frac{C_N}{1 + |\mathbf{x} - \mathbf{y}|^N} . \qquad (64)$$

Using (60) we can rewrite (55) in the following way (in the $\phi = 0$ case for simplicity)

$$e^{\mathcal{W}(A,0)} = \int P(d\psi^{\leq 0}) \int P(d\psi^{(1)}) e^{-\mathcal{V}(\psi^{(0)} + \psi^{(1)}) + B(A, \psi^{(\leq 0)} + \psi^{(1)})} , \qquad (65)$$

where $P(d\psi^{\leq 0})$ and $P(d\psi^{(1)})$ have respectively propagator $\sum_{\varepsilon=\pm} e^{i\varepsilon p_F(x-y)} g_\varepsilon^{(\leq 0)}$ $(\mathbf{x} - \mathbf{y})$ and $g^{(1)}(\mathbf{x} - \mathbf{y})$. The integration of the "high energy" degrees of freedom can be done using the *invariance of exponential* property

$$\int P(d\psi) e^{X(\psi+\phi)} = \exp\left[\sum_{n=0}^{\infty} \frac{1}{n!} \mathcal{E}^T(X(\cdot + \phi; n))\right] = e^{X'(\phi)} , \qquad (66)$$

where $\mathcal{E}^T(X(\cdot + \phi; n))$ are the fermionic *truncated expectation*, which are given by the sum over all the possible *connected* Feynman graphs (see Fig. 1). By using (66) we get:

$$e^{\mathcal{W}(A,0)} = \int P(d\psi^{\leq 0}) e^{-\mathcal{V}^{(0)}(\psi^{(0)}) + B^{(0)}(A, \psi^{(\leq 0)})} , \qquad (67)$$

where $\mathcal{V}^{(0)}(\psi^{(\leq 0)})$ and $B^{(0)}(A, \psi^{(\leq 0)})$ is sum of monomials of any degree multiplied by suitable kernels; in particular $\mathcal{V}^{(0)}(\psi^{(\leq 0)})$ is given by:

$$\mathcal{V}^{(0)}(\psi^{(0)}) = \sum_{n=2}^{\infty} W_n^{(0)}(\mathbf{x}_1, \ldots, \mathbf{x}_n) \prod_{i=1}^{n} \psi_{\mathbf{x}_i,\varepsilon_i}^{\sigma_i} \qquad (68)$$

and a similar expression holds for $B^{(0)}(A, \psi^{(\leq 0)})$ with kernels $W_{n,m}^{(0)}$ with n ψ and m A fields.

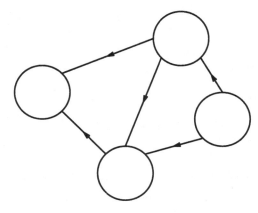

Fig. 1.   An example of connected Feynman graph contributing to $\mathcal{E}^T$.

The kernels $W_{n,m}^{(0)}$ are *analytic functions* of λ and ν in a complex disk, and we describe briefly how such a *nonperturbative* property is obtained. According to (66), the kernels $W_n^{(0)}$ can be written as power series in λ, ν, and each order k can be written as sum of connected Feynman graphs with λ and ν vertices and propagators $g^{(1)}(\mathbf{x})$; by (64), it is easy to check that each Feynman graph at order k is bounded by $k!^{-1}[\max(\lambda, \nu)]^k C^k$, for a suitable constant C. This is however not enough for proving the convergence of the expansion for $W_n^{(0)}$, as the number of graphs contributing to order k are $O(k!^2)$ so that the k-order contribution is bounded by $k![\max(\lambda, \nu)]^k C^k$. The presence of such $k!$ apparently prevent the convergence of the series. While in the case of bosonic models such factorial is really there, in the case of fermionic systems cancellations between Feynman diagrams (ultimately related to the fermionic anticommutativity properties) have the effect that the real final bound is much better and convergence of the series for $W_n^{(0)}$ can be finally achieved.[40] Indeed the fermionic expectation can be written as a *determinant*

$$\int P(d\psi)\psi_{\mathbf{x}_1}^{-(1)} \cdots \psi_{\mathbf{x}_n}^{-(1)} \psi_{\mathbf{y}_1}^{+(1)} \cdots \psi_{\mathbf{y}_n}^{+(1)} = \mathrm{Det}\, G \tag{69}$$

where $G_{i,j} = g^{(1)}(\mathbf{x}_i - \mathbf{y}_i)$; writing $g^{(1)}(\mathbf{x}_i - \mathbf{y}_i)$ in the form

$$g^{(1)}(\mathbf{x} - \mathbf{y}) = \int d\mathbf{z} A^{(1)}(\mathbf{x} - \mathbf{z})\bar{B}^{(1)}(\mathbf{y} - \mathbf{z}) = (A, B) \tag{70}$$

then, if (A, A) and $(B, B) \leq C$ by the *Gram inequality*

$$|\mathrm{Det}\, G| \leq C^n. \tag{71}$$

Note that, even if $\mathrm{Det}\, G$ is expressed by $n!$ terms, in the final bound no $n!$ appear. This fact is however still not enough; in order to integrate over the coordinates one needs to exploit the decay properties of the propagator (64), which are lost in the

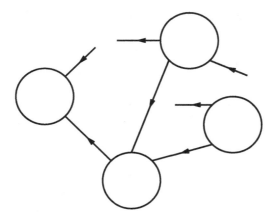

Fig. 2. One of the term in the Battle–Brydges–Federbush formula.

bound (71). One needs to extract from the truncated expectation a minimal set of propagators ensuring the connection between vertices (which allow the integration over coordinates) leaving all the other fields grouped in determinants, so that they can be bounded by the Gram bound. This is made possible by the Battle–Brydges–Federbush formula[41] (see, for instance, again Ref. 39 for a derivation); the truncated expectation is written not as sum of Feynman diagrams, but sum of *trees* of propagators times determinants (represented in the Fig. 2 as uncoupled lines). The trees of propagators are used to perform the integration over the coordinates and the determinants are bounded by Gram bounds, so that the convergence of the expansion for the kernels $W_n^{(0)}$ for λ, ν not too large follows.

The integration of the high energy fields $\psi^{(1)}$ says that the generating function on the model (34), namely (55), can be *exactly* rewritten as (67), that is in terms of two kinds of fermions with a band-width cut-off around the two Fermi points. *The Tomonaga approximation corresponds to neglecting in $\mathcal{V}^{(0)}$ the irrelevant terms with $n \geq 4$ in $\mathcal{V}^{(0)}$ and in replacing $\cos(k' + \pm p_F) - \cos p_F$ with $\pm(\sin p_F)k'$*; the exact RG analysis below allows to take such terms into account. It is natural to continue the analysis of (67) again integrating the momenta closer and closer to the Fermi points. We can write:

$$\chi(k - \varepsilon p_F, k_0) = \sum_{j=-\infty}^{0} f_j(\mathbf{k}') \tag{72}$$

where $\mathbf{k}' = \mathbf{k} - \varepsilon \mathbf{p}_F$, $f_j(\mathbf{k}')$ (see Fig. 3) is a compact support function with support in $t_0\gamma^{j-1} \leq |\mathbf{k}'|_{T^1} \leq t_0\gamma^{j+1}$, $\mathbf{p}_F = (0, p_F)$ and $t_0, \gamma > 1$ are suitable constants. We can write the propagator $\hat{g}_\varepsilon^{(\leq 0)}$ as:

$$\hat{g}_\varepsilon^{(\leq 0)}(\mathbf{k}') = \sum_{j=-\infty}^{0} \hat{g}_\varepsilon^{(j)}(\mathbf{k}'), \tag{73}$$

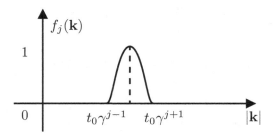

Fig. 3. The function $f_j(\mathbf{k})$.

where $\hat{g}_\varepsilon^{(\le 0)}(\mathbf{k}')$ is a propagator living at momentum scales $O(\gamma^j)$ close one of the two Fermi points. The above decomposition of the propagator induced an analogous decomposition of the fermionic fields as sum of fields $\psi^{(\le 0)} = \sum_{h=-\infty}^0 \psi^{(h)}$.

After the integration of the fields $\psi_\varepsilon^{(0)}$, $\psi_\varepsilon^{(-1)}, \ldots, \psi_\varepsilon^{(h+1)}$ by the reiterate application of (66) we get

$$e^{\mathcal{W}_{L,\beta}(A,0)} = \int P_{Z_{h-1},\delta_{h-1}}(d\psi^{(\le h)}) e^{\mathcal{V}^{(h)}(\sqrt{Z_{h-1}}\psi^{(\le h)}) + \mathcal{B}^{(h)}(\sqrt{Z_{h-1}}\psi^{(\le h)}, A)}, \qquad (74)$$

where $P_{Z_h,\delta_h}(d\psi^{(\le h)})$ is the *Gaussian Grassmann integration* with propagator

$$\hat{g}_\varepsilon^{(\le h)}(\mathbf{k}') = \frac{\chi_h(\mathbf{k}')}{Z_h} \frac{1}{-ik_0 + (1+\delta_h)(\cos p_F - \cos(k' + \varepsilon p_F))} \qquad (75)$$

with $\chi_h(\mathbf{k}')$ a smooth compact support function nonvanishing for $|\mathbf{k}'| \le t_0\gamma^h$, Z_h is the wavefunction renormalization and $v_h = \sin p_F(1+\delta_h)$ is the effective Fermi velocity; finally the *effective interaction* $\mathcal{V}^{(h)}(\psi)$ is a sum over monomials in the Grassmann variables

$$\mathcal{V}^{(h)}(\psi) = \gamma^h v_h F_\nu^{(h)} + \lambda_h F_\lambda^{(h)} + R_h, \qquad (76)$$

where

$$\begin{aligned} F_\nu^{(h)} &= \sum_{\varepsilon = \pm} \int d\mathbf{x} \psi_{\mathbf{x},\varepsilon}^{(\le h)+} \psi_{\mathbf{x},\varepsilon}^{(\le h)-}, \\ F_\lambda^{(\le h)} &= \int d\mathbf{x} \psi_{\mathbf{x},+}^{(\le h)+} \psi_{\mathbf{x},-}^{(\le h)+} \psi_{\mathbf{x},+}^{(\le h)-} \psi_{\mathbf{x},-}^{(\le h)-} \end{aligned} \qquad (77)$$

and $R^{(h)}$ is sum over monomials with n ψ fields with *negative* scaling dimension D, defined by:

$$D = 2 - \frac{n}{2}. \qquad (78)$$

In the same way (for $A^{(1)} = 0$ for definiteness)

$$
\begin{aligned}
&\mathcal{B}^{(h)}(\sqrt{Z_{h-1}}\psi^{(\leq h)}, A) \\
&= \sum_{\varepsilon} \left[\int d\mathbf{x} Z^{(1)}_{h-1} A^{(0)}_{\mathbf{x}} \psi^{+(\leq h)}_{\mathbf{x},\varepsilon} \psi^{-(\leq h)}_{\mathbf{x},\varepsilon} + Z^{(+)}_{h-1} A^{(0)}_{\mathbf{x}} e^{2i\varepsilon p_F x} \psi^{+(\leq h)}_{\mathbf{x},\varepsilon} \psi^{-(\leq h)}_{\mathbf{x},-\varepsilon} \right] \\
&\quad + \bar{R}_h \,,
\end{aligned}
\tag{79}
$$

and $\bar{R}_h$ is sum over monomials with n ψ fields and m A fields with negative scaling dimension $D = 2 - (n/2) - m$. A similar expression holds setting $A^{(0)} = 0$, and the corresponding renormalization constants will be called $Z^{(2)}_{h-1}, \hat{Z}^{(+)}_{h-1}$.

It is natural the interpretation of λ_h in (77) as the *effective coupling* of the model and of ν_h is the renormalization of the chemical potential. In the RG language, the terms with positive or vanishing dimension are called *relevant* or *marginal* terms, respectively; the terms in R_h or $\bar{R}_h$, having *negative* scaling dimension, are *irrelevant*. The kernels $W^{(h)}_{n,m}$ are expressed by expansions in the effective couplings λ_j, ν_j, $j = 1, 0, -1, \ldots, h$ which are convergent, by the Gram bound, *provided that* λ_j, ν_j are not too large; from them an expansion for the correlations and the other physical observables in terms of λ_j, ν_j is obtained. Such renormalized expansions in the effective couplings λ_j, ν_j *should not* be confused with the naive perturbative expansion of the correlations in λ, ν; while the latter is plagued by infrared divergences the renormalized one is finite and even convergent provided that λ_j, ν_j are not too large uniformly in j. One has then to analyze the flow of the effective parameters.

Note first that ν_j is a relevant coupling and it is possible to choose the parameter ν, by a fixed point argument, so that $\nu_j = O(\gamma^j)$; one can also prove, due to symmetry cancellations in the flow equation, that $\delta_j \to_{j \to -\infty} \delta_{-\infty} = O(\lambda)$. In order to study the flow of λ_j one first notices that the single scale propagator can be decomposed, for $h \leq 0$, as

$$
g^{(h)}_{\rho}(\mathbf{x}, \mathbf{y}) = g^{(h)}_{T,\varepsilon}(\mathbf{x}, \mathbf{y}) + r^{(h)}_{\varepsilon}(\mathbf{x}, \mathbf{y}) \,,
\tag{80}
$$

where

$$
g^{(h)}_{T,\varepsilon}(\mathbf{x}, \mathbf{y}) = \frac{1}{L^2} \sum_{\mathbf{k}} e^{-i\mathbf{k}(\mathbf{x}-\mathbf{y})} \frac{1}{Z_h} \frac{1}{-ik_0 + \varepsilon v_F k} \,,
\tag{81}
$$

with $v_F = \sin p_F (1 + \delta_{-\infty})$ and for any N

$$
|r^{(h)}_{\varepsilon}(\mathbf{x} - \mathbf{y})| \leq C_N \frac{\gamma^{2h}}{1 + (\gamma^h |\mathbf{x} - \mathbf{y}|^N)} \,.
\tag{82}
$$

Note that $g^{(h)}_{T,\varepsilon}(\mathbf{x}, \mathbf{y})$ verifies bound similar to (82) with γ^{2h} in the numerator replaced by γ^h; this means that we can rewrite the propagator as a Dirac propagator at scale h plus a correction $r^{(h)}$ whose size is $O(\gamma^{2h})$ instead of $O(\gamma^h)$, that is much smaller.

The above decomposition induces a similar decomposition in the flow equation; we can write

$$\lambda_{j-1} = \lambda_j + \beta_\lambda^{(j)}(\lambda_j; \ldots; \lambda) + O(\lambda^2 \gamma^j) \,, \tag{83}$$

where in the last term we have included all the *irrelevant* contributions, containing at least an $r_\varepsilon^{(k)}$ or a ν_k or a $\mathcal{R}\mathcal{V}^{(0)}$, and it is asymptotically vanishing for dimensional reasons. The flow of λ_j is then driven by $\beta_\lambda^{(j)}$ which is expressed by a convergent series with each term sum of $O(1)$ renormalized Feynman graphs. A lowest order computation shows that such graphs cancel out; this is true to all orders (but of course it cannot be checked by an explicit computation) as a consequence of the following nonperturbative property, called *vanishing of the beta function*

$$|\beta_\lambda^{(j)}(\lambda_j, \ldots, \lambda_j)| \leq C\lambda_j^2 \gamma^j \,. \tag{84}$$

We will briefly resume the main steps of the proof of (84), which is the key of all the RG analysis, in the following section. Assuming (84), it follows that the sequence λ_j converges, as $j \to -\infty$, to an analytic function

$$\lambda_{-\infty}(\lambda) = \lambda[\hat{v}(0) - \hat{v}(2p_F)] + O(\lambda^2) \tag{85}$$

such that $|\lambda_j - \lambda_{-\infty}| \leq C\lambda^2 \gamma^j$. The effective wavefunction renormalization Z_j verifies the following equation

$$\frac{Z_{j-1}}{Z_j} = 1 + \beta_z^{(j)}(\lambda_j, \ldots, \lambda_0) + O(\lambda \gamma^j) \,, \tag{86}$$

where again the last term in the r.h.s. includes the irrelevant terms and

$$Z_j \sim A_j(\lambda)\gamma^{\eta j} \,, \quad \eta = \log_\gamma(1 + \beta_z^{(-\infty)}(\lambda_{-\infty}, \ldots, \lambda_{-\infty}) \tag{87}$$

and $A_j \to_{j \to -\infty} A$ as $O(\lambda \gamma^j)$. Note that the exponent η is written as power series in the variable $\lambda_{-\infty}/v_F$ and it depends only from $\beta_z^{(j)}$ and not from the irrelevant term; on the contrary $A(\lambda)$ depends also on the irrelevant part of the flow equation. Similarly, $Z_j^{(i)}$, $i = 1, 2$ appearing in (79)

$$\frac{Z_{j-1}^{(i)}}{Z_j} = 1 + \beta_1^{(j)}(\lambda_j, \ldots, \lambda_0) + O(\lambda \gamma^j) \tag{88}$$

and $\beta_1^{(j)}$ is asymptotically vanishing

$$|\beta_\pm^{(j)}(\lambda_j, \ldots, \lambda_j)| \leq C\lambda^2 \gamma^j \tag{89}$$

with

$$Z_j^{(1)} \sim A_{1,j} 2^{\eta j} \,, \quad Z_j^{(2)} \sim A_{2,j} 2^{\eta j} \,. \tag{90}$$

Note that, while the exponents of $Z_j^{(1)}$ and $Z_j^{(2)}$ are the same (and equal to η), the amplitudes A_1 and A_2 are *different*. Finally $Z_j^{(2)} \sim \gamma^{\eta_2 j}$ with again η_2 is written as power series in the variable $\lambda_{-\infty}/v_F$.

By taking derivatives with respect to the external fields $\phi^{\pm}$ we get that the 2-point function can be written as:

$$\langle a_{\mathbf{x}}^- a_{\mathbf{0}}^+ \rangle = \left[\sum_{h=-\infty}^{0} \sum_{\varepsilon=\pm} e^{i\varepsilon p_F(x-y)} g_\varepsilon^{(h)}(\mathbf{x}-\mathbf{y}) + g^{(1)}(\mathbf{x}-\mathbf{y}) \right] (1 + O(\lambda)). \tag{91}$$

The above expression says that the 2-point function is sum of single scale propagators, each one with its wavefunction renormalization Z_h and an effective fermi velocity $v_h = \sin p_F (1 + \delta_h)$. From the above expression one can derive the large distance decay

$$\langle a_{\mathbf{x}}^- a_{\mathbf{0}}^+ \rangle_T \sim g_0(\mathbf{x}) \frac{1 + \lambda f(\lambda)}{(x_0^2 + v_F^2 x^2)^{(\eta/2)}}, \tag{92}$$

where $f(\lambda)$ is a bounded function, $\eta = b\lambda^2 + O(\lambda^3)$, with $b > 0$, and

$$g_0(\mathbf{x}) = \sum_{\varepsilon=\pm} \frac{e^{i\varepsilon p_F x}}{-ix_0 + \varepsilon v_F x}, \tag{93}$$

$$v_F = \sin p_F + O(\lambda), \quad p_F = \cos^{-1}(\mu + \lambda) + O(\lambda). \tag{94}$$

In momentum space, the meaning of (91) is even more transparent; it simply says, due to the compact support properties of the propagator, that for momenta close to the Ferm points $\mathbf{k}' \sim t_0 \gamma^j$, $\mathbf{k} = \mathbf{k}' + \varepsilon \mathbf{p}_F$, $j \leq 0$ the 2-point function is essentially given by $|\mathbf{k}'|^\eta/(-ik_0 + \varepsilon v_F k')$, while for momenta far from the Fermi point the 2-point function is essentially given by $\hat{g}^{(1)}(\mathbf{k})$, that is it has the same behavior of the noninteracting one. A similar expansion holds for the density–density correlation which is given by:

$$\langle \rho_{\mathbf{x}} \rho_0 \rangle_T \sim \cos(2p_F x) \Omega^{3,a}(\mathbf{x}) + \Omega^{3,b}(\mathbf{x}), \tag{95}$$

$$\Omega^{3,a}(\mathbf{x}) = \frac{1 + A_1(\mathbf{x})}{2\pi^2 [x^2 + (v_F x_0)^2]^{X_+}}, \tag{96}$$

$$\Omega^{3,b}(\mathbf{x}) = \frac{1}{2\pi^2 [x^2 + (v_F x_0)^2]} \left\{ \frac{x_0^2 - (x/v_F)^2}{x^2 + (v_F x_0)^2} + A_2(\mathbf{x}) \right\}, \tag{97}$$

with $|A_1(\mathbf{x})|, |A_2(\mathbf{x})| \leq C|\lambda|$ and $X_+ = 1 - a_1\lambda + O(\lambda^2)$ with

$$X_+ = 1 - \eta_+ + \eta = 1 - a_1\lambda + O(\lambda^2) \tag{98}$$

and

$$a_1 = [\hat{v}(0) - \hat{v}(2p_F)]/(\pi \sin p_F) \tag{99}$$

Note that while the exponent of the oscillating part is anomalous, the exponent of the nonoscillating part is equal to the noninteracting one, as a consequence of the vanishing of the Beta function (89); note also that the above result is in agreement reduces to (0.47) when $\hat{v}(k) = e^{ik}$ $p_F = \pi/2$, in agreement with the exact solution of the XXZ model. Finally the Cooper pair density correlation, that is the correlation of the operator $\rho_\mathbf{x}^c = a_\mathbf{x}^+ a_{\mathbf{x}'}^+ + a_\mathbf{x}^- a_{\mathbf{x}'}^-$, $\mathbf{x}' = (x+1, x_0)$, behaves as:

$$\langle \rho_\mathbf{x}^c \rho_\mathbf{0}^c \rangle_T \sim \frac{1 + A_3(\mathbf{x})}{2\pi^2 (x^2 + v_F^2 x_0^2)^{X_-}}, \tag{100}$$

with $X_- = 1 + a_1\lambda + O(\lambda^2)$, a_1 being the same constant appearing in the first order of X_+.

The above formulas give a rather explicit expression for the correlations can be found. They were derived assuming the vanishing of the beta function (84), (89), whose proof will be recalled below.

5. Emerging Symmetries and Vanishing of Beta Function

The above RG analysis relies on two nonperturbative properties, namely (84) and (89); they can be checked at lowest orders in the renormalized expansion, but the cancellations are so complex that an explicit check at all orders is essentially impossible. The cancellations expressed by (84) and (89) are related to gauge symmetries which are however only asymptotic and not exact in the model (34). The key observation is to introduce a continuum model, related to the scaling limit of the lattice model (34), regularized by a nonlocal interaction and an ultraviolet γ^N, $N \gg 1$ and infrared γ^h, $h \ll 0$ cut-offs, to be finally removed. The beta function of this continuum model coincides, up to asymptotically vanishing $O(\lambda\gamma^j)$ terms, with the one of the model (34); we will then show that gauge symmetries imply the vanishing of the Beta function of this model and consequently of the model (34). This strategy provides a rigorous way of implementing the ideas of *emerging symmetries* in the model (34).

We consider the following Grassmann integral, which is the generating function of a system of massless Dirac fermions in the continuum with a nonlocal current–current interaction.

$$e^{W(J,\phi)} = \int P(d\psi) e^{V^{(N)}(\psi) + \int d\mathbf{x} J_{\mu,\mathbf{x}} j_{\mu,\mathbf{x}} + \sum_{\varepsilon=\pm} [\psi_{\mathbf{x},\varepsilon}^+ \phi_{\mathbf{x},\varepsilon}^- + \psi_{\mathbf{x},\varepsilon}^- \phi_{\mathbf{x},\varepsilon}^+]} \tag{101}$$

with $\mathbf{x} = (x_0, x) \in \mathbb{R}^2$, $\psi, \bar{\psi}$ are Euclidean $d = 1 + 1$ spinors, $\bar{\psi} = \psi^+ \gamma_0$, $\psi^\pm = (\psi_+^\pm, \psi_-^\pm)$, $P(d\psi^{(\leq N)})$ is the fermionic Gaussian integration with propagator

$$g^{(h,N)}(\mathbf{k}) = i \frac{\chi_{h,N}(\mathbf{k})}{\mathbf{k}} \tag{102}$$

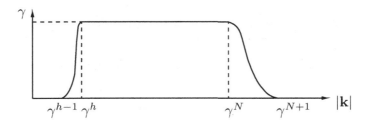

Fig. 4. The cut-off function $\chi_{h,N}(\mathbf{k})$.

$\not{k} = \gamma_0 k_0 + c\gamma_1 k_1$, which in components appear to be equal to $g^{(\leq N)}_{T,\varepsilon}(\mathbf{x})$ (81), and

$$V^{(N)}(\psi) = \frac{1}{4}\tilde{\lambda}_\infty \int d\mathbf{x} d\mathbf{y} v(\mathbf{x} - \mathbf{y}) j_\mu(\mathbf{x}) j_\mu(\mathbf{y}) \tag{103}$$

with $j_\mu(\mathbf{x}) = \bar{\psi}_\mathbf{x} \gamma_\mu \psi_\mathbf{x}$ and $v(\mathbf{x} - \mathbf{y})$ a short range symmetric interaction $\hat{v}(0) = 1$; moreover

$$\gamma_0 = \begin{pmatrix} 0 & 1 \\ 1 & 0 \end{pmatrix}, \quad \gamma_1 = \begin{pmatrix} 0 & -i \\ i & 0 \end{pmatrix}$$

Finally the cut-off function $\chi_{h,N}(\mathbf{k})$ is $= 1$ for $\gamma^{h-1} \leq \sqrt{k_0^2 + c^2 k^2} \leq \gamma^{N+1}$ and zero otherwise (see Fig. 4). γ^N plays the role of an ultraviolet cut-off, and at the end the limit $N \to \infty$ must be taken, while γ^h acts as an infrared cut-off. Note that the propagator of the ψ_ε fields is

$$g^{(h,N)}_\varepsilon(\mathbf{k}) = \frac{\chi_{h,N}(\mathbf{k})}{-ik_0 + c\varepsilon k} = \frac{\chi_{h,N}(\mathbf{k})}{D_\varepsilon(\mathbf{k})}, \tag{104}$$

where $D_\varepsilon(\mathbf{k}) = -ik_0 + \varepsilon c k$ and the interaction is:

$$V^{(N)}(\psi) = \tilde{\lambda}_\infty \int d\mathbf{x} d\mathbf{y} v(\mathbf{x} - \mathbf{y}) \psi^+_{+,\mathbf{x}} \psi^-_{+,\mathbf{x}} \psi^+_{-,\mathbf{y}} \psi^-_{-,\mathbf{y}} \tag{105}$$

Therefore if we replace $v(\mathbf{x} - \mathbf{y})$ with $v(x - y)\delta(x_0 - y_0)$ and we replace $\chi_N(\mathbf{k})$ with $\chi_{h,N}(k)$ we obtain, in the limit $-h, N \to \infty$, the Luttinger model.

Contrary to what happens in the model (34), in which the lattice imposes an ultraviolet cut-off, the model (101) poses either an infrared and an ultraviolet problem. The fields is written as $\psi = \sum_{j=h}^{N} \psi^{(j)}$, where $\psi^{(j)}$ has propagator with support in $\gamma^{j-1} \leq \sqrt{k_0^2 + c^2 k^2} \leq \gamma^{j+1}$. According to power counting based on dimensional considerations, also in the ultraviolet region ($h > 0$) the scaling dimension is $2 - n/2 - m$, but one can take advantage from the nonlocality of the potential to improve the scaling dimension and show that indeed the dimension is always negative; in other words the theory is *superrenormalizable* in the ultraviolet while is renormalizable in the infrared region. The integration of the ultraviolet region is discussed in detail in Refs. 34 and 35 and is related to a previous analysis in Ref. 29 for the Yukawa$_2$

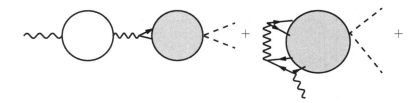

Fig. 5. Decomposition of $W_{2,1}^{(k)}$: the gray blobs represent respectively $W_{2,1}^{(k)}$ and $W_{2,2}^{(k)}$, the paired wiggly lines represent v, the paired line $g^{(k,N)}$.

model. The main idea of the power counting improvement can be grasped from the following picture Fig. 5. Let us consider the kernels $W_{2,1}^{(k)}$ with $k > 0$, with one external fields J and two fermions fields ψ; the scaling dimension is apparently vanishing by power counting. Such kernels can be decomposed (without loosing the determinant bounds which are essential for convergence) in a sum of two kind of terms, as shown in Fig. 5.

Note that the scaling dimension $2 - n/2 - m$ in the ultraviolet region is obtained integration over all the interactions $v(\mathbf{x} - \mathbf{y})$; however the terms represented by the second graph in Fig. 4 remain connected cutting the wiggly line representing the nonlocal interaction $v(\mathbf{x} - \mathbf{y})$, so that we can simply bound v by a constant and use one of the two propagators to perform the integral over the coordinates; this gives a gain γ^{-2k} in the bound, so that the dimension becomes negative for such terms. More explicitly, the second term in Fig. 4 can be bounded by:

$$C|\tilde{\lambda}_\infty||v|_{L^\infty}|W_{2,2}^{(k)}|_{L^1} \sum_{k \leq i' \leq j \leq i \leq N} |g^{(j)}|_{L^1}|g^{(i)}|_{L^1}|g^{(i')}|_{L^\infty} \leq C\tilde{\lambda}_\infty^2 \gamma^{-2k} \qquad (106)$$

On the other hand the first kind of terms in Fig. 4 becomes disconnected cutting the internal wiggly line; however the local part of the first term is vanishing by symmetry as:

$$\int d\mathbf{x}[g_\varepsilon^{(k,N)}(\mathbf{x} - \mathbf{z})]^2 = 0 \qquad (107)$$

so that the bound in the first term has an extra γ^{-k}. A similar power counting improvement can be repeated for the other terms to show that the dimension is always negative in the ultraviolet region. Note that in this analysis is essential that the interaction is short ranged, and it cannot be repeated for a local interaction in which $|v|_{L^\infty}$ is not bounded.

After the integration of the ultraviolet fields $\psi^{(N)}, \psi^{(N-1)}, \ldots, \psi^{(0)}$, we get a Grassmann integral very similar to (67), with the difference that Grassmann fields have propagators $g_{T,\varepsilon}^{(h,0)}(\mathbf{x}-\mathbf{y})$ (81) with c replacing v_F; the integration of the infrared scales is similar to the one for the model (34) but by symmetry $\nu_j = \delta_j = 0$, $\tilde{Z}_j^{(1)} = \tilde{Z}_j^{(2)}$ and the single scale propagator is given by $g_{T,\varepsilon}^{(h)}(\mathbf{x} - \mathbf{y})$ with v_F replaced by c. In the chain model, as we have seen, Lorentz symmetry is broken so that $\delta_j \neq 0$ (the velocity is renormalized) and $Z_{h-1}^{(1)}/Z_{h-1}^{(2)} = 1 + O(\lambda)$ (ρ and J are not component of the relativistic current). The effective coupling verify a flow equation

$$\tilde{\lambda}_{j-1} = \tilde{\lambda}_j + \beta_\lambda^{(j)}(\tilde{\lambda}_j, \dots, \tilde{\lambda}_0) + O(\tilde{\lambda}_\infty^2 \gamma^{\vartheta j}), \tag{108}$$

where $\beta_\lambda^{(j)}(\tilde{\lambda}_j, \dots, \tilde{\lambda}_j)$ is the *same* as function of $\tilde{\lambda}_j$ of the one appearing in (83) for the model (101), provided that we choose $c = v_F$; the same is true for $\beta_1^{(j)}$ (88). Moreover

$$\tilde{Z}_j \sim \gamma^{\eta j}, \quad \eta = \log_\gamma(1 + \beta_z^{(-\infty)}(\tilde{\lambda}_{-\infty}, \dots, \tilde{\lambda}_{-\infty})) \tag{109}$$

where $\beta_z^{(-\infty)}$ is the *same* as appearing in (86).

We have then seen that the dominant part of the beta function for the effective coupling $\tilde{\lambda}_j$ and for $\tilde{Z}_j^1/\tilde{Z}_j$ coincides with the one of the chain model (34). We will now take advantage from the fact that the model (101) verify extra symmetries, which at the end will lead to the proof of (84) and (89). Note first that, denoting $\langle \cdots \rangle_{h,N}$ the averages in the model (101) with infrared cut-off γ^h and ultraviolet cut-off γ^N, for momenta $\mathbf{k} \sim \gamma^h$ we get, if $D_\varepsilon(\mathbf{k}) = -ik_0 + \varepsilon c k$

$$\langle \hat{\rho}_{\mathbf{p}} \hat{\psi}_{2\mathbf{k},\varepsilon}^- \hat{\psi}_{-\mathbf{k},\varepsilon}^+ \rangle_{h,N} = -\frac{\tilde{Z}_h^{(1)}}{\tilde{Z}_h^2 D_\varepsilon(\mathbf{k})^2}(1 + O(\tilde{\lambda}_h^2)) \tag{110}$$

$$\langle \hat{\psi}_{\mathbf{k},\varepsilon}^- \hat{\psi}_{\mathbf{k},\varepsilon}^+ \rangle_{h,N} = \frac{1}{\tilde{Z}_h D_\varepsilon(\mathbf{k})}(1 + O(\tilde{\lambda}_h^2)) \tag{111}$$

$$\langle \hat{\psi}_{\mathbf{k},+}^- \hat{\psi}_{-\mathbf{k},+}^+ \hat{\psi}_{\mathbf{k},-}^- \hat{\psi}_{-\mathbf{k},-}^+ \rangle_{h,N} = \frac{-\tilde{\lambda}_h + O(\tilde{\lambda}_h^2)}{\tilde{Z}_h^2 D_+(\mathbf{k})^2 D_-(\mathbf{k})^2} \tag{112}$$

From the above equations we see that the value of the effective renormalizations $\tilde{Z}_h^{(1)}, \tilde{Z}_h$ and of the effective coupling $\tilde{\lambda}_h$ are related to the value of the two, three and four point functions with an infrared scale γ^h, computed for momenta close to the infrared cut-off scale. On the other hand, the value of $\tilde{Z}_k^{(1)}, \tilde{Z}_k, \tilde{\lambda}_k$ for $k \geq h$ in a theory with infrared cut-off at scale γ^h are the same as in a theory with *no* infrared cut-off, as a consequence of the compact support properties of the single scale propagators. Therefore, we can consider a sequence of models varying the value of the infrared cut-off scale; if we can derive a set of Ward Identities relating the correlations and we compute them at the infrared scale, we get, from (110), (111), (112), relations between the effective renormalizations and couplings. In order to do that one has to face a well known problem, which is present in any Wilsonian RG approach; the momentum cut-off breaks the local symmetries and produces additional terms in the Ward Identities.

In order to understand the idea let us start considering the functional integral (101) in which the cut-off function $\chi_{h,N}$ is replaced by one; the Grassmann integral is therefore only formal (to be well defined it requires a regularization) but let us, for the moment, ignore this fact. We can now perform the change of variables

$$\psi^{\pm}_{\mathbf{x},\varepsilon} \to e^{\pm i\alpha_{\varepsilon,\mathbf{x}}}\psi^{\pm}_{\mathbf{x},\varepsilon} \tag{113}$$

and perform a derivative with respect to $\alpha_{\mathbf{x},\varepsilon}$ and to the external fields we get the following Ward Identity, if $D_{\varepsilon}(\mathbf{p}) = -ip_0 + \varepsilon cp$

$$D_{\varepsilon}(\mathbf{p})\langle\hat{\rho}_{\mathbf{p},\varepsilon}\hat{\psi}^{-}_{\mathbf{k}-\mathbf{p},\varepsilon'}\hat{\psi}^{+}_{\mathbf{k},\varepsilon'}\rangle = \delta_{\varepsilon,\varepsilon'}\left[\langle\hat{\psi}^{-}_{\mathbf{k}-\mathbf{p},\varepsilon'}\hat{\psi}^{+}_{\mathbf{k}-\mathbf{p},\varepsilon'}\rangle - \langle\hat{\psi}^{-}_{\mathbf{k},\varepsilon'}\hat{\psi}^{+}_{\mathbf{k},\varepsilon'}\rangle\right] \tag{114}$$

This relation says that the two and three point function are *not* independent and it suggests, using (110), (111), that $\tilde{Z}^{(1)}_h \sim \tilde{Z}_h$, that is the current renormalization and the wavefunction renormalization diverges with the same exponent. However, the above Ward Identity is not true in presence of the cut-off function $\chi_{h,N}$.

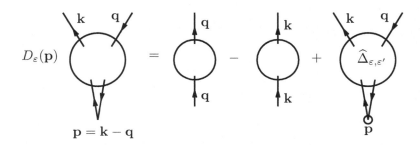

Fig. 6.   Graphical representation of the Ward Identity (115).

If we repeat the above computation in presence of the cut-off function $\chi_{h,N}$ we get that the Ward Identity acquires a correction term; the true Ward Identity for the model (101) is:

$$D_{\varepsilon}(\mathbf{p})\langle\hat{\rho}_{\mathbf{p},\varepsilon}\hat{\psi}^{-}_{\mathbf{k}-\mathbf{p},\varepsilon'}\hat{\psi}^{+}_{\mathbf{k},\varepsilon'}\rangle_{h,N} = \delta_{\varepsilon,\varepsilon'}\left[\langle\hat{\psi}^{-}_{\mathbf{k}-\mathbf{p},\varepsilon'}\hat{\psi}^{+}_{\mathbf{k}-\mathbf{p},\varepsilon'}\rangle_{h,N} - \langle\hat{\psi}^{-}_{\mathbf{k},\varepsilon'}\hat{\psi}^{+}_{\mathbf{k},\varepsilon'}\rangle_{h,N}\right]$$
$$+ \hat{\Delta}_{\varepsilon,\varepsilon'}(\mathbf{p},\mathbf{k}) \tag{115}$$

with

$$\hat{\Delta}_{\varepsilon,\varepsilon'}(\mathbf{p},\mathbf{k}) = \frac{1}{L^2}\sum_{\mathbf{k}'} C_{h,N}(\mathbf{k}',\mathbf{k}'-\mathbf{p})\langle\hat{\psi}^{+}_{\mathbf{k}',\varepsilon}\hat{\psi}^{-}_{\mathbf{k}'-\mathbf{p},\varepsilon};\hat{\psi}^{-}_{\mathbf{k},\varepsilon'}\hat{\psi}^{+}_{\mathbf{k}-\mathbf{p},\varepsilon'}\rangle, \tag{116}$$

where

$$C_{h,N}(\mathbf{k},\mathbf{k}-\mathbf{p}) = (\chi^{-1}_{h,N}(\mathbf{k}-\mathbf{p})-1)D_{\varepsilon}(\mathbf{k}-\mathbf{p}) - (\chi^{-1}_{h,N}(\mathbf{k})-1)D_{\varepsilon}(\mathbf{k}). \tag{117}$$

With respect to the (114), the presence of the infrared and ultraviolet cut off produce the extra term $\Delta_{\varepsilon\varepsilon'}(\mathbf{p},\mathbf{k})$ in (115); its presence can be easily checked at lowest order in perturbation theory starting from the following (trivial) identity for the propagator (104)

$$D_\varepsilon(\mathbf{p})\frac{\chi_{h,N}(\mathbf{k})}{D_\varepsilon(\mathbf{k})}\frac{\chi_{h,N}(\mathbf{k}-\mathbf{p})}{D_\varepsilon(\mathbf{k}-\mathbf{p})} = \frac{\chi_{h,N}(\mathbf{k}-\mathbf{p})}{D_\varepsilon(\mathbf{k}-\mathbf{p})} - \frac{\chi_{h,N}(\mathbf{k})}{D_\varepsilon(\mathbf{k})}$$
$$+ C_{h,N}(\mathbf{k},\mathbf{k}-\mathbf{p})\frac{\chi_{h,N}(\mathbf{k})}{D_\varepsilon(\mathbf{k})}\frac{\chi_{h,N}(\mathbf{k}-\mathbf{p})}{D_\varepsilon(\mathbf{k}-\mathbf{p})}. \quad (118)$$

Following the previous formal computation, one could be tempted to conclude that $\Delta_{\varepsilon,\varepsilon'}$ is vanishing in the limit $-h, N \to \infty$; this conclusion is however false and the following formula holds, proved in Ref. 34 and 35

$$\Delta_{\varepsilon,\varepsilon'}(\mathbf{p},\mathbf{k}) = \frac{\tilde{\lambda}_\infty}{4\pi c}v(\mathbf{p})D_{-\varepsilon}(\mathbf{p})\langle\hat{\rho}_{\mathbf{p},-\varepsilon}\hat{\psi}^-_{\mathbf{k}-\mathbf{p},\varepsilon'}\hat{\psi}^+_{\mathbf{k},\varepsilon'}\rangle_{h,N} + D_\varepsilon(\mathbf{p})R_{\varepsilon,\varepsilon'}(\mathbf{p};\mathbf{k}) \quad (119)$$

and, for $h \le 0$, N large enough and $|\mathbf{p}|, |\mathbf{k}| \sim \gamma^h$

$$|R_{\varepsilon,\varepsilon'}(\mathbf{p};\mathbf{k})| \le C|\tilde{\lambda}_\infty|\frac{\gamma^{-2h}}{Z_h}. \quad (120)$$

Again the proof of (119) and (120) is technical, but the main ideas can be understood qualitatively. The term $\Delta_{\varepsilon,\varepsilon'}$ can be written as the derivative of a Grasmmann integral, similar to (101) but in which $\int d\mathbf{x}J_{\mu,\mathbf{x}}j_{\mu,\mathbf{x}}$ is replaced by:

$$\int d\mathbf{k}d\mathbf{p}J_{\mathbf{p}}C_{h,N}(\mathbf{k},\mathbf{k}+\mathbf{p})\hat{\psi}^+_{\mathbf{k},\varepsilon}\hat{\psi}^-_{\mathbf{k}+\mathbf{p},\varepsilon}. \quad (121)$$

The analysis of the ultraviolet scales is similar to the one of (101); the terms with two external fermionic fields and a J external field $\tilde{W}^{(h)}_{2,1}$ are again dimensionally marginal and one needs a dimensional improvement produced by the decomposition. The second term in Fig. 7 represents the terms which remain connected cutting the wiggly line representing the nonlocal interaction $v(\mathbf{x}-\mathbf{y})$, so that we can simply bound v by a constant and use one of the two propagators to perform the integral over the coordinates; this gives a gain γ^{-2k} in the bound, so that the dimension

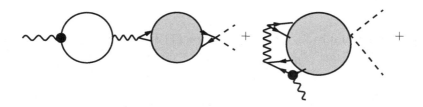

Fig. 7.   Decomposition of $\tilde{W}^{(k)}_{1,2}$.

becomes negative for such terms. There is however an extra gain due to the fact that necessarily at least one of the two fields in (121) must have scale N; this is a consequence of the definition (117) ($1 - \chi_N$ has to be nonvanishing). Therefore, the second term in Fig. 7 can be bounded by:

$$C|\tilde{\lambda}_\infty||v|_{L^\infty}|\tilde{W}_{2,2}^{(k)}|_{L^1} \sum_{k \le i' \le j \le i \le N} |g^{(N)}|_{L^1}|g^{(i)}|_{L^1}|g^{(i')}|_{L^\infty} \le C\tilde{\lambda}_\infty^2 \gamma^{-N-k} \qquad (122)$$

so that its contribution is now *vanishing* as $N \to \infty$. Another crucial difference with respect to the analysis in (0.4) is the fact that the local part of the bubble in the first term is nonvanishing and given by:

$$\lim_{N \to \infty} \tilde{\lambda}_\infty \int \frac{d\mathbf{k}}{(2\pi)^2} C_N(\mathbf{k}, \mathbf{k} - \mathbf{p}) g_\varepsilon^{(\le N)}(\mathbf{k}) g_\varepsilon^{(\le N)}(\mathbf{k} - \mathbf{p})|_{\mathbf{p}=0}$$

$$= -\frac{\tilde{\lambda}_\infty}{4\pi c} D_{-\varepsilon}(\mathbf{p}) \int_0^\infty d\rho \chi_0'(\rho)$$

$$= \frac{\tilde{\lambda}_\infty}{4\pi c} D_{-\varepsilon}(\mathbf{p}). \qquad (123)$$

This explain the presence of the first term of the r.h.s. of (119).

By inserting (119) in (115), computing the resulting expression at the cut-off scale γ^h and using (110) and (111) we get that:

$$\frac{\tilde{Z}_h^{(1)}}{\tilde{Z}_h} = 1 + O(\bar{\lambda}_h^2). \qquad (124)$$

Comparing this with (88) we get, by a contradiction argument, the asymptotic vanishing of the beta function (89).

A similar argument can be repeated for proving (84). One writes a Schwinger–Dyson equation for the model (101), that is:

$$-\langle \psi_{\mathbf{k}_1,+}^- \psi_{\mathbf{k}_2,+}^+ \psi_{\mathbf{k}_3,-}^- \psi_{\mathbf{k}_4,-}^+ \rangle_{h,N}$$

$$= \tilde{\lambda}_\infty \hat{g}_-^{(h,N)}(\mathbf{k}_4) \Big[v(\mathbf{k}_1 - \mathbf{k}_2) \langle \psi_{\mathbf{k}_1,+}^- \psi_{\mathbf{k}_2,+}^+ \rho_{\mathbf{k}_1-\mathbf{k}_2,+} \rangle_{h,N} \langle \psi_{\mathbf{k}_3,-}^- \psi_{\mathbf{k}_3,-}^+ \rangle_{h,N}$$

$$+ \frac{1}{L^2} \sum_{\mathbf{p}} \tilde{\lambda}_\infty \hat{v}(\mathbf{p}) \langle \rho_{\mathbf{p},+} \psi_{\mathbf{k}_1,+}^- \psi_{\mathbf{k}_2,+}^+ \psi_{\mathbf{k}_3,-}^- \psi_{\mathbf{k}_4-\mathbf{p},-}^+ \rangle_{h,N} \Big]. \qquad (125)$$

The l.h.s is proportional to λ_h when $\mathbf{k}_i \sim \gamma^h$ by (112) and one writes in the first term in the r.h.s. the function $\langle \psi_{\mathbf{k}_1,+}^- \psi_{\mathbf{k}_2,+}^+ \rho_{\mathbf{k}_1-\mathbf{k}_2,+} \rangle$ in terms of the 2-point function by using the Ward Identity (119) and (115). Similarly we can use for the second term in the r.h.s. a Ward Identity for $\langle \rho_{\mathbf{p},+} \psi_{\mathbf{k}_1,+}^- \psi_{\mathbf{k}_2,+}^+ \psi_{\mathbf{k}_3,-}^- \psi_{\mathbf{k}_4-\mathbf{p},-}^+ \rangle_{h,N}$; now one gets a term involving the integral of the correction R in the Ward Identity and the main technical problem, discussed in Refs. 33 and 43, is to bound such a term uniformly in h; the conclusion is that the r.h.s. is proportional to $\tilde{\lambda}_\infty + O(\tilde{\lambda}_\infty^2)$ and this implies the vanishing of the beta function in (108) and consequently (84); therefore $\tilde{\lambda}_{-\infty}$ is analytic in $\tilde{\lambda}_\infty$ and

$$\tilde{\lambda}_{-\infty} = \tilde{\lambda}_\infty + O(\tilde{\lambda}_\infty^2). \tag{126}$$

6. Ward Identities and Anomalies

In the previous section we have considered a fermionic model (101) with linear dispersion relation, a momentum infrared and ultraviolet cut-off and a nonlocal interaction; the infrared behavior of this model is the same as the one of the chain model (34) but it verifies extra symmetries allowing to deduce Ward Identities. The momentum cut-off produces corrections with respect to the naïve Ward Identities which are not small at all, and are not vanishing even when cut-offs are removed. Indeed (115) and (119) acquire a simpler form in the limit $-h, N \to \infty$ at fixed momenta where they can be rewritten as

$$D_\varepsilon(\mathbf{p})\langle \hat{\rho}_{\mathbf{p},\varepsilon} \hat{\psi}^-_{\mathbf{k}-\mathbf{p},\varepsilon} \hat{\psi}^+_{\mathbf{k},\varepsilon}\rangle = \langle \hat{\psi}^-_{\mathbf{k}-\mathbf{p},\varepsilon} \hat{\psi}^+_{\mathbf{k}-\mathbf{p},\varepsilon}\rangle - \langle \hat{\psi}^-_{\mathbf{k},\varepsilon} \hat{\psi}^+_{\mathbf{k},\varepsilon}\rangle$$
$$+ \frac{\tilde{\lambda}_\infty}{4\pi c} D_{-\varepsilon}(\mathbf{p})\langle \hat{\rho}_{\mathbf{p},-\varepsilon} \hat{\psi}^-_{\mathbf{k}-\mathbf{p},\varepsilon} \hat{\psi}^+_{\mathbf{k},\varepsilon}\rangle \tag{127}$$

$$D_{-\varepsilon}(\mathbf{p})\langle \hat{\rho}_{\mathbf{p},-\varepsilon} \hat{\psi}^-_{\mathbf{k}-\mathbf{p},\varepsilon} \hat{\psi}^+_{\mathbf{k},\varepsilon}\rangle = \frac{\tilde{\lambda}_\infty}{4\pi c} D_\varepsilon(\mathbf{p})\langle \hat{\rho}_{\mathbf{p},\varepsilon} \hat{\psi}^-_{\mathbf{k}-\mathbf{p},\varepsilon} \hat{\psi}^+_{\mathbf{k},\varepsilon}\rangle$$

or, using the relativistic notation $j_\mu = \bar{\psi}\gamma_\mu\psi$ $j_{\mu,5} = \bar{\psi}\gamma_\mu\gamma_5\psi$ and using that $j_0 = -ij_{5,1} = \rho_1 + \rho_{-1}$, $j_1 = ij_{5,0} = i(\rho_1 - \rho_{-1})$ and calling $\mathbf{p}_\mu = (\omega, cp)$

$$-i\mathbf{p}_\mu\langle j_{\mu,\mathbf{p}} \psi_{\mathbf{k}} \psi^+_{\mathbf{k}+\mathbf{p}}\rangle = A[\langle \psi_{\mathbf{k}} \psi^+_{\mathbf{k}}\rangle - \langle \psi_{\mathbf{k}+\mathbf{p}} \psi^+_{\mathbf{k}+\mathbf{p}}\rangle]$$
$$-i\mathbf{p}_\mu\langle j_{5,\mu,\mathbf{p}} \psi_{\mathbf{k}} \psi^+_{\mathbf{k}+\mathbf{p}}\rangle = \bar{A}\gamma_5[\langle \psi_{\mathbf{k}} \psi^+_{\mathbf{k}}\rangle - \langle \psi_{\mathbf{k}+\mathbf{p}} \psi^+_{\mathbf{k}+\mathbf{p}}\rangle] \tag{128}$$

with

$$A(\mathbf{p}) = \frac{1}{1 - \tau\hat{v}(\mathbf{p})}, \quad \bar{A}(\mathbf{p}) = \frac{1}{1 + \tau\hat{v}(\mathbf{p})} \tag{129}$$

and

$$\tau = \frac{\tilde{\lambda}_\infty}{4\pi c}. \tag{130}$$

In the same way the Ward Identities for the density operators in the model (101) are, in the limit of removed cut-off $-h, N \to \infty$

$$D_+(\mathbf{p})\langle \rho_{\mathbf{p},+} \rho_{-\mathbf{p},+}\rangle - \tau D_-(\mathbf{p})\langle \rho_{\mathbf{p},-} \rho_{-\mathbf{p},+}\rangle = \frac{1}{4\pi c} D_-(\mathbf{p})$$
$$D_-(\mathbf{p})\langle \rho_{\mathbf{p},-} \rho_{-\mathbf{p},+}\rangle - \tau D_+(\mathbf{p})\langle \rho_{\mathbf{p},+} \rho_{-\mathbf{p},+}\rangle = 0 \tag{131}$$

from which we get

$$\langle j^{0,\mathbf{P}} j^{0,-\mathbf{P}} \rangle = -\frac{1}{4\pi c} \frac{1}{1-\tau^2} \left[\frac{D_-(\mathbf{p})}{D_+(\mathbf{p})} + \frac{D_+(\mathbf{p})}{D_-(\mathbf{p})} - 2\tau \right]$$

$$\langle j^{1\mathbf{P}} j^{1-\mathbf{P}} \rangle = -\frac{1}{4\pi c} \frac{1}{1-\tau^2} \left[\frac{D_-(\mathbf{p})}{D_+(\mathbf{p})} + \frac{D_+(\mathbf{p})}{D_-(\mathbf{p})} + 2\tau \right].$$

$$(132)$$

Finally we can write the Schwinger-function for the 2-point function in the model (101)

$$\langle \psi^-_{\mathbf{k},\varepsilon} \psi^+_{\mathbf{k},\varepsilon} \rangle_{h,N} = g^{[h,N]}_\omega(\mathbf{k})$$

$$+ \tilde{\lambda}_\infty g^{[h,N]}_\varepsilon(\mathbf{k}) \frac{1}{L^2} \sum_{\mathbf{p}} \hat{v}(\mathbf{p}) \langle \rho_{\mathbf{p},-\varepsilon} \psi^-_{\mathbf{k},-} \psi^+_{\mathbf{k}-\mathbf{p},-} \rangle_{h,N} \Bigg]$$

$$(133)$$

We can now insert in the last term the Ward Identity (115) and (119); it turns out that the integral of the correction is vanishing so that in the limit $-h, N \to \infty$

$$D_\varepsilon(\mathbf{k}) \langle \hat{\psi}^-_{\mathbf{k},\varepsilon} \hat{\psi}^+_{\mathbf{k},\varepsilon} \rangle = 1 + \tilde{\lambda}_\infty \int \frac{d\mathbf{p}}{(2\pi)^2} [A(\mathbf{p}) - \bar{A}(\mathbf{p})] \frac{v(\mathbf{p})}{D_{-\varepsilon}(\mathbf{p})} \langle \hat{\psi}^-_{\mathbf{k}-\mathbf{p},\varepsilon} \hat{\psi}^+_{\mathbf{k}-\mathbf{p},\varepsilon} \rangle \quad (134)$$

Passing in Fourier transform we get a PDE whose solution is given by:

$$\langle \psi^-_{\mathbf{x}} \psi^+_{\mathbf{y}} \rangle \sim g_\varepsilon(\mathbf{x} - \mathbf{y}) \frac{1}{|\mathbf{x} - \mathbf{y}|^\eta} \quad (135)$$

with

$$\eta = \frac{\tilde{\lambda}_\infty}{4\pi c} \left[\frac{1}{1-\tau} - \frac{1}{1+\tau} \right]. \quad (136)$$

In the same way we can consider the limit of removed cut-off of (125) and we get:

$$\langle \psi^+_{\mathbf{x},\varepsilon} \psi^-_{\mathbf{x},-\varepsilon} \psi^+_{\mathbf{y},-\varepsilon} \psi^-_{\mathbf{y},\varepsilon} \rangle_T \sim \frac{C}{|\mathbf{x} - \mathbf{y}|^{2X_+}} \quad (137)$$

and

$$\langle \psi^+_{\mathbf{x},\varepsilon} \psi^+_{\mathbf{x},-\varepsilon} \psi^-_{\mathbf{y},-\varepsilon} \psi^-_{\mathbf{y},\varepsilon} \rangle_T \sim \frac{C}{|\mathbf{x} - \mathbf{y}|^{2X_-}}, \quad (138)$$

so that:

$$X_+ = 1 - \bar{A}(0)(\lambda_\infty/2\pi c) = 1 - \frac{(\tilde{\lambda}_\infty/2\pi c)}{1 + (\tilde{\lambda}_\infty/4\pi c)}$$

$$X_- = 1 + A(0)(\lambda_\infty/2\pi c) = 1 + \frac{(\tilde{\lambda}_\infty/2\pi c)}{1 - (\tilde{\lambda}_\infty/4\pi c)};$$

$$(139)$$

The exponents in the model (101) verify the relations (48), even if the exponents are of course different with respect to the Luttinger model ones (21).

The Grassmann integral (101) (with no infrared cut-off) can be rewritten, via an Hubbard–Stratonovich transformation, in the following way if $\tilde{\lambda}_\infty = e^2$

$$\mathcal{W}_N(J, \phi) = \log \int P(d\psi^{(\leq N)})P(dA)e^{\int d\mathbf{x}[e\bar{\psi}_\mathbf{x}(A_{\mu,\mathbf{x}}\gamma_\mu)\psi_\mathbf{x} + J_{\mu,\mathbf{x}}A_{\mu,\mathbf{x}} + \phi_\mathbf{x}\bar{\psi}_\mathbf{x} + \bar{\phi}_\mathbf{x}\psi_\mathbf{x}]},$$

(140)

in which the bosonic propagator is given by $\langle A_{\mu,\mathbf{x}} A_{\nu,\mathbf{y}} \rangle = \delta_{\mu,\nu} v(\mathbf{x} - \mathbf{y})$. Therefore, the model (101) can be equivalently written as a QFT model in $d = 1 + 1$ in which massless Dirac fermions interact with a vector boson field. The WI (130) can be rewritten as:

$$-i\gamma_\mu \mathbf{p}_\mu \langle j_{5,\mu,\mathbf{p}} \psi_{\mathbf{k},\omega} \bar{\psi}_{\mathbf{k}+\mathbf{p}} \rangle = [\langle \psi_\mathbf{k} \bar{\psi}_\mathbf{k} \rangle - \langle \psi_{\mathbf{k}+\mathbf{p}} \bar{\psi}_{\mathbf{k}+\mathbf{p}}^- \rangle] + \frac{\tau}{e}\varepsilon_{\mu,\nu}\langle A_{\nu,\mathbf{p}}\psi_{\mathbf{k},\omega}\bar{\psi}_{\mathbf{k}+\mathbf{p}} \rangle,$$

(141)

where we have used that:

$$\varepsilon_{\mu,\nu}\langle A_{\nu,\mathbf{p}}\psi_{\mathbf{k},\omega}\bar{\psi}_{\mathbf{k}+\mathbf{p}} \rangle = ev(\mathbf{p})\langle j_{5,\mu,\mathbf{p}}\psi_{\mathbf{k},\omega}\bar{\psi}_{\mathbf{k}+\mathbf{p}} \rangle,$$

(142)

where $\tau/e = e/4\pi c$ is the *chiral anomaly*. The anomaly is linear in e, a property called *anomaly nonrenormalization*; it was first proved by Adler and Bardeen as an identity in perturbation theory[45] in QED_4 and (141) is the form it acquires in $d = 1+1$, see Ref. 46. The results in Refs. 45 and 46 are however true *order by order* in the expansion, while (141) is a nonperturbative statement (it makes no use of the loop cancellation or on any graphs arguments) obtained by a rigorous analysis of Grassmann integrals, and its validity relies on (119).

The model (101) can be also considered a regularization of the massless Thirring model[4]; more exactly, if $v(x)$ is a regularized delta and $\lim_{K\to\infty} v(x) = \delta(x)$, the Thirring model can be constructed taking the limits $K, N \to \infty$. Such limit gives a nontrivial limit only if we perform the change of variables $\psi \to \sqrt{Z}\psi$ and we choose Z diverging in the limit $K, N \to \infty$. If the limit $K \to \infty$ is removed after the limit $N \to \infty$ one has to choose $Z = \gamma^{-\eta K}$ and it is found[34,43] that the WI is given by (128), (129), (130) with $v(p)$ replaced by one. This Ward Identity, obtained by the analysis of the Grassman integral (101), coincides with the one derived by Johnson[4] postulating the validity of anomalous commutation rules

$$[\psi(x), j^0(y)] = a\delta(x - y)\psi(y), \quad [\psi(x), j_5^0(y)] = \bar{a}\delta(x - y)\psi(y) \quad (143)$$

and fixing the value of $a, \bar{a}$ by a self consistence argument.

On the other hand, if one performs the limit in the opposite order, that is the $N \to \infty$ limit is taken after $K \to \infty$ choosing $Z = \gamma^{-\eta N}$ one finds again the Ward

Identity (128), (129) but now τ is given by[44]:

$$\tau = \frac{\tilde{\lambda}_\infty}{4\pi c} + b\tilde{\lambda}_\infty^2 + O(\tilde{\lambda}_\infty^3) \tag{144}$$

that is the anomaly acquires higher orders corrections. The validity of the anomaly nonrenormalization in the Thirring model depends then on how the cut-offs are removed. Having higher orders correction in the anomaly would prevent to get simple expressions of the exponents like (137), (138) and would prevent the possibility of checking the scaling relations.

The Ward Identities (128) for the lattice model (34) are different with respect to the Luttinger model ones (31), (32). However a RG analysis similar to the one for the model (101) could be repeated also for the Luttinger model; in such case, the cut-off $\chi_N(\mathbf{k})$ should be replaced by a cut-off involving only the spatial momenta, that is $\chi_N(k)$ depending only from the spatial momentum. The ultraviolet fields can be integrated[47] and the Ward Identities acquire a correction given by (115) with

$$\lim_{-h,N\to\infty} \Delta_{\varepsilon,\varepsilon'}(\mathbf{p};\mathbf{k}) = \frac{\tilde{\lambda}_\infty}{2\pi}\hat{v}(p)p\langle\rho_{\mathbf{p},-\varepsilon}\hat{\psi}_{\mathbf{k}-\mathbf{p},\varepsilon'}^-\hat{\psi}_{\mathbf{k},\varepsilon'}^+\rangle\,. \tag{145}$$

The computation is analogue to (123) with $\chi_N(\mathbf{k})$ replaced by $\chi_N(k)$. Therefore (127) is replaced by:

$$D_\varepsilon(\mathbf{p})\langle\hat{\rho}_{\mathbf{p},\varepsilon}\hat{\psi}_{\mathbf{k}-\mathbf{p},\varepsilon}^-\hat{\psi}_{\mathbf{k},\varepsilon}^+\rangle = \langle\hat{\psi}_{\mathbf{k}-\mathbf{p},\varepsilon}^-\hat{\psi}_{\mathbf{k}-\mathbf{p},\varepsilon}^+\rangle - \langle\hat{\psi}_{\mathbf{k},\varepsilon}^-\hat{\psi}_{\mathbf{k},\varepsilon}^+\rangle$$
$$+ \frac{\tilde{\lambda}_\infty}{2\pi}p\langle\hat{\rho}_{\mathbf{p},-\varepsilon}\hat{\psi}_{\mathbf{k}-\mathbf{p},\varepsilon}^-\hat{\psi}_{\mathbf{k},\varepsilon}^+\rangle \tag{146}$$

$$D_{-\varepsilon}(\mathbf{p})\langle\hat{\rho}_{\mathbf{p},-\varepsilon}\hat{\psi}_{\mathbf{k}-\mathbf{p},\varepsilon}^-\hat{\psi}_{\mathbf{k},\varepsilon}^+\rangle = \frac{\tilde{\lambda}_\infty}{2\pi}p\langle\hat{\rho}_{\mathbf{p},\varepsilon}\hat{\psi}_{\mathbf{k}-\mathbf{p},\varepsilon}^-\hat{\psi}_{\mathbf{k},\varepsilon}^+\rangle$$

from which (31), (32) are recovered. In this case the anomaly produces two different velocities in the Ward Identity, related to charge and current excitations; on the contrary, this phenomenon cannot appear in (101) due to Lorentz invariance, and the anomaly produces the asymmetry of the renormalization of the current and chiral current $A \neq \bar{A}$.

7. Proof of the Luttinger Liquid Relations in Nonsolvable Models

The conclusion of the above analysis is that the exponents of the chain model (34) can be written in terms of *convergent* expansions; such series allow to compute the indices with arbitrary precision (with explicit computation of lowest orders and a rigorous bounds of the rest), but their extreme complexity makes impossible the explicit verification of the Luttinger liquid relations (48) and (49) from them.

We have seen that the critical exponents η, X_+, X_- can be represented as power series in the variable $\lambda_{-\infty}/v_F$ and we can choose $\tilde{\lambda}_\infty(\lambda)$ and $c = v_F(\lambda)$ in the model

(101) so that, by using (126) and (85)

$$\lambda_{-\infty}(\lambda) = \tilde{\lambda}_{-\infty}(\tilde{\lambda}_{\infty}) \tag{147}$$

As a consequence of (109), *with this choice of c, $\tilde{\lambda}_{\infty}$ the exponents of the model (101) and (34) are the same.* In addition it is a corollary of the RG analysis the validity of rigorous relations between the correlations of the model (34) and (101); that is, for $|\mathbf{k}|, |\mathbf{k} + \mathbf{p}| \leq \kappa$ (in the l.h.s are correlations for the chain model (34) and in the r.h.s. for the model (101))

$$\langle \hat{\rho}_{\mathbf{p}} \hat{a}^{+}_{\mathbf{k}+\varepsilon\mathbf{p}_F} \hat{a}^{-}_{\mathbf{k}+\varepsilon\mathbf{p}+\mathbf{p}_F} \rangle = \frac{Z^{(3)}}{Z^2} \langle \hat{j}_{0,\mathbf{p}} \hat{\psi}^{+}_{\mathbf{k},\varepsilon} \hat{\psi}^{-}_{\mathbf{k}+\mathbf{p},\varepsilon} \rangle (1 + r_1)$$

$$\langle \hat{J}_{\mathbf{p}} \hat{a}^{+}_{\mathbf{k}+\varepsilon\mathbf{p}_F} \hat{a}^{-}_{\mathbf{k}+\mathbf{p}+\mathbf{p}^{\omega}_F} \rangle = \frac{\tilde{Z}^{(3)}}{Z^2} \langle \hat{j}_{1,\mathbf{p}} \hat{\psi}^{+}_{\mathbf{k},\varepsilon} \hat{\psi}^{-}_{\mathbf{k}+\mathbf{p},\varepsilon} \rangle (1 + r_2) \tag{148}$$

with $|r_1|, |r_2| \leq C\kappa^{\vartheta}$ where

$$\frac{\tilde{Z}^{(3)}}{Z^{(3)} \sin p_F} = 1 + 2a_1\lambda + O(\lambda^2) \tag{149}$$

with

$$a_1 = \frac{1}{2\pi v_F}[\hat{v}(0) - \hat{v}(2p_F)]. \tag{150}$$

In order to prove (148) we note that the models (34) and (101) differ by *irrelevant terms* in the infrared region, and we can tune the (finite) value of the wavefunction renormalization and of the vertex or current renormalization at scale $h = 0$ so that the corresponding running renormalizations are asymptotically coinciding as $h \to -\infty$. Note also that $Z^{(3)}$ and $\tilde{Z}^{(3)}$ are different, as a consequence of the fact that in the lattice model (34) Lorentz symmetry is broken by the irrelevant terms; this can be explicitly checked at lowest order computing the two graphs in Fig. 8 for the lattice model (34).

Similarly the current–current or the density–density renormalizations are related by the following relations:

$$\langle \hat{\rho}_{\mathbf{p}} \hat{\rho}_{-\mathbf{p}} \rangle = \left[\frac{Z^{(3)}}{Z} \right]^2 \langle \hat{j}_{0,\mathbf{p}} \hat{j}_{0,-\mathbf{p}} \rangle + \hat{A}_{\rho,\rho}(\mathbf{p})$$

$$\langle \hat{J}_{\mathbf{p}} \hat{J}_{-\mathbf{p}} \rangle = \left[\frac{\tilde{Z}^{(3)}}{Z} \right]^2 \langle \hat{j}_{1,\mathbf{p}} \hat{j}_{i,-\mathbf{p}} \rangle + \hat{A}_{j,j}(\mathbf{p}) \tag{151}$$

with

$$|A_{\rho,\rho}(\mathbf{x})|, |A_{j,j}(\mathbf{x})| \leq \frac{C}{|\mathbf{x}|^{2+\vartheta}}. \tag{152}$$

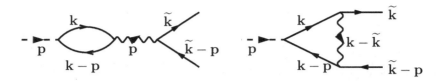

Fig. 8.   The first order contributions to $Z^{(3)}$ and $\tilde{Z}^{(3)}$, giving the value is $\pm a_1$.

Therefore $A_{\rho,\rho}(\mathbf{p})$, $A_{j,j}(\mathbf{p})$ are *continuous* (and with a nonvanishing limit in $\mathbf{p} = 0$), while $\langle \hat{\rho}_{\mathbf{p}} \hat{\rho}_{-\mathbf{p}} \rangle$, $\langle \hat{J}_{\mathbf{p}} \hat{J}_{-\mathbf{p}} \rangle$ are *not* continuous; this regularity improvement is a consequence of the fact that such terms are generated by the presence of the irrelevant terms.

With the above choice of $\tilde{\lambda}_\infty$ and $c = v_F$ the exponents of the chain model (34) coincides with the exponents of (101), which are expressed by (136) and (139), so that the expoents verify Luttinger liquid relations (48) with:

$$K = \frac{1 - (\tilde{\lambda}_\infty / 4\pi v_F)}{1 + (\tilde{\lambda}_\infty / 4\pi v_F)}. \tag{153}$$

and

$$\tilde{\lambda}_\infty = \lambda[\hat{v}(0) - \hat{v}(2p_F)] + O(\lambda^2), \quad v_F = \sin p_F + O(\lambda) \tag{154}$$

depending from all the model details. It remains to prove (49). Note first that the Ward Identies of the model (101), namely (129), implies the following Ward Identities for the model (34), for $|\mathbf{k}|, |\mathbf{k} + \mathbf{p}| \le \kappa$, through (148)

$$-ip_0 \langle \hat{\rho}_{\mathbf{p}} \hat{a}^+_{\mathbf{k}+\varepsilon\mathbf{p}_F} \hat{a}^-_{\mathbf{k}+\mathbf{p}+\varepsilon\mathbf{p}_F} \rangle + \varepsilon p v_F \frac{Z^{(3)}}{\tilde{Z}^{(3)}} \langle \hat{J}_{\mathbf{p}} \hat{a}^+_{\mathbf{k}+\varepsilon\mathbf{p}_F} \hat{a}^-_{\mathbf{k}+\mathbf{p}+\varepsilon\mathbf{p}_F} \rangle$$

$$= \frac{Z^{(3)}}{Z} \frac{1}{(1-\tau)} [\langle \hat{a}^+_{\mathbf{k}+\varepsilon\mathbf{p}_F} \hat{a}^-_{\mathbf{k}+\varepsilon\mathbf{p}_F} \rangle - \langle \hat{a}^+_{\mathbf{k}+\mathbf{p}+\varepsilon\mathbf{p}_F} \hat{a}^-_{\mathbf{k}+\mathbf{p}+\varepsilon\mathbf{p}_F} \rangle](1 + O(\kappa^\vartheta)) \tag{155}$$

and

$$-ip_0 \langle \hat{J}_{\mathbf{p}} \hat{a}^+_{\mathbf{k}} \hat{a}^-_{\mathbf{k}+\mathbf{p}} \rangle + \varepsilon p v_F \frac{\tilde{Z}^{(3)}}{Z^{(3)}} \langle \hat{\rho}_{\mathbf{p}} \hat{a}^+_{\mathbf{k}} \hat{a}^-_{\mathbf{k}+\mathbf{p}} \rangle$$

$$= \frac{\tilde{Z}^{(3)}}{Z} \frac{1}{(1+\tau)} [\langle \hat{a}^+_{\mathbf{k}} \hat{a}^-_{\mathbf{k}} \rangle - \langle \hat{a}^+_{\mathbf{k}+\mathbf{p}} \hat{a}^-_{\mathbf{k}+\mathbf{p}} \rangle](1 + O(\kappa^\vartheta)).$$

The first Ward Identity must coincide with the exact Ward Identity found for the model (34), obtained by the continuity equation; we see then that the finite renormalizations are not independent but are related by the following exact relations

$$\frac{Z^{(3)}}{(1-\tau)Z} = 1, \quad v_F \frac{Z^{(3)}}{\tilde{Z}^{(3)}} = 1 \tag{156}$$

from which

$$K = 1 - \frac{1}{\pi v_F}[\hat{v}(0) - \hat{v}(2p_F)]\lambda + O(\lambda^2)\,. \tag{157}$$

By (156) and calling $J_\mathbf{p} = v_F j_\mathbf{p}$ we get:

$$-ip_0\langle\hat{\rho}_\mathbf{p}\hat{a}^+_{\mathbf{k}+\varepsilon\mathbf{p}_F}\hat{a}^-_{\mathbf{k}+\mathbf{p}+\varepsilon\mathbf{p}_F}\rangle + \varepsilon p v_F\langle\hat{j}_\mathbf{p}\hat{a}^+_{\mathbf{k}+\varepsilon\mathbf{p}_F}\hat{a}^-_{\mathbf{k}+\mathbf{p}+\varepsilon\mathbf{p}_F}\rangle$$
$$= [\langle\hat{a}^+_{\mathbf{k}+\varepsilon\mathbf{p}_F}\hat{a}^-_{\mathbf{k}+\varepsilon\mathbf{p}_F}\rangle - \langle\hat{a}^+_{\mathbf{k}+\mathbf{p}+\varepsilon\mathbf{p}_F}\hat{a}^-_{\mathbf{k}+\mathbf{p}+\varepsilon\mathbf{p}_F}\rangle](1 + O(\kappa^\vartheta)) \tag{158}$$

and

$$-ip_0\langle\hat{j}_\mathbf{p}\hat{a}^+_\mathbf{k}\hat{a}^-_{\mathbf{k}+\mathbf{p}}\rangle + \varepsilon p v_F\langle\hat{\rho}_\mathbf{p}\hat{a}^+_\mathbf{k}\hat{a}^-_{\mathbf{k}+\mathbf{p}}\rangle$$
$$= K[\langle\hat{a}^+_\mathbf{k}\hat{a}^-_\mathbf{k}\rangle - \langle\hat{a}^+_{\mathbf{k}+\mathbf{p}}\hat{a}^-_{\mathbf{k}+\mathbf{p}}\rangle](1 + O(\kappa^\vartheta))\,. \tag{159}$$

The first Ward Identity (158) for the chain model (34) was derived by the continuity equation, while the second follows from *emerging* chiral symmetry and cannot be derived directly by the symmetries of the Hamiltonian. From (132) and (151) we get, if $D_\varepsilon(\mathbf{p}) = -i\omega + \varepsilon v_F p$

$$\langle\hat{\rho}_\mathbf{p}\hat{\rho}_{-\mathbf{p}}\rangle = \left[\frac{Z^{(3)}}{Z}\right]^2 \frac{1}{4\pi v_F} \frac{1}{1-\tau^2}\left[\frac{D_-(\mathbf{p})}{D_+(\mathbf{p})} + \frac{D_+(\mathbf{p})}{D_-(\mathbf{p})} - 2\tau\right] + \hat{A}_{\rho,\rho}(\mathbf{p})\,. \tag{160}$$

As we noticed $\hat{A}_{\rho,\rho}(\mathbf{p})$ is continuous and its value at $\mathbf{p} = 0$ is determined by the condition (43), that is $\langle\hat{\rho}_\mathbf{p}\hat{\rho}_{-\mathbf{p}}\rangle_{0,p} = 0$; with this value of $\hat{A}_{\rho,\rho}(0)$ we get:

$$\langle\hat{\rho}_\mathbf{p}\hat{\rho}_{-\mathbf{p}}\rangle = \left[\frac{Z^{(3)}}{Z}\right]^2 \frac{1}{4\pi v_F} \frac{1}{1-\tau^2} \frac{v_F^2 p^2}{\omega^2 + v_F^2 p^2} + O(p)\,. \tag{161}$$

Finally, using (153) and (156) we get:

$$\langle\hat{\rho}_\mathbf{p}\hat{\rho}_{-\mathbf{p}}\rangle = \frac{K}{\pi v_F}\frac{v_F^2 p^2}{\omega^2 + v_F^2 p^2} + O(p) \tag{162}$$

and the conclusion is that:

$$\kappa = \lim_{p\to 0}\lim_{\omega\to 0}\langle\hat{\rho}_\mathbf{p}\hat{\rho}_{-\mathbf{p}}\rangle = \frac{K}{\pi v_F}\,. \tag{163}$$

In the same way we get:

$$\langle\hat{J}_\mathbf{p}\hat{J}_{-\mathbf{p}}\rangle = \left[\frac{\tilde{Z}^{(3)}}{Z}\right]^2 \frac{1}{4\pi v_F} \frac{1}{1-\tau^2}\left[\frac{D_-(\mathbf{p})}{D_+(\mathbf{p})} + \frac{D_+(\mathbf{p})}{D_-(\mathbf{p})} + 2\tau\right] + \hat{A}_{j,j}(\mathbf{p}) \tag{164}$$

and using (153) and (156) we get:

$$D(\mathbf{p}) = \frac{Kv_F}{\pi} \frac{\omega^2}{\omega^2 + v_F^2 p^2} + O(\omega) \tag{165}$$

so that:

$$D = \lim_{\omega \to 0} \lim_{p \to 0} D(\mathbf{p}) = \frac{Kv_F}{\pi}. \tag{166}$$

This concludes the proof of (48), (49) for the a generic nonsolvable chain model (34) and coupling not too large.

8. Conclusions

We have reviewed a recent approach to one-dimensional systems of interacting fermions based on exact RG and Constructive QFT methods. The exponents and several physical quantities can be written in terms of *convergent* expansions and several Luttinger liquid relations are rigorously established, via a combination of regularity results and *emerging* Ward Identites related to asymptotic symmetries. Such relations are true in the Luttinger model, whose linear dispersion relation allows an exact solution, but their validity in more realistic fermionic models with nonlinear dispersion relation (as lattice or Jellium models), even if widely accepted, was unproven except in certain particular solvable cases. The approach reviewed above proves the validity of the Luttinger liquid conjecture, in the Euclidean region and for coupling not too large (the analytic estimate of the convergence radius could be improved by computer assisted analysis, as for KAM series), for generic 1D systems solvable or not solvable; we have focused to spinless lattice fermions for definiteness, but the proof holds for continuum Jellium models or spinning Hubbard models with repulsive interaction in the nonhalf filled band case. A similar approach has been applied also to classical bidimensional spin models like the Ashkin–Teller or the Eight vertex or any models of coupled spins with a quartic interaction,[44,48] where a proof of several of the Kadanoff relations between critical exponents[9] has been achieved. An important feature of this approach is that the irrelevant terms are fully taken into account: the analysis is performed at Euclidean times but the irrelevant terms are expected to play an even more important role at real times. Therefore, it would be important to extend such exact methods to the computation of the dynamic responses at real frequencies.

References

1. S. Tomonaga, *Prog. Theor. Phys.* **5**, 544 (1950).
2. J. M. Luttinger, *J. Math. Phys.* **4**, 1154 (1963).
3. W. Thirring, *Ann. Phys.* **3**, 91 (1958).

4. K. Johnson, *Nuovo Cimento* **20**, 773 (1961).
5. D. Mattis and E. Lieb, *J. Math. Phys.* **6**, 304 (1965).
6. D. M. Haldane, *Phys. Rev. Lett.* **45**, 1358 (1980).
7. D. M. Haldane, *J. Phys. C* **14**, 2585 (1981).
8. A. Luther and I. Peschel, *Phys. Rev. B* **12**, 3908 (1975).
9. L. P. Kadanoff, *Phys. Rev. Lett.* **39**, 903 (1977).
10. A. W. Overhauser, *Physica* **1** 307 (1965).
11. A. Theunmann, *Phys. Rev. B* **15**, 4524 (1977).
12. G. Benfatto, G. Gallavotti and V. Mastropietro, *Phys. Rev. B* **45**(10), 5468 (1991).
13. V. Mastropietro, *Nuovo Cimento* **109**(1), 1 (1993).
14. W. Metzner and C. Di Castro, *Phys. Rev. B* **47**, 16107 (1993).
15. T. D. Schultz, D. C. Mattis and E. H. Lieb, *Rev. Mod. Phys.* **36**, 856 (1964).
16. D. Sutherland, *J. Math. Phys.* **11**, 3183 (1970).
17. R. J. Baxter, *Exactly Solved Models in Statistical Mechanics* (Academic Press Inc., London, 1989).
18. C. N. Yang and C. P. Yang, *Phys. Rev.* **147**, 303 (1966).
19. G. G. Santos *Phys. Rev. B* **46**, 14217 (1992).
20. A. A. Abrikosov, L. P. Gorkov and I. E. Dzylaloshinski, *Methods of Quantum Field Theory in Statistical Physics* (Prentice Hall, 1963).
21. I. E. Dzylaloshinski and A. I. Larkin, *Sov. Phys. JETP* **38**(1), 202 (1974).
22. J. Solyom, *Adv. Phys.* **28**(2), 209 (1979).
23. K. G. Wilson, *Phys. Rev. B* **4**, 3174 (1971).
24. J. Polchinski, *Nucl. Phys. B* **231**, 269 (1984).
25. G. Gallavotti, *Rev. Mod. Phys.* **57**, 471 (1985).
26. G. Benfatto and G. Gallavotti, *J. Stat. Phys.* **59**, 541 (1990).
27. K. Gawedzki and A. Kupiainen, *Commun. Math. Phys.* **102**, 1 (1985).
28. J. Feldman *et al.*, *Commun. Math. Phys.* **103**, 67 (1986).
29. A. Lesniewski, *Commun. Math. Phys.* **108**, 437 (1987).
30. G. Benfatto *et al.*, *Commun. Math. Phys.* **160**, 93 (1994).
31. G. Benfatto and V. Mastropietro, *Rev. Math. Phys.* **13**, 1323 (2001).
32. G. Benfatto and V. Mastropietro, *Commun. Math. Phys.* **231**(1), 97 (2002).
33. G. Benfatto and V. Mastropietro, *Commun. Math. Phys.* **258**, 609 (2005).
34. V. Mastropietro, *J. Math. Phys.* **48**, 022302 (2007).
35. G. Benfatto, P. Falco and V. Mastropietro, *Commun. Math. Phys.* **292**(2), 569 (2009).
36. G. Benfatto, P. Falco and V. Mastropietro, *Phys. Rev. Lett.* **104**, 075701 (2010).
37. G. Benfatto and V. Mastropietro, *J. Stat. Phys.* **138**(6), 1084 (2010).
38. G. Benfatto and V. Mastropietro, *J. Stat. Phys.* **142**(2), 251 (2010).
39. V. Mastropietro, *Nonperturbative Renormalization* (World Scientific, 2008).
40. E. R. Caianiello, *Nuovo Cimento* **10**, 1634 (1960).
41. G. Battle and P. Federbush, *Ann. Phys.* **142**, 95 (1982); D. Brydges, A short course on cluster expansions, eds. H. Les, K. Osterwalder and R. Stora (North Holland Press, 1986).
42. K. Osterwalder and R. Stora (eds.), *A Short Course on Cluster Expansions* (North Holland Press, 1986).
43. V. Mastropietro, *J. Phys. A* **40**, 10349 (2007).
44. G. Benfatto, P. Falco and V. Mastropietro, *Commun. Math. Phys.* **273**, 67 (2007).
45. S. L. Adler and W. A. Bardeen, *Phys. Rev.* **182**, 1517 (1969).
46. H. Georgi and J. M. Rawls, *Phys. Rev. D* **3**, 874 (1971).
47. G. Gentile and B. Scoppola, *Commun. Math. Phys.* **154**, 135 (1993).
48. V. Mastropietro, *Commun. Math. Phys.* **244**, 595 (2004).

THE LUTTINGER LIQUID AND INTEGRABLE MODELS*

J. SIRKER

Department of Physics and Research Center OPTIMAS,
Technical University Kaiserslautern,
D-67663 Kaiserslautern, Germany

Many fundamental one-dimensional lattice models such as the Heisenberg or the Hubbard model are integrable. For these microscopic models, parameters in the Luttinger liquid theory can often be fixed and parameter-free results at low energies for many physical quantities such as dynamical correlation functions obtained where exact results are still out of reach. Quantum integrable models thus provide an important testing ground for low-energy Luttinger liquid physics. They are, furthermore, also very interesting in their own right and show, for example, peculiar transport and thermalization properties. The consequences of the conservation laws leading to integrability for the structure of the low-energy effective theory have, however, not fully been explored yet. I will discuss the connection between integrability and Luttinger liquid theory here, using the anisotropic Heisenberg model as an example. In particular, I will review the methods which allow to fix free parameters in the Luttinger model with the help of the Bethe ansatz solution. As applications, parameter-free results for the susceptibility in the presence of nonmagnetic impurities, for spin transport, and for the spin-lattice relaxation rate are discussed.

Keywords: Luttinger liquids; integrable models; conservation laws.

1.   Introduction

The Tomonaga–Luttinger liquid[1–4] is believed to describe the low-energy properties of gapless one-dimensional interacting electron systems irrespective of the precise nature of the microscopic Hamiltonian. This *universality* can be understood in a renormalization group sense as irrelevance of band curvature and additional interaction terms which might arise when deriving this low-energy effective theory from a microscopic model. Similar to the important role Onsager's exact solution[5] of the two-dimensional Ising model has played in establishing and confirming general renormalization group theory, exactly solvable one-dimensional quantum models have been crucial for the development of Luttinger liquid theory.

Integrable models are, furthermore, also interesting in their own right and a number of almost ideal realizations are known today. One example are cuprate spin chains such as Sr_2CuO_3 whose magnetic properties are well described by the integrable one-dimensional Heisenberg model.[6–14] Furthermore, cold atomic gases represent quantum systems which are to a high degree isolated from the surroundings and whose Hamiltonians are easily tunable. This makes it possible to use them as

---

*This article first appeared in International Journal of Modern Physics B, Vol. 26, No. 22 (2012).

quantum simulators to study almost perfect realizations of integrable systems such as the Lieb–Liniger model[15,16] or the fermionic Hubbard model.[17]

In Sec. 2 I will discuss quantum integrability with a particular emphasis on Bethe ansatz integrable models and outline possible effects on transport and the thermalization of closed quantum systems. In the rest of the paper, I will then concentrate on the anisotropic Heisenberg (or XXZ) model as one specific example for a Bethe ansatz integrable model. In Sec. 3 I describe how the Luttinger model, including leading irrelevant operators, can be obtained from this microscopic model by using bosonization techniques. In Sec. 4 I then briefly outline important aspects of the Bethe ansatz solution. In Sec. 5 it is shown that a comparison of the results of Secs. 3 and 4 allows to fix parameters in the Luttinger liquid theory for the XXZ model. Applications of the parameter-free low-energy effective theory to calculate various properties of spin chains are considered in Sec. 6. This includes the calculation of susceptibilities in the presence of nonmagnetic impurities, and results for spin transport and NMR relaxation rates. The final section is devoted to a brief summary and some conclusions.

2. Quantum Integrability

A classical system with Hamilton function $\mathcal{H}$ and phase space dimension $2N$ is integrable if it has N constants of motion $\mathcal{Q}_n$ with

$$\{\mathcal{H}, \mathcal{Q}_n\} = 0 \quad \text{and} \quad \{\mathcal{Q}_i, \mathcal{Q}_j\} = 0 \quad \text{if} \quad i \neq j \,. \tag{1}$$

Here $\{\cdot, \cdot\}$ denotes the Poisson bracket. Quantum integrability, on the other hand, is much harder to define precisely, see, for example, Ref. 18. In this regard it is important to note that every quantum system in the thermodynamic limit, irrespective of integrability, has infinitely many conservation laws

$$[H, |E_n\rangle\langle E_n|] = 0 \,, \tag{2}$$

where H is the Hamiltonian with eigenstates $|E_n\rangle$ and $[\cdot, \cdot]$ denotes the commutator. Apart from these *nonlocal* conservation laws a quantum system can have local conservation laws given by

$$\mathcal{Q}_n = \sum_j q_{n,j} \quad \text{or} \quad \mathcal{Q}_n = \int dx \, q_n(x) \,, \tag{3}$$

where $q_{n,j}$ is a density operator acting on n neighboring sites in the case of a lattice model while for a continuum model $q_n(x)$ is a fully local density operator. A generic example for a local conservation law is the Hamiltonian itself for models with short range interactions. In Bethe ansatz integrable models, a whole set of such local conservation laws does exist which can be obtained from the transfer matrix of the corresponding two-dimensional classical model by taking successive derivatives of the transfer matrix τ with respect to the spectral parameter λ

$$\mathcal{Q}_n \propto \frac{\partial^n}{\partial \lambda^n} \ln \tau(\lambda)|_{\lambda=\xi} \,. \tag{4}$$

Here ξ is the spectral parameter at which the transfer matrix is evaluated. These conserved quantities are directly related to the existence of so-called R-matrices which fulfill the Yang–Baxter equations and from which the transfer matrices $\tau(\lambda)$ can be constructed.[19]

For a low-energy effective theory describing such an integrable model, we have to demand — at least in principle — that the low-energy Hamiltonian H_{eff} also fulfills

$$[H_{\mathrm{eff}}, \mathcal{Q}_n] = 0 \,. \tag{5}$$

This corresponds to a fine-tuning of parameters in the Luttinger model. In particular, it might mean that certain terms which are not forbidden by general symmetry considerations have to vanish. Such a program has not fully been explored yet; in Sec. 5.2 we will see, as an example, that the conserved quantity $\mathcal{Q}_3$ for the XXZ model does indeed prevent certain terms from occurring in the low-energy theory.

2.1. Consequences for transport

Local conservation laws can have a dramatic effect on the transport properties.[20] This can be easily understood as follows. We can always define a local current density j_l by making use of the continuity equation

$$\frac{\partial}{\partial t}\rho_l + j_l - j_{l-1} = 0 \,, \tag{6}$$

where ρ_l is the density at site l. The current itself is then given by $\mathcal{J} = \sum_l j_l$. This current could be, for example, an electric, spin or thermal current. Conserved quantities Q_n can now prevent a current from decaying completely leading to ballistic transport and a *finite Drude weight*

$$D(T) \equiv \lim_{t\to\infty} \lim_{L\to\infty} \frac{1}{2LT}\langle \mathcal{J}(t)\mathcal{J}(0)\rangle \geq \lim_{L\to\infty} \frac{1}{2LT} \sum_n \frac{\langle \mathcal{J}Q_n\rangle^2}{\langle Q_n^2\rangle} \,. \tag{7}$$

Here Q_n can denote a local or nonlocal conserved quantity, L is the length of the system, and T the temperature. The second relation in Eq. (7) is the Mazur inequality[21] which becomes an equality if all conservation laws, local and non-local, are included.[22,23] In order to obtain a possible nonzero Drude weight at finite temperatures within a Luttinger model description, the relevant conservation laws have to be taken into account explicitly. One way to achieve this is discussed in Sec. 6.2. Importantly, one expects that only local or pseudo-local conservation laws $\mathcal{Q}_n$ with $\langle \mathcal{J}Q_n\rangle \neq 0$ can give rise to a finite bound in Eq. (7) so that $D(T > 0) \neq 0$ is characteristic for an integrable model.

2.2. *Consequences for thermalization in closed systems*

Additional local conservation laws can also have a profound impact on a possible thermalization of a closed quantum system. Imagine that we prepare an initial state $|\Psi(0)\rangle$ and follow the unitary time evolution of this state under an integrable Hamiltonian. One says that a closed quantum system in the thermodynamic limit has thermalized if for any *local* observable $\mathcal{O}$ the limit

$$\mathcal{O}_\infty = \lim_{t\to\infty} \langle \Psi(t)|\mathcal{O}|\Psi(t)\rangle \tag{8}$$

is well-defined and time independent and can also be expressed as an ensemble average

$$\mathcal{O}_\infty \equiv \mathrm{Tr}(\mathcal{O}\rho) \tag{9}$$

with an appropriately chosen density matrix ρ. Note that even for a generic closed quantum system temperature T is not defined by an external bath but rather by the energy of the initial state

$$\langle \Psi(0)|H|\Psi(0)\rangle = \underbrace{\mathrm{Tr}(He^{-H/T})/Z}_{\mathrm{Tr}(\rho_c H)} \tag{10}$$

with T acting as a Lagrange multiplier and Z being the partition function. For an integrable model we have to demand that the relation (10) also holds if we replace $H \to \mathcal{Q}_n$ and the canonical density matrix ρ_c by the density matrix[24]

$$\rho = \frac{1}{Z}\exp\left(-\sum_j \lambda_j \mathcal{Q}_j\right), \tag{11}$$

which now contains a Lagrange multiplier λ_j for *each* of the locally conserved quantities. The existence of additional local conservation laws therefore severely restricts a possible thermalization of the system leading to additional constraints which are incorporated by the Lagrange multipliers in Eq. (11). Experimental indications for such constraints have been seen in realizations of the Lieb–Liniger model in ultracold gases.[16]

3. Low-Energy Description of the *XXZ* Model

In the following sections, we want to concentrate on one of the simplest integrable lattice models, the *XXZ* model

$$H = J \sum_{j=1}^{N(N-1)} \left[-\frac{1}{2}(c_j^\dagger c_{j+1} + \mathrm{h.c.}) - h\left(c_j^\dagger c_j - \frac{1}{2}\right) + \Delta\left(n_j - \frac{1}{2}\right)\left(n_{j+1} - \frac{1}{2}\right) \right]. \tag{12}$$

Here c $(c^\dagger)$ annihilates (creates) a spinless fermion, J gives the energy scale, Δ characterizes the nearest neighbor density–density interaction with the density operator $n_j = c_j^\dagger c_j$. N is the number of sites and the boundary conditions might be either periodic (sum runs up to N with $c_{N+1}^{(\dagger)} \equiv c_1^{(\dagger)}$) or open (sum runs only up to $N-1$). h acts as a chemical potential. With the help of the Jordan–Wigner transformation

$$S_j^z \to n_j - \frac{1}{2}, \quad S_j^+ \to (-1)^j c_j^\dagger e^{i\pi\phi_j}, \quad S_j^- \to (-1)^j c_j e^{-i\pi\phi_j}, \tag{13}$$

where $\phi_j = \sum_{l=1}^{j-1} n_l$ we can also express this model in terms of spin-1/2 operators

$$H = J \sum_{j=1}^{N(N-1)} [S_j^x S_{j+1}^x + S_j^y S_{j+1}^y + \Delta S_j^z S_{j+1}^z - hS_j^z]. \tag{14}$$

For both kinds of boundary conditions the *XXZ* model is integrable by Bethe ansatz.[17,25–28] The Luttinger liquid approach is applicable in the critical regime which is given by $-1 < \Delta \leq 1$ for $h = 0$. In general, the range of anisotropies for which the model is critical depends on the applied magnetic field h. In the free fermion case, the model (12) is easily solved by Fourier transform leading to

$$H_0 = \sum_p \epsilon_p c_p^\dagger c_p \quad \text{with} \quad \epsilon_p = -J(\cos p + h). \tag{15}$$

where we have set the lattice constant $a = 1$. The allowed momenta are given by $p = 2\pi n/N$ with $n = 0, \ldots, N-1$ for periodic boundary conditions (PBCs) or $p = \pi n/(N+1)$, $n = 1, \ldots, N$ for open boundary conditions (OBCs). In Sec. 4 we will briefly discuss the Bethe ansatz solution of this model for OBCs.

Let us first revisit the derivation of an effective low-energy description, the Luttinger theory, by bosonization following Refs. 4, 29 and 30. First, we replace the fermionic operators in the continuum limit by two fields $\psi_{R,L}$ defined near the two Fermi points $\pm k_F = \pm \arccos(-h)$:

$$c_j \to \psi(x) = e^{ik_F x} \psi_R(x) + e^{-ik_F x} \psi_L(x). \tag{16}$$

In a second step, we use standard Abelian bosonization to write the fermion fields as

$$\psi_{R,L}(x) \sim \frac{1}{\sqrt{2\pi\alpha}} e^{-i\sqrt{2\pi}\phi_{R,L}(x)}, \tag{17}$$

where $\alpha \sim k_F^{-1}$ is a short distance cutoff. Instead of working with the left and right components $\phi_{R,L}$ we can define a bosonic field $\tilde{\phi}$ and its dual field $\tilde{\theta}$ by

$$\tilde{\phi} = \frac{\phi_L - \phi_R}{\sqrt{2}}, \quad \tilde{\theta} = \frac{\phi_L + \phi_R}{\sqrt{2}}, \tag{18}$$

which satisfy the standard bosonic commutation rule $[\tilde{\phi}(x), \partial_{x'}\tilde{\theta}(x')] = i\delta(x - x')$.

If we bosonize the kinetic energy term of Eq. (12) keeping only the lowest order we obtain

$$H_0^{\text{kin}} = iv_F \int_0^L dx (: \psi_R^\dagger \partial_x \psi_R : - : \psi_L^\dagger \partial_x \psi_L :)$$

$$= \frac{v_F}{2} \int_0^L dx [(\partial_x \phi_R)^2 + (\partial_x \phi_L)^2], \tag{19}$$

where $::$ denotes normal ordering and $v_F = J \sin k_F$ is the Fermi velocity. This approximation corresponds to a linearization of the dispersion ϵ_p at the Fermi points $\pm k_F$. In this case the bosonic model is quadratic in $\partial_x \phi_{R,L}$. Corrections to the kinetic energy appear due to band curvature. Including these curvature terms, we can write the expansion of the dispersion ϵ_p near the two Fermi points as

$$\epsilon_k^{R,L} \approx \pm v_F k + \frac{k^2}{2M} \mp \frac{\gamma k^3}{6} + \cdots, \tag{20}$$

where $k \equiv p \mp k_F$ for the right or left movers, respectively, $M = (J \cos k_F)^{-1}$ is the effective mass and $\gamma = J \sin k_F$. Note that the inverse mass M^{-1} vanishes in the particle-hole symmetric case $h = 0$. In this case, the curvature correction is cubic in momentum. Bosonization of the k^2-term leads to a correction cubic in $\partial_x \phi_{R,L}$, whereas the term cubic in momentum gives a quartic correction in terms of the bosonic fields. Cubic and quartic terms in the bosonic operators will also arise from the interaction term in Eq. (12). The scaling dimension of these terms is 3 and 4 respectively so that they are formally irrelevant. The interaction will, however, also produce additional marginal terms, quadratic in the bosonic fields, which together with (19) lead to the exactly solvable Luttinger model

$$H_{\text{LL}} = \frac{v_F}{2} \int dx \left\{ \left(1 + \frac{g_4}{2\pi v_F}\right) [(\partial_x \phi_R)^2 + (\partial_x \phi_L)^2] - \frac{g_2}{\pi v_F} \partial_x \phi_L \partial_x \phi_R \right\}. \tag{21}$$

Here $g_2 = g_4 = 2J\Delta[1 - \cos(2k_F)] = 4J\Delta \sin^2 k_F$ are interaction parameters. The Hamiltonian (21) can be rewritten in the form

$$H_{\text{LL}} = \frac{1}{2} \int dx \left[vK(\partial_x \tilde{\theta})^2 + \frac{v}{K}(\partial_x \tilde{\phi})^2 \right], \tag{22}$$

where v (the renormalized velocity) and K (the Luttinger parameter) are given by

$$v = v_F \sqrt{\left(1 + \frac{g_4}{2\pi v_F}\right)^2 - \left(\frac{g_2}{2\pi v_F}\right)^2} \approx v_F \left(1 + \frac{2\Delta}{\pi} \sin k_F\right), \qquad (23)$$

$$K = \sqrt{\frac{1 + \dfrac{g_4}{2\pi v_F} - \dfrac{g_2}{2\pi v_F}}{1 + \dfrac{g_4}{2\pi v_F} + \dfrac{g_2}{2\pi v_F}}} \approx 1 - \frac{2\Delta}{\pi} \sin k_F . \qquad (24)$$

Expressions (23) and (24) are approximations valid in the limit $|\Delta| \ll 1$. In Sec. 5 we will review how these parameters in the Luttinger liquid Hamiltonian can be fixed exactly for arbitrary interaction strengths $-1 < \Delta \leq 1$ using the Bethe ansatz solution.

The Luttinger parameter in the Hamiltonian (22) can be absorbed by performing a canonical transformation that rescales the fields in the form $\tilde{\phi} \to \sqrt{K}\phi$ and $\tilde{\theta} \to \theta/\sqrt{K}$ leading to

$$H_{\mathrm{LL}} = \frac{v}{2} \int dx [(\partial_x \theta)^2 + (\partial_x \phi)^2] . \qquad (25)$$

We can also define the right and left components of these rescaled bosonic fields by

$$\varphi_{R,L} = \frac{\theta \mp \phi}{\sqrt{2}} . \qquad (26)$$

These are related to $\phi_{R,L}$ by a Bogoliubov transformation.

3.1. Irrelevant operators in the finite field case

The leading irrelevant operators stem from the k^2-term in Eq. (20) and give rise to dimension three operators $\sim (\partial_x \phi_{R,L})^3$. Similar terms will also arise by bosonizing the interaction term. Instead of deriving these terms from the microscopic Hamiltonian, we can introduce them phenomenologically by considering the symmetries of the problem. In particular, the low-energy effective theory has to be symmetric under the parity transformation $\phi_L \to \phi_R$, $\phi_R \to \phi_L$, and $x \to -x$. We therefore can parametrize these terms as[30]

$$\delta H = \frac{\sqrt{2\pi}}{6} \int dx \{ \eta_- [(\partial_x \varphi_L)^3 - (\partial_x \varphi_R)^3]$$
$$+ \eta_+ [(\partial_x \varphi_L)^2 \partial_x \varphi_R - (\partial_x \varphi_R)^2 \partial_x \varphi_L] \} . \qquad (27)$$

We will see in Sec. 5 that we can relate the amplitudes $\eta_\pm$ to quantities which are known from the exact solution. The derivation of these terms starting from the microscopic Hamiltonian, on the other hand, would only allow us to obtain the coupling constants to first-order in Δ with[30]

$$\eta_- \approx \frac{1}{M}\left(1 + \frac{2\Delta}{\pi}\sin k_F\right), \quad \eta_+ \approx -\frac{3\Delta}{\pi M}\sin k_F\,. \tag{28}$$

From this expansion we see that (a) both terms vanish in the limit $h \to 0$ where $M^{-1} \to 0$, and (b) that the term mixing right and left movers parametrized by η_+ is only present in the interacting case, $\Delta \neq 0$.

3.2. Irrelevant operators for zero field

In the particle-hole symmetric case, $h = 0$, the first correction to the linear dispersion relation is cubic in momentum, see Eq. (20). Instead of bosonizing this term starting from the microscopic Hamiltonian (12) we again introduce the corresponding terms in the bosonic model based on symmetry arguments. The dimension four operators allowed by symmetry can be parametrized as

$$\delta\mathcal{H} = \frac{\pi\zeta_-}{12}[: (\partial_x\varphi_R)^2 :: (\partial_x\varphi_R)^2 : + : (\partial_x\varphi_L)^2 :: (\partial_x\varphi_L)^2 :]$$

$$+ \frac{\pi\zeta_+}{2}[: (\partial_x\varphi_R)^2 :: (\partial_x\varphi_L)^2 :] + \pi\zeta_3[: (\partial_x\varphi_R)^3 :: \partial_x\varphi_L :$$

$$+ : (\partial_x\varphi_L)^3 :: \partial_x\varphi_R :]\,. \tag{29}$$

The explicit bosonization of the corresponding band curvature and interaction terms yields the coupling constants again only to lowest lowest order

$$\zeta_- \approx -J\left(1 + \frac{\Delta}{\pi}\right), \quad \zeta_+ \approx -\frac{\Delta J}{\pi}, \quad \zeta_3 = 0\,. \tag{30}$$

In addition, the Umklapp scattering term $\delta\mathcal{H}_U \sim e^{4ik_F x}\Psi_R^\dagger(x)\Psi_L(x)\Psi_R^\dagger(x+1)$ $\Psi_L(x+1) + \text{h.c.}$ is commensurate in this case, $4k_F = 2\pi$, and therefore has to be kept in the low-energy effective theory. Bosonizing this term leads to

$$\delta\mathcal{H}_U = \lambda\cos(4\sqrt{\pi K}\phi)\,, \tag{31}$$

and to lowest order in Δ we have $\lambda = J\Delta/(2\pi^2)$. For OBC, there is also an irrelevant boundary operator allowed

$$\delta\mathcal{H}_B \sim (\delta(x) + \delta(L))(\partial_x\phi)^2\,. \tag{32}$$

Finally, we want to consider the Luttinger model (25) with an additional small magnetic field δh added, ignoring the irrelevant terms

$$H = \frac{v}{2}\int_0^L\left[(\partial_x\phi)^2 + (\partial_x\theta)^2 - \frac{2}{v}\sqrt{\frac{K}{\pi}}\delta h\partial_x\phi\right]\,. \tag{33}$$

By performing a shift in the boson field

$$\phi \to \phi + \sqrt{\frac{K}{\pi}} \frac{x}{v} \delta h \tag{34}$$

we return to the quadratic Hamiltonian (25) with an additional constant shift $-LK(\delta h)^2/(2\pi v)$. The bulk susceptibility per site is therefore given by

$$\chi_{\text{bulk}} = K/\pi v \,. \tag{35}$$

This result does not only hold for $h = 0$ but also for any finite field h_0 at which we want to calculate χ with $K = K(\Delta, h_0)$ and $v = v(\Delta, h_0)$.

4. The Bethe Ansatz Solution

To exactly solve the interacting system for PBC or OBC, one can use the coordinate Bethe ansatz.[17,25–28] Here we want to review very briefly some of the essential results needed to fix the parameters in the Luttinger model and refer the reader to Refs. 28, 31–33 for a more detailed discussion. The coordinate Bethe ansatz starts from the fully polarized state ("the vacuum") and one derives coupled eigenvalue equations $H|M\rangle = E|M\rangle$ for states $|M\rangle$ with M spins flipped. The eigenenergies can then be written as

$$E = J \sum_{j=1}^{M} \cos k_j + J\Delta \left(\frac{N-1}{4} - M \right). \tag{36}$$

The structure is similar to the noninteracting case, however, the momenta k_j are shifted from their positions for $\Delta = 0$. They can be determined from a set of coupled nonlinear equations. In the thermodynamic limit, a single integral equation for the density of roots $\rho(x)$ is obtained which parametrizes the allowed momenta

$$\vartheta(x, \gamma) + \frac{1}{2N} [\vartheta(x, \gamma) + \vartheta(x, \pi - \gamma) + \vartheta(x, 2\gamma)]$$

$$= \rho(x) + \int_{-B}^{B} \vartheta(x - y, 2\gamma) \rho(y) dy \tag{37}$$

where

$$\vartheta(x, \gamma) = \frac{1}{\pi} \frac{\sin \gamma}{\cosh 2x - \cos \gamma} \tag{38}$$

and we have set $\Delta = \cos \gamma$. Equation (37) is the integral equation for the XXZ chain with OBC in the thermodynamic limit, including the boundary correction $\sim \mathcal{O}(1/N)$. Omitting the $1/N$ correction this is the standard integral equation for PBC.[28] The integral equation contains an unknown parameter B and an unknown function $\rho(x)$. It can be solved analytically by Fourier transform for $B = \infty$ and one finds (ignoring the $1/N$ correction)

$$\rho(x) = \frac{1}{2\gamma \cosh \pi x/\gamma}. \tag{39}$$

The magnetization per site m and the ground state energy per site e for general B are given by

$$m = 1/2 - \int_{-B}^{B} \rho(x)dx + 1/(2N), \tag{40}$$
$$e = -hs^z - \frac{J \sin \gamma}{2} \int_{-B}^{B} \vartheta(x, \gamma)\rho(x)dx + \frac{J}{4}\left(\cos \gamma + \frac{2 - \cos \gamma}{N}\right).$$

Inserting Eq. (39) into Eq. (40) one finds that $B = \infty$ corresponds to $m = 0$, i.e., to the case of zero magnetic field $h = 0$. In general, $B = B(h)$ and the dependence on magnetic field has to be determined numerically. This can be achieved by using the stationarity condition

$$\frac{\partial}{\partial B}[e(h) - e(h = 0)] = 0. \tag{41}$$

5. Fixing Parameters of the Luttinger Model Using Integrability

One of the main motivations to apply Luttinger liquid theory to integrable models is that parameters in the Luttinger liquid theory such as velocities, Luttinger parameters, coupling constants of irrelevant operators and prefactors of correlation functions which usually are nonuniversal and therefore unknown, can often be determined in the case of an integrable model.[31,34–37] This makes it possible to obtain parameter-free results at low energies. The general idea is to calculate static observables at zero or finite temperatures exactly using the Bethe ansatz and to compare with results obtained within Luttinger theory. In the following we briefly review this method to obtain the velocity and Luttinger liquid parameter, Sec. 5.1, and the coupling constants for band curvature and Umklapp terms for the XXZ model, Sec. 5.2.

5.1. *Velocity and Luttinger liquid parameter*

To obtain the velocity of elementary excitations, we have to consider the change in energy when replacing the ground state distribution of roots by a distribution which contains an excitation near the Fermi points. In the case of finite magnetic field, the obtained Bethe ansatz equations can only be solved numerically. Here we want to restrict ourselves to the zero field case. Replacing the ground state distribution (39) in the expression for the energy (40) by the distribution containing an excitation

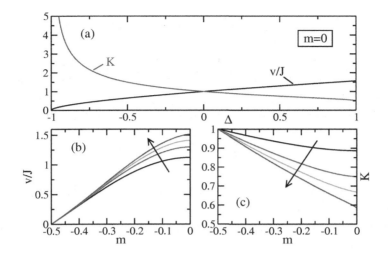

Fig. 1. (a) Velocity v/J and Luttinger parameter K for magnetization $m = 0$ as a function of Δ. (b) Velocity and (c) Luttinger parameter as a function of magnetization m for $\Delta = 0.2, 0.5, 0.7, 0.9$ (in arrow direction).

gives the energy in terms of the momentum change.[28] This allows to read off the spin velocity

$$v = \left. \frac{\partial E}{\partial k} \right|_{k=k_F} = \frac{J\pi}{2} \frac{\sin \gamma}{\gamma} = \frac{J\pi}{2} \frac{\sqrt{1 - \Delta^2}}{\arccos \Delta} \, . \tag{42}$$

The spin velocity therefore increases from $v_F = J$ (remember that we have set $a = 1$) at the free fermion point to $v = J\pi/2$ at the isotropic antiferromagnetic Heisenberg point. Conversely, the velocity vanishes, as expected, for $\Delta \to -1$ corresponding to the isotropic ferromagnet.

To determine the Luttinger parameter K it is easiest to calculate the bulk susceptibility $\chi_{\text{bulk}}(h) = \partial m / \partial h$ using Eq. (40). To do so, $B(h)$ is required. For finite magnetic fields this again requires a numerical solution. For infinitesimal fields, on the other hand, $B(h)$ can be determined analytically[28,32,33] and

$$\chi_{\text{bulk}} = \frac{1}{J} \frac{\gamma}{(\pi - \gamma)\pi \sin \gamma} = \frac{1}{2v(\pi - \gamma)} \tag{43}$$

is obtained. Comparing with Eq. (35) we find

$$K = \frac{\pi}{2(\pi - \gamma)} = \frac{\pi}{2(\pi - \arccos \Delta)} \, . \tag{44}$$

Therefore $K = 1$ for the free fermion model and $K = 1/2$ at the isotropic antiferromagnetic Heisenberg point. Equations (42) and (44) agree to first order with the expressions (24). The velocity and Luttinger parameter, both for zero and finite fields, as obtained from the Bethe ansatz solution, are shown in Fig. 1.

5.2. *Coupling constants of irrelevant operators*

Next, we want to review how the coupling constants $\zeta_{\pm,3}$, Eq. (29), and λ, Eq. (31), can be fixed in the zero field case, and the coupling constants $\eta_\pm$, Eq. (27), for finite field. The zero field case has been first considered by Lukyanov[31] and analytical formulas for the coupling constants have been obtained. The finite field case has been treated in Refs. 30 and 38 leading to formulas which require a numerical solution of the Bethe ansatz equations.

5.2.1. *The zero field case*

The simplest way to determine the Umklapp scattering amplitude λ is to consider an open chain with a small magnetic field added.[33] In the low-energy description this means that we have to consider (33) with the Umklapp term (31) added. We can then again perform the shift (34). This brings us back to the standard Luttinger liquid Hamiltonian (25) and the magnetic field now appears in the Umklapp term (31). In first-order perturbation theory in Umklapp scattering we then find the following boundary correction to the ground state energy[39]

$$E_U^{(1)} = \lambda \int_0^\infty dx \left\langle \cos\left(4\sqrt{\pi K}\phi + \frac{4Khx}{v}\right)\right\rangle_0 . \tag{45}$$

Here $\langle\cdots\rangle_0$ denotes the correlation function calculated for the free boson model. For PBC this correlation function would vanish, however, for OBC we obtain

$$E_U^{(1)} = \lambda \int_0^\infty dx \frac{\cos\left(\dfrac{4Khx}{v}\right)}{(2x)^{4K}} . \tag{46}$$

Note that this is a $1/N$ correction to the ground state energy per site $e = E/N$. By partial integration we can split of the convergent part and find

$$E_U^{(1,\text{conv})} = -\lambda(2K)^{4K}\Gamma(-4K)\sin(2K\pi)\left(\frac{h}{v}\right)^{4K-1} . \tag{47}$$

At the same time, we can apply the Bethe ansatz to analytically calculate the so-called boundary susceptibility χ_B given by $\chi = \chi_{\text{bulk}} + \chi_B/N + \mathcal{O}(N^2)$ to leading orders in h.[32,33] The amplitude λ of the Umklapp term can now be found by comparing the exact result for χ_B with Eq. (47). This leads to

$$\lambda = \frac{2K\Gamma(2K)\sin\pi/2K}{\pi\Gamma(2-2K)}\left[\frac{\Gamma\left(1+\dfrac{1}{4K-2}\right)}{2\sqrt{\pi}\Gamma\left(1+\dfrac{K}{2K-1}\right)}\right]^{4K-2} . \tag{48}$$

In Ref. 31 this result has been obtained first by calculating the bulk correction to the ground state energy. Note, however, that this requires second-order perturbation theory in the Umklapp scattering. Particular care has to be taken when considering the isotropic antiferromagnet, $\Delta = 1$. In this case, Umklapp scattering becomes marginally irrelevant and λ has to be replaced by a running coupling constant which depends on the length scale where the system is considered. In general, both the length of the system and temperature will be of importance and the running coupling constant $g(L, v/T)$ can be introduced by the replacements $K \to (1 + g)/2$ and $\lambda \to -g/4$. An explicit solution of the renormalization group equations for g is only possible if one of those two length scales dominates. In the thermodynamic limit, for example, this scale will be set by temperature alone and one finds[31]

$$1/g + \ln(g)/2 = \ln(T_0/T) \tag{49}$$

with $T_0 = \sqrt{\pi/2}e^{1/4+\tilde{\gamma}}$ where $\tilde{\gamma}$ is the Euler constant. The scale T_0 has again been fixed by comparing with the Bethe ansatz result for the bulk susceptibility χ_{bulk} in the isotropic case.

Here integrability has been used to fix a coupling constant. The conservation laws underlying integrability discussed in Sec. 2 can, however, have an even more profound effect.[13,14] For the XXZ model the first of the nontrivial conserved quantities is the energy current $J^E = \mathcal{Q}_3$ given by

$$J^E = J^2 \sum_j [S^y_{j-1} S^z_j S^x_{j+1} - S^x_{j-1} S^z_j S^y_{j+1} + \Delta(S^x_{j-1} S^y_j S^z_{j+1} - S^z_{j-1} S^y_j S^x_{j+1})$$
$$+ \Delta(S^z_{j-1} S^x_j S^y_{j+1} - S^y_{j-1} S^x_j S^z_{j+1})] \, . \tag{50}$$

The latter is defined by the continuity equation of the energy density at zero field

$$j^E_{j+1} - j^E_j = -\partial_t \mathcal{H}_j = i[\mathcal{H}_j, H] \, , \tag{51}$$

where $H = \sum_j \mathcal{H}_j$ is the Hamiltonian (14) with $h = 0$, PBC and $J^E = \sum_j j^E_j$. The energy current operator for the Luttinger model can be obtained from (51) by taking the continuum limit. This leads to

$$J^E_0 = \int dx \, j^E_0(x) = \frac{v^2}{2} \int dx \, [(\partial_x \varphi_R)^2 - (\partial_x \varphi_L)^2] = -v^2 \int dx \, \partial_x \phi \partial_x \theta \, . \tag{52}$$

This operator is conserved, i.e., $[J^E_0, H_{\text{LL}}] = 0$. The irrelevant operators (29) lead to a correction of the energy current which can again be calculated using the continuity equation (51). To first-order one finds

$$\delta J^E = \pi v \int dx \left\{ \frac{\zeta_-}{3} [(\partial_x \varphi_R)^4 - (\partial_x \varphi_L)^4] + 2\zeta_3 [(\partial_x \varphi_R)^3 \partial_x \varphi_L - (\partial_x \varphi_L)^3 \partial_x \varphi_R] \right\} \, . \tag{53}$$

For $J^E = J^E_0 + \delta J^E$ to be conserved as required by integrability, we have to require that $[J^E, H] = [J^E_0 + \delta J^E, H_{\text{LL}} + \delta H] = 0$ up to the considered order. Since

$[J_E^0, H] = [J_0^E, H_{\text{LL}} + \delta H] = 0$ this implies that $[\delta J^E, H_{\text{LL}}] = 0$. The ζ_--term in Eq. (53) does not mix right and left movers and therefore obviously commutes with H_{LL}. The ζ_3-term, on the other hand, does mix the two modes and therefore does not commute with H_{LL}. Integrability therefore implies that $\zeta_3 = 0$. We see that apart from determining the precise values of coupling constants in the low-energy effective theory, there is a more fundamental consequence: integrability corresponds to a fine tuning of the coupling constants such that the local conservation laws are fulfilled. In particular, terms which are in general allowed by symmetry might be absent.

We are left with only two amplitudes, $\zeta_\pm$, for the dimension four operators. Let us briefly review how they can be fixed as well. Using Eq. (26) we can express both terms by the boson field ϕ and the dual field θ. Now performing again the shift (34) for a small applied magnetic field δh we find a first-order correction to the ground state energy per site

$$e_{\zeta_\pm}^{(1)} = (\zeta_- + 3\zeta_+) \frac{K^2(\delta h)^4}{24\pi v^4}. \tag{54}$$

The $(\delta h)^4$-term in the ground state energy can also be calculated analytically by Bethe ansatz.[31-33] One finds that the result consists of two distinct, additive, contributions. One of those vanishes at the free fermion point and is therefore associated with the ζ_+-term in the low-energy effective theory which mixes right and left movers. The other term then determines ζ_- leading to[31]

$$\zeta_- = -\frac{v}{4\pi K} \frac{\Gamma\left(\frac{6K}{4K-2}\right)\Gamma^3\left(\frac{1}{4K-2}\right)}{\Gamma\left(\frac{3}{4K-2}\right)\Gamma^3\left(\frac{K}{2K-1}\right)}, \quad \zeta_+ = -\frac{v}{2\pi}\tan\left(\frac{\pi K}{2K-1}\right). \tag{55}$$

The dependence on anisotropy of all three coupling constants $\lambda, \zeta_\pm$ is shown in Fig. 2. Note that the amplitude ζ_+ diverges for $K = (2n+1)/4n$ with $n \in \mathbb{N}$. Corrections to observables calculated in perturbation theory in the irrelevant operators are, however, usually finite. What happens is that at these special points the scaling dimensions of different irrelevant operators coincide and the two diverging amplitudes "conspire" to produce a finite result. This point has been discussed in detail in Ref. 33 using the susceptibility as an example.

5.2.2. *The finite field case*

For a finite magnetic field ($B < \infty$), the Bethe ansatz integral equation cannot be solved analytically. However the amplitudes $\eta_\pm$ can be related to changes in the Luttinger parameter and velocity when changing the field.[30,38] This allows for an accurate numerical determination of these parameters.

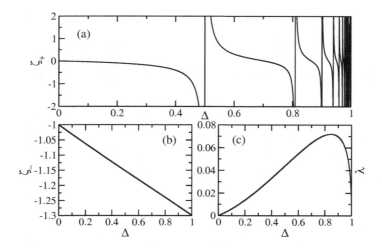

Fig. 2. (a) ζ_+ as function of anisotropy Δ. The amplitude diverges whenever the argument of the tan-function in Eq. (55) is $\pm\pi/2 (\mathrm{mod}\, 2\pi)$. (b) ζ_- and (c) λ, Eq. (48) as a function of Δ.

The basic idea is again quite simple. We consider the Luttinger liquid Hamiltonian at some finite magnetic field h_0. This means that our left and right modes live near Fermi points $\pm k_F \neq \pm\pi/2$. Now we apply an additional small magnetic field δh which we can take care of by the boson shift (34). If we now calculate the free energy we obtain

$$f = -\frac{\pi T^2}{6v(h)} - \underbrace{\frac{K(h)}{2\pi v(h)}}_{=\chi(h)/2}(J\delta h)^2 \tag{56}$$

with $h = h_0 + \delta h$. The interaction parameters $K(h)$ and $v(h)$ can be determined numerically as described in Sec. 4. Now we can expand (56) in δh and obtain to lowest order

$$\delta f = \frac{\pi T^2}{6v^2(h_0)}\frac{\partial v}{\partial h}\bigg|_{h=h_0}\delta h\,; \quad \delta\chi = \frac{K(h_0)}{\pi v(h_0)}\left[\frac{1}{K}\frac{\partial K}{\partial h}\bigg|_{h=h_0} - \frac{1}{v}\frac{\partial v}{\partial h}\bigg|_{h=h_0}\right]\delta h\,. \tag{57}$$

These corrections have to stem from the dimension three operators (27). The second approach therefore is to keep $K = K(h = h_0)$, $v = v(h = h_0)$ fixed and to perform the shift (34) also in (27). Calculating again the free energy by standard techniques we now find

$$\delta f = (3\eta_- - \eta_+)\frac{\pi\sqrt{K}J\delta h T^2}{18v^3(h_0)} + (\eta_- - \eta_+)\frac{K^{3/2}(J\delta h)^3}{6\pi v^3(h_0)}\,;$$

$$\delta\chi = (\eta_+ - \eta_-)\frac{K^{3/2}J}{\pi v^3(h_0)}\delta h\,. \tag{58}$$

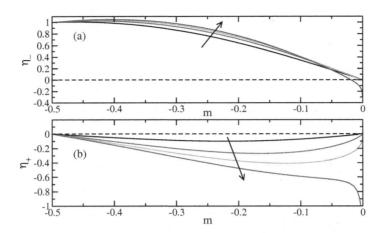

Fig. 3. (a) η_- as a function of magnetization m for $\Delta = 0.2, 0.5, 0.7, 0.9$ (in arrow direction). (b) Same for η_+. Note that $\eta_\pm \to -\infty$ for $m \to 0$ in the case $\Delta = 0.9$ leading to a sign change of η_-.

A comparison of (57) and (58) yields two equations from which one obtains[30,38]

$$\eta_- = \frac{v}{K^{1/2}} \frac{\partial v}{\partial h} + \frac{v^2}{2K^{3/2}} \frac{\partial K}{\partial h}, \quad \eta_+ = \frac{3v^2}{2K^{3/2}} \frac{\partial K}{\partial h}. \tag{59}$$

From this result a few general conclusions can be drawn. For a free model, η_+ should vanish because a mixing of right and left movers is then impossible. This is indeed the case since K does not change when applying a field in this case. Furthermore, η_+ will also be absent for models such as the Calogero–Sutherland model where K remains independent of the applied field even in the interacting case. The η_--term, on the other hand, is already present in a noninteracting system due to band curvature with $\eta_- = M^{-1}$ for $\Delta = 0$. The result (59) of course also agrees with the expansion for small Δ, Eq. (28).

The parameters $\eta_\pm$ are shown in Fig. 3 for different anisotropies Δ as a function of magnetization m. While $\eta_\pm \to 0$ for $m \to 0$ if $K > 5/8$ as expected from the weak coupling expansion (28) both parameters diverge, on the other hand, for $K < 5/8$ ($\Delta > \cos(\pi/5) \approx 0.81$). This behavior is discussed in more detail in Ref. 14 and we remind the reader that divergencies also occur in the amplitude ζ_+ for $h = 0$.

6. Applications

We now want to consider a few examples where the low-energy effective theory has been used to obtain parameter-free results for several important observables.

6.1. Impurities, Friedel oscillations and nuclear magnetic resonance

One of the best known realizations of the spin-1/2 antiferromagnetic Heisenberg chain is the cuprate Sr_2CuO_3.[6] In this system excess oxygen dopes holes into the chain which seem to be basically immobile.[7,40] Effectively, this leads to randomly distributed nonmagnetic impurities which cut the spin chain into finite segments. The magnetic properties are therefore determined by an ensemble of finite chains of random length N with OBC.

In a chain with OBC, translational invariance is broken leading to a position dependent local susceptibility

$$\chi_j = \frac{\partial}{\partial h}\langle S_j^z \rangle_{h=0} = \frac{1}{T}\langle S_j^z S_{\text{tot}}^z \rangle_{h=0}\,, \tag{60}$$

where T is the temperature and $S_{\text{tot}}^z = \sum_j S_j^z$. In order to calculate χ_j in the low-energy limit, we can express the spin operator in terms of the bosonic field

$$S_j^z \approx \sqrt{\frac{K}{\pi}}\partial_x \Phi + c(-1)^j \cos\sqrt{4\pi K}\Phi\,. \tag{61}$$

Here the prefactor of the uniform part is fixed by the condition $\sum_j S_j^z = S_{\text{tot}}^z$.[4,29] The amplitude of the alternating part, on the other hand, can be fixed with the help of the Bethe ansatz solution. The techniques required are, however, much more involved than the ones reviewed in the previous section. In particular, one finds that $c = \sqrt{A_z/2}$ with A_z as given in Eq. (4.3) of Ref. 34.

Using Eq. (61) we can write $\chi_j = \chi^{\text{uni}} + (-1)^j \chi_j^{\text{st}}$. The uniform part χ^{uni} for the Luttinger model is given by

$$\chi^{\text{uni}} = -\frac{\partial^2 f}{\partial h^2}\bigg|_{h=0} = \frac{1}{LT}\frac{\sum_{S_z} S_z^2 \exp\left[-\frac{\pi v}{2KLT}S_z^2\right]}{\sum_{S_z} \exp\left[-\frac{\pi v}{2KLT}S_z^2\right]}\,. \tag{62}$$

For $LT/v \to 0$ and L even $\chi_s \sim (2/LT)\exp[-\pi v/2KLT]$ whereas for L odd $\chi_s \sim (4LT)^{-1}$. For $LT/v \to \infty$ the thermodynamic limit result (35) is recovered. Note that this zeroth order result is position independent and shows scaling with LT. Corrections to scaling occur due to the irrelevant bulk and boundary operators. For $0 \le \Delta \le 1$ the leading bulk irrelevant operator is due to Umklapp scattering (31). This leads to a first-order correction in the free energy

$$\delta f_1 = \frac{\lambda}{L}\int_0^L dx \langle \cos(4\sqrt{\pi K}\phi)\rangle_{S_z} \exp(-8\pi K\langle \phi\phi\rangle_{\text{osc}})\,, \tag{63}$$

where we have used the mode expansion for OBC

$$\phi(x,t) = \sqrt{\frac{\pi}{16K}} + \sqrt{\frac{\pi}{K}}S_{\text{tot}}^z \frac{x}{L} + \sum_{n=1}^{\infty}\frac{\sin(\pi n x/L)}{\sqrt{\pi n}}(e^{-i\pi n\frac{vt}{L}}a_n + e^{i\pi n\frac{vt}{L}}a_n^\dagger)\,, \tag{64}$$

to split the expectation value of the Umklapp operator into an S_z (zero mode) and an oscillator part. Furthermore, we have used the cumulant theorem for the oscillator part. It is now straightforward, although a bit tedious, to evaluate the two parts of

(63). From this the correction to the uniform part of the susceptibility in first-order in Umklapp scattering can readily be obtained.[11]

The boundary operator (32) yields a further correction[11]

$$\delta\chi_2^{\text{uni}} = \frac{\pi v b}{2KT^2 L^3} \left[\frac{\sum\limits_{S_z} S_z^4 e^{-\frac{\pi v S_z^2}{2KLT}}}{\sum\limits_{S_z} e^{-\frac{\pi v S_z^2}{2KLT}}} - \frac{\left(\sum\limits_{S_z} S_z^2 e^{-\frac{\pi v S_z^2}{2KLT}} \right)^2}{\left(\sum\limits_{S_z} e^{-\frac{\pi v S_z^2}{2KLT}} \right)^2} \right]. \tag{65}$$

In the thermodynamic limit, Eq. (65) reduces to $\delta\chi_2^{\text{uni}} \to Kb/(\pi v L)$. The field theory result in this limit can be compared with the calculation of the boundary susceptibility based on the Bethe ansatz[32,33] and the proportionality constant b can be fixed

$$b = 2^{-1/2} \sin[\pi K/(4K-2)]/\cos[\pi/(8K-4)]. \tag{66}$$

To first-order in Umklapp scattering and in the dimension three boundary operator a parameter-free result for χ^{uni} can therefore be obtained.

The alternating part of the susceptibility (60) can be written as $\chi_j^{\text{st}} = (c/T)\exp(-2\pi K\langle\phi\phi\rangle_{\text{osc}})\langle\cos\sqrt{4\pi K}\phi\rangle_{S_z}$ where we have again split the correlation function into an oscillator and a zero mode part using the mode expansion. The calculation is now completely analogous to the calculation of the correction (63) leading to

$$\chi_j^{\text{st}} = -\left(\frac{\pi}{N+1}\right)^K \frac{\eta^{3K}(e^{-\frac{\pi v}{TL}})}{\theta_1^K\left(\frac{\pi j}{N+1}, e^{-\frac{\pi v}{2TL}}\right)} \frac{\sum\limits_m m\sin[2\pi m j/(N+1)]e^{-\pi v m^2/(2KLT)}}{\sum\limits_m e^{-\pi v m^2/(2KLT)}}. \tag{67}$$

Here $\eta(q)$ is the Dedekind eta function, and $\theta_1(u,q)$ the elliptic theta function of the first kind. In the thermodynamic limit, $L = Na \to \infty$, where we have reintroduced the lattice constant a for clarity, we can simplify our result and obtain

$$\chi_j^{\text{st}} = \frac{2cK}{v} \frac{x}{\left[\frac{v}{\pi T} \sinh\left(\frac{2\pi Tx}{v}\right) \right]^K}. \tag{68}$$

with $x = ja$. This agrees for the isotropic Heisenberg case, $K = 1/2$, with the result in Ref. 41. In Fig. 4 the parameter-free formula for $\chi_j = \chi^{\text{uni}} + (-1)^j\chi_j^{\text{st}}$ is compared to Quantum Monte Carlo data.[12]

The position dependent susceptibility is directly measured as Knight shift in NMR. The hyperfine interaction couples nuclear and electron spins and the Knight shift of the nuclear resonance frequency for a chain segment of length N is given by $K_j^{(N)} = (\gamma_e/\gamma_n)\sum_{j'} A^{j-j'}\chi_{j'}^{(N)}$, where γ_e (γ_n) is the electron (nuclear) gyromagnetic

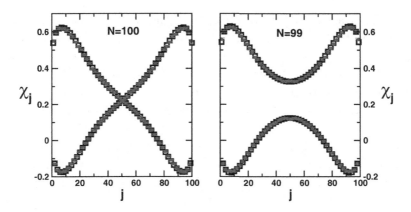

Fig. 4. Local susceptibility for a finite open XXZ chain with length N, $\Delta = 0.3$ at temperature $T/J = 0.02$. The squares denote the result of the parameter-free field theory formula, the circles are results obtained by Quantum Monte Carlo calculations.[12]

ratio, respectively. The hyperfine interaction is short ranged so that usually only A^0 and $A^{\pm 1}$ matter. For a random distribution of nonmagnetic impurities within a chain the NMR spectrum reflects the distribution of Knight shifts for an ensemble of spin chains with random lengths. This leads to rather complicated NMR spectra[7,40] whose properties can be fully understood using the parameter-free results for the susceptibility discussed above.[12]

6.2. *The spin-lattice relaxation rate and transport*

The spin current (or particle current in the fermionic language) for the XXZ model is defined by

$$\mathcal{J} = -\frac{iJ}{2} \sum_l (S_l^+ S_{l+1}^- - S_{l+1}^+ S_l^-) \approx -\sqrt{\frac{K}{\pi}} \int dx\, \partial_x \theta\,. \tag{69}$$

Whether or not the integrable XXZ model supports ballistic transport at finite temperatures has been the topic of a long-standing debate.[13,14,20,42–44] As discussed in Sec. 2.1 ballistic transport is signaled by a nonzero Drude weight and related to the part of the current which cannot decay due to conservation laws, see Eq. (7). For finite magnetic field, the Mazur inequality indeed immediately yields a nonzero Drude weight.[20] In this case the conserved energy current (52) becomes

$$\tilde{\mathcal{J}}_0^E = -v^2 \int dx\, \partial_x \theta \partial_x \phi - hv\sqrt{\frac{K}{\pi}} \int dx\, \partial_x \theta = \mathcal{J}_0^E + h\mathcal{J} \tag{70}$$

where we have used again the shift in the boson field, Eq. (34). The equal time correlations in (7) can now be evaluated for the Luttinger model (25) and[14]

$$D \geq D_{\text{Mazur}} = \frac{1}{2TL} \frac{\langle \mathcal{J}\tilde{\mathcal{J}}_0^E \rangle^2}{\langle (\tilde{\mathcal{J}}_0^E)^2 \rangle} = \frac{vK/2\pi}{1 + \dfrac{\pi^2}{3K}\left(\dfrac{T}{h}\right)^2} \quad (T, h \ll J)\,. \tag{71}$$

For $T/h \to 0$ the Mazur bound obtained from the overlap with $\tilde{\mathcal{J}}_0^E$ saturates the exact zero temperature Drude weight $D(T = 0) = vK/2\pi$.[45] Furthermore, one can also use the Bethe ansatz to calculate the Mazur bound D_{Mazur} exactly. The obtained result agrees with (71) up to temperatures of order J.[14]

For zero magnetic field, however, the overlap between all local conserved quantities $\mathcal{Q}_n$ of the XXZ model which can be constructed from the transfer matrix (4) and the current $\mathcal{J}$ vanishes, because the $\mathcal{Q}_n$ are even under particle-hole transformations while $\mathcal{J}$ is odd. Recently, a quantity — not related to the conserved quantities obtained from the Bethe ansatz solution — has been constructed for an open XXZ chain which is conserved up to boundary terms.[46] This quantity seems to protect part of the current in the thermodynamic limit, a view which appears to be supported by new numerical data.[47] As in the finite field case, the correct picture therefore seems to be that at finite temperatures a diffusive and a ballistic transport channel coexist.[13] For temperatures $T/J \in [0.2, 0.5]$ and $h = 0$, the Drude weight, however, seems to be much suppressed compared to its zero temperature value known exactly from Bethe ansatz. Furthermore, the Drude weight seems to vanish completely at finite temperatures in the isotropic case, $\Delta = 1$. We therefore ignore a protected part of the current for now and first concentrate on the diffusive channel. The corrections due to a possible conserved part of the current are discussed at the end of this section.

From an experimental point of view, it is of great interest to understand the intrinsic mechanism which leads to spin diffusion in the XXZ model. Spin diffusion has directly been observed in the spin-lattice relaxation rate $1/T_1$ measured in NMR experiments as a magnetic field dependence $1/T_1 \sim (1/\sqrt{h})$.[8,48] Both the spin-lattice relaxation rate and the conductivity can be calculated within the Luttinger model from the retarded boson propagator

$$\langle \phi\phi \rangle^{\text{ret}}(q, \omega) = \frac{v}{\omega^2 - v^2 q^2 - \Pi^{\text{ret}}(q, \omega)} \,. \tag{72}$$

For $\Pi^{\text{ret}}(q, \omega) = 0$ this is just the free boson propagator. In the zero field case, the leading irrelevant operator is the Umklapp term (31) and we will concentrate here on calculating the self-energy $\Pi^{\text{ret}}(q, \omega)$ in first-order in this perturbation. Further contributions to the self-energy will stem from the band curvature terms and are discussed in Ref. 14. Note, however, that only Umklapp scattering can give the bosons a finite lifetime and thus lead to diffusive transport. The calculation of the correlation function (72) in first-order in Umklapp scattering is straightforward.[49] We find[13,14]

$$\Pi^{\text{ret}}(q, \omega) \approx -2i\gamma\omega \,, \tag{73}$$

where the decay rate γ is given by

$$2\gamma = 8\pi K \lambda^2 \sin(4\pi K) \left(\frac{2\pi}{v}\right)^{8K-2} \Gamma(1/2 - 2K)\Gamma(2K)$$

$$\times \frac{B(2K, 1-4K)}{\sqrt{\pi}2^{4K+1}} \cot(2\pi K)T^{8K-3} \tag{74}$$

in the anisotropic case and by

$$2\gamma = \pi g^2 T \tag{75}$$

in the isotropic case, $\Delta = 1$. The running coupling constant $g = g(T)$ is determined by (49). The Kubo formula directly relates the conductivity $\sigma(q, \omega)$ to the calculated bosonic Green's function

$$\sigma(q, \omega) = \frac{K}{\pi} i\omega \langle \phi\phi \rangle^{ret}(q, \omega) \tag{76}$$

and due to the finite relaxation rate γ at finite temperatures one find a Lorentzian for the real part of the conductivity

$$\sigma'(q = 0, \omega) = \frac{vK}{\pi} \frac{2\gamma}{\omega^2 + (2\gamma)^2} . \tag{77}$$

At the same time, also the spin-lattice relaxation rate can be expressed by the same bosonic Green's function

$$\frac{1}{T_1} \approx -\frac{2KT}{\pi\omega_e} \int \frac{q^2 dq}{2\pi} |A(q)|^2 \text{Im}\langle\phi\phi\rangle_{\text{ret}}(q, \omega_e) . \tag{78}$$

Here $\omega_e = \mu_B h$ is the electron magnetic resonance frequency.[14] If the hyperfine coupling form factor $A(q)$ picks out the $q \sim 0$ contributions of the integral (78) as in the oxygen NMR experiment on Sr_2CuO_3 in Ref. 8 then $1/T_1$ and the conductivity $\sigma'(q \sim 0, \omega)$ are directly related. The experimentally considered spin chains are almost isotropic, $\Delta \approx 1$. Using the parameter-free result (75) we obtain the diffusive behavior

$$\frac{1}{T_1 T} \sim \sqrt{\frac{\gamma(T)}{\omega_e}} \sim \sqrt{\frac{T/\ln^2(J/T)}{\omega_e}} . \tag{79}$$

The only free parameters remaining depend on microscopic details of the considered compound. Both the exchange constant J and the hyperfine coupling constant $A(q)$ can be fixed by analyzing the susceptibility and performing a $K - \chi$ analysis respectively.[6,8,40] Integrability therefore makes it possible to obtain an analytical result for the spin-lattice relaxation rate which includes the intrinsic relaxation processes. The comparison of the result (79) with experiment as shown in Fig. 5 demonstrates that Umklapp scattering seems to be the dominant source for relaxation and that other contributions, e.g., due to electron-phonon scattering are small.

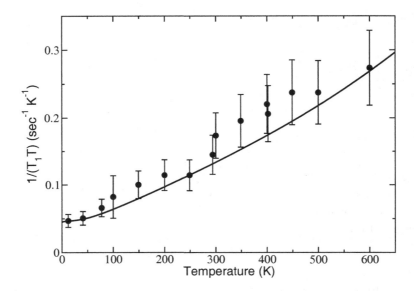

Fig. 5. Spin-lattice relaxation rate $1/T_1$ for Sr_2CuO_3. Experimental data (symbols) are compared to the field theoretical result. The only free parameters are the exchange constant $J \sim 1800$ K and the hyperfine coupling tensor which have been determined experimentally. A more detailed analysis is presented in Ref. 14.

Finally, we have to discuss the possible conservation of part of the current even at zero magnetic field. The quantity constructed in Ref. 46 contains terms active over several lattice sites for $0 < \Delta < 1$. It is therefore nontrivial to bosonize this term and to consider the consequences of its (almost) conservation for the low-energy effective theory in a similar spirit to our discussion of J^E in Sec. 5.2. An alternative method discussed in Refs. 13, 14 and 50 is to use a memory matrix formalism to calculate the self-energy (72) in the presence of conservation laws. Instead of Eq. (73) the self-energy then reads

$$\Pi(q = 0, \omega) \approx \frac{-2i\gamma\omega}{1 + 2i\gamma y/\omega} \quad \text{with} \quad y = \frac{\langle \mathcal{J}Q \rangle^2}{\langle \mathcal{J}^2 \rangle \langle Q^2 \rangle - \langle \mathcal{J}Q \rangle^2}, \tag{80}$$

where Q is the conserved quantity with $\langle \mathcal{J}Q \rangle \neq 0$. As a consequence, the current–current correlation function at large times is now given by

$$\frac{1}{L} \langle \mathcal{J}(t)\mathcal{J} \rangle = \frac{KvT}{\pi(1 + y)} [y + \exp(-2\gamma(1 + y)t)], \tag{81}$$

where the first term is proportional to the Drude weight and the second term describes the diffusive part. For $\Delta = 1$ the Drude weight seems to be zero at finite temperatures[46] so that our theoretical analysis of $1/T_1$ is not affected. In any case, the spin chains in Sr_2CuO_3 do not represent an integrable system and there is therefore certainly no ballistic channel. It is known, for example, that in this compound also a weak next-nearest neighbor coupling $J_2 \sim 0.1J$ exists. While this coupling destroys integrability it will only lead to a weak renormalization of the Umklapp amplitude λ.[29] It is thus legitimate to use the results for the integrable model to analyze the experimental data for $1/T_1$.

7. Conclusions

Integrable gapless one-dimensional quantum models allow to test many predictions of Luttinger liquid theory. As such they have been vital in confirming the universal applicability of the latter. On the other hand, Luttinger liquid theory has also helped to understand integrable quantum models better. Except for the simplest integrable systems, such as free bosonic or fermionic particles and to some extent also Calogero–Sutherland type models, an exact calculation of correlation functions for arbitrary distances and times has not been achieved yet. A combination of Luttinger liquid theory and integrability then often allows to obtain a more complete picture. For Bethe ansatz integrable systems such as the Lieb–Liniger Bose gas, the (anisotropic) Heisenberg, the Hubbard or the supersymmetric t–J model it is possible to fix the velocity of the collective excitations v and the Luttinger parameter K as a function of density and interaction strength. For the Luttinger model, (dynamical) correlations can then be easily calculated.

Often one can go even one step further and determine the amplitudes of leading irrelevant operators acting as corrections to the Luttinger Hamiltonian as well as amplitudes of correlation functions, exactly. This article is by no means a complete review of all the results which have been obtained in this field. Instead, I have used the XXZ model as an example and have summarized how v, K as well as the amplitudes of the leading irrelevant band curvature terms and Umklapp scattering can be determined. As an application, I have shown that this allows to derive parameter-free results for the local susceptibility in open Heisenberg chains and quantitatively explains Knight shift spectra which have been investigated by nuclear magnetic resonance for compounds such as Sr_2CuO_3. As a second application, I have presented results for the relaxation rate of the Luttinger liquid bosons at finite temperatures in first-order in Umklapp scattering. From this a parameter-free formula for the spin-lattice relaxation rate $1/T_1$ can be obtained which is in excellent agreement with experiment. The same bosonic correlation function and the same relaxation rate determine, on the other hand, also the particle transport. In general, the XXZ model has coexisting ballistic and diffusive transport channels. The relative weight of each of these channels can be determined by combining the Luttinger model — keeping the dangerously irrelevant Umklapp term with known amplitude — with a memory-matrix calculation.

There are many more interesting results which have not been covered in this review. One of the perhaps most fascinating recent developments is the so-called nonlinear Luttinger liquid theory which allows to calculate dynamic response functions while taking band curvature into account.[51] This has lead to the discovery of new power laws near edge singularities. For integrable models the exponents of these power laws can be determined exactly.[52] A comprehensive overview about these developments has been given in a recent excellent review.[53]

Acknowledgments

I acknowledge support by the excellence graduate school MAINZ and the collaborative research center SFB/TR49.

References

1. J. M. Luttinger, *J. Math. Phys.* **4**, 1154 (1963).
2. D. C. Mattis and E. H. Lieb, *J. Math. Phys.* **6**, 304 (1965).
3. A. Luther and I. Peschel, *Phys. Rev. B* **12**, 3908 (1975).
4. T. Giamarchi, *Quantum Physics in One Dimension* (Clarendon Press, Oxford, 2004).
5. L. Onsager, *Phys. Rev.* **65**, 117 (1944).
6. N. Motoyama, H. Eisaki and S. Uchida, *Phys. Rev. Lett.* **76**, 3212 (1996).
7. M. Takigawa *et al.*, *Phys. Rev. B* **55**, 14129 (1997).
8. K. R. Thurber *et al.*, *Phys. Rev. Lett.* **87**, 247202 (2001).
9. S. Eggert, *Phys. Rev. B* **53**, 5116 (1996).
10. J. Sirker *et al.*, *Phys. Rev. Lett.* **98**, 137205 (2007).
11. J. Sirker *et al.*, *J. Stat. Mech. Theory Exp.* P02015 (2008).
12. J. Sirker and N. Laflorencie, *Europhys. Lett.* **86**, 57004 (2009).
13. J. Sirker, R. G. Pereira and I. Affleck, *Phys. Rev. Lett.* **103**, 216602 (2009).
14. J. Sirker, R. G. Pereira and I. Affleck, *Phys. Rev. B* **83**, 035115 (2011).
15. E. H. Lieb and W. Liniger, *Phys. Rev.* **130**, 1605 (1963).
16. T. Kinoshita, T. Wenger and D. S. Weiss, *Nature* **440**, 900 (2006).
17. C. N. Yang and C. P. Yang, *Phys. Rev.* **150**, 321 (1966).
18. J.-S. Caux and J. Mossel, *J. Stat. Mech. Theory Exp.* P02023 (2011).
19. F. H. L. Essler *et al.*, *The One-Dimensional Hubbard Model* (Cambridge University Press, Cambridge, 2005).
20. X. Zotos, F. Naef and P. Prelovšek, *Phys. Rev. B* **55**, 11029 (1997).
21. P. Mazur, *Physica* **43**, 533 (1969).
22. M. Suzuki, *Physica* **51**, 277 (1971).
23. P. Jung and A. Rosch, *Phys. Rev. B* **75**, 245104 (2007).
24. J. M. Deutsch, *Phys. Rev. A* **43**, 2046 (1991).
25. H. Bethe, *Z. Phys.* **71**, 205 (1931).
26. E. K. Sklyanin, *J. Phys. A* **21**, 2375 (1988).
27. F. C. Alcaraz *et al.*, *J. Phys. A* **20**, 6397 (1987).
28. M. Takahashi, *Thermodynamics of One-Dimensional Solvable Problems* (Cambridge University Press, Cambridge, UK, 1999).
29. S. Eggert and I. Affleck, *Phys. Rev. B* **46**, 10866 (1992).
30. R. G. Pereira *et al.*, *J. Stat. Mech. Theory Exp.* P08022 (2007).
31. S. Lukyanov, *Nucl. Phys. B* **522**, 533 (1998).
32. M. Bortz and J. Sirker, *J. Phys. A: Math. Gen.* **38**, 5957 (2005).
33. J. Sirker and M. Bortz, *J. Stat. Mech. Theory Exp.* P01007 (2006).
34. S. Lukyanov and V. Terras, *Nucl. Phys. B* **654**, 323 (2003).
35. N. Kitanine, J. M. Maillet and V. Terras, *Nucl. Phys. B* **554**, 647 (1999).

36. N. Kitanine *et al.*, *Nucl. Phys. B* **567**, 554 (2000).

37. N. Kitanine *et al.*, *J. Stat. Mech. Theory Exp.* P05028 (2011).

38. R. G. Pereira *et al.*, *Phys. Rev. Lett.* **96**, 257202 (2006).

39. S. Fujimoto and S. Eggert, *Phys. Rev. Lett.* **92**, 037206 (2004).

40. J. P. Boucher and M. Takigawa, *Phys. Rev. B* **62**, 367 (2000).

41. S. Eggert and I. Affleck, *Phys. Rev. Lett.* **75**, 934 (1995).

42. J. V. Alvarez and C. Gros, *Phys. Rev. Lett.* **88**, 077203 (2002).

43. X. Zotos, *Phys. Rev. Lett.* **82**, 1764 (1999).

44. F. Heidrich-Meisner *et al.*, *Phys. Rev. B* **66**, 140406(R) (2002).

45. B. S. Shastry and B. Sutherland, *Phys. Rev. Lett.* **65**, 243 (1990).

46. T. Prosen, *Phys. Rev. Lett.* **106**, 217206 (2011).

47. C. Karrasch, J. H. Bardarson and J. E. Moore, *Phys. Rev. Lett.* **108**, 227206 (2012).

48. F. L. Pratt *et al.*, *Phys. Rev. Lett.* **96**, 247203 (2006).

49. M. Oshikawa and I. Affleck, *Phys. Rev. B* **65**, 134410 (2002).

50. A. Rosch and N. Andrei, *Phys. Rev. Lett.* **85**, 1092 (2000).

51. A. Imambekov and L. I. Glazman, *Science* **323**, 228 (2009).

52. R. G. Pereira, S. R. White and I. Affleck, *Phys. Rev. Lett.* **100**, 027206 (2008).

53. A. Imambekov, T. L. Schmidt and L. I. Glazman, arXiv:1110.1374 (2011).

LONG TIME CORRELATIONS OF NONLINEAR LUTTINGER LIQUIDS*

RODRIGO G. PEREIRA

Instituto de Física de São Carlos, Universidade de São Paulo,
C.P. 369 São Carlos, SP, 13560-970, Brazil
rpereira@ifsc.usp.br

An overview is given of the limitations of Luttinger liquid theory in describing the real time equilibrium dynamics of critical one-dimensional systems with nonlinear dispersion relation. After exposing the singularities of perturbation theory in band curvature effects that break the Lorentz invariance of the Tomonaga–Luttinger model, the origin of high frequency oscillations in the long time behaviour of correlation functions is discussed. The notion that correlations decay exponentially at finite temperature is challenged by the effects of diffusion in the density–density correlation due to umklapp scattering in lattice models.

Keywords: Luttinger liquid; band curvature; dynamical responses.

1. Introduction

When Haldane coined the name "Luttinger liquid" (LL),[1] the point was to emphasize the universality of the theory beyond the exactly solvable model studied by Tomonaga and Luttinger.[2,3] Indeed, the mapping from interacting fermions to noninteracting bosons that renders the Tomonaga–Luttinger (TL) model solvable depends crucially on the approximation of a linear dispersion relation for low energy excitations. However, the thermodynamic properties predicted by LL theory are asymptotically exact in the low energy limit for generic critical one-dimensional systems because perturbations associated with band curvature are irrelevant in the renormalization group sense. This explains why LL behavior is observed in so many different systems, such as quantum wires, carbon nanotubes, spin chain compounds and cold atoms in optical lattices.[4]

Neglecting irrelevant perturbations, the elementary excitation of the one-component TL model is a free boson with massless relativistic dispersion $\omega_q = v|q|$. As a result, the model is Lorentz invariant: Its correlation functions are preserved by continuous rotations between space and Euclidean time. More precisely, the TL model is invariant under local conformal transformations in $(1 + 1)$ dimensions. This allows one to bring in the arsenal of conformal field theory (CFT), with central charge $c = 1$, to compute correlation functions.[5] Conformal invariance dictates that for large distances x and long real times t the ground state correlation function for

*This article first appeared in International Journal of Modern Physics B, Vol. 26, No. 22 (2012).

a given field $\Phi(x)$ decays as a power law

$$\langle \Phi(x,t)\Phi^\dagger(0,0)\rangle \sim \frac{1}{(x-vt)^{2\Delta_+}(x+vt)^{2\Delta_-}}, \tag{1}$$

where Δ_+ and Δ_- are conformal dimensions. The conformal dimensions of physical operators are determined by the Luttinger parameter K of the TL model. Both the velocity v and the Luttinger parameter K can be extracted from the finite size spectrum in general, or calculated exactly for integrable models in particular.[6] The result can be generalized to models with more than one gapless degree of freedom, such as the Hubbard model away from half-filling.[7] Using the conformal mapping to the cylinder geometry with compactified Euclidean time direction, one can predict that at finite temperatures correlation functions decay exponentially in both x and t.[8]

Perhaps due to the remarkable success of CFT methods in LL physics, the validity of Eq. (1) is often overstated. The truth is, for any model in the LL universality class where the dispersion relation is not exactly linear, *the CFT result for time-dependent correlation functions does not give the correct long time behavior* for real time $t > |x|/v$, i.e., inside the light cone. While there are examples of gapless one-dimensional systems to which CFT techniques clearly do not apply — for instance spinful fermions in the spin-incoherent LL regime[9] or the ferromagnetic Bose gas[10] — the general reason for the breakdown of LL theory in real time dynamics is the effect of band curvature. Although formally irrelevant, perturbations to the TL model that take the form of boson decay processes generate singular contributions to dynamical correlation functions. The solution to this quandary gave birth to the subject of "nonlinear LLs"[11] (see Ref. 12 for a detailed review). Some predictions of the field theory for nonlinear LLs have been recently confirmed by an exact form factor approach.[12]

The purpose of this chapter is to provide an overview of the breakdown of LL theory in equilibrium dynamics and discuss its consequences for the long time behavior of correlation functions in critical one-dimensional systems. We will mainly focus on two aspects: the contribution of high energy modes to correlation functions at zero temperature and the diffusive contribution due to umklapp processes which dominates the long time tail at finite temperatures.

2. Breakdown of LL Theory by Band Curvature Effects

Fermi liquid theory breaks down in one dimension because scattering between two disconnected Fermi points always leads to singularities in the particle-hole and particle–particle channels, making quasiparticles unstable as the *fermion* self-energy diverges.[14] Similarly, LL theory fails to describe dynamical response functions because the *boson* self-energy due to band curvature terms is singular. However, the singularity in this case is connected with the macroscopic degeneracy of states comprised of multiple bosons with the same chirality, which is an artifact of the linear dispersion approximation in the TL model.

Consider the TL model for spinless fermions (following the standard g-ology notation)[4]

$$H = \sum_{r=R,L} \int dx \left[v_F : \psi_r^\dagger(-ir\partial_x)\psi_r : + \frac{g_2}{2} : (\psi_r^\dagger\psi_r)(\psi_{-r}^\dagger\psi_{-r}) : + \frac{g_4}{2} : (\psi_r^\dagger\psi_r)^2 : \right].$$
(2)

Here $r = R, L = \pm$ denotes right and left movers, defined from single-particle states with momentum around $\pm k_F$, and :: refers to normal ordering with respect to the free fermion ground state. The fermion field operator reads $\Psi(x) \approx e^{ik_F x}\psi_R(x) + e^{-ik_F x}\psi_L(x)$. Bosonization[4] maps the fermionic model to the Gaussian model

$$H = \sum_{r=R,L} \int dx \frac{v}{2} [: (\partial_x\varphi_R)^2 : + : (\partial_x\varphi_L)^2 :],$$
(3)

where the normal ordering is with respect to the vacuum of bosons and the chiral bosonic fields obey the commutation relation $[\varphi_r(x), \partial_{x'}\varphi_{r'}(x')] = -ir\delta_{r,r'}\delta(x - x')$. For g_2, $g_4 \ll v_F$, the renormalized velocity is given approximately by $v \approx v_F - g_4/(2\pi) + \mathcal{O}(g^2)$. In the notation used here, the chiral fermion fields are bosonized as

$$\psi_r(x) \approx (2\pi\alpha)^{-1/2} \exp[-i\sqrt{2\pi}(\lambda\varphi_r + \bar{\lambda}\varphi_{-r})],$$
(4)

with $\lambda = (\sqrt{K} + 1/\sqrt{K})/2$ and $\bar{\lambda} = (-\sqrt{K} + 1/\sqrt{K})/2$. Here α is a short-distance cutoff and K is the Luttinger parameter given approximately by $K \approx 1 - g_2/(2\pi v_F) + \mathcal{O}(g^2)$. Within the TL model, the single fermion Green's function at zero temperature is given by $G(x,t) = -i\langle\Psi(x,t)\Psi^\dagger(0,0)\rangle = e^{ik_F x}G_R(x,t) + e^{-ik_F x}G_L(x,t)$, where the Green's functions for the chiral fermions read[15]

$$G_r(x,t) = -i\langle\psi_r(x,t)\psi_r^\dagger(0,0)\rangle = (r/2\pi)(x - rvt)^{-\lambda^2}(x + rvt)^{-\bar{\lambda}^2}.$$
(5)

The result in Eq. (5) has the form of Eq. (1) with conformal dimensions $\Delta_\pm = 1/8(K + (1/K) \pm 2r)$. Taking the Fourier transform with the proper time ordering prescription one finds that the particle addition part of the single fermion spectral function does not have a quasiparticle peak; instead it behaves as a power law above a threshold energy, $A(k,\omega) \sim k^{\bar{\lambda}^2}(\omega - v\delta k)^{-2+\lambda^2}$ for $\delta k = k - k_F \ll k_F$.[16,17] Importantly, the exponent in the spectral function is directly related to the singularities of $G_r(x,t)$ in Eq. (5) along the light cone $x = \pm vt$. For $g_2, g_4 \ll v_F$, the interaction-dependent exponent can also be obtained by resumming logarithmic divergences in the perturbation theory in the fermion description,[18] but the advantage of bosonization is that the fermion interactions are treated easily by the rescaling of the bosonic fields leading to Eq. (4).

What if we perturb the TL model with band curvature effects, which are indeed present in any real system? Let us assume, as usual, that we are allowed to truncate the Hilbert space to low energy states around the Fermi points, but now we add a parabolic term in the dispersion[1]

$$\delta H = -\frac{1}{2m} \int dx \sum_{r=R,L} : \psi_r^\dagger \partial_x^2 \psi_r : , \tag{6}$$

where m is the effective mass at the Fermi level. After bosonizing and performing the Bogoliubov transformation that diagonalizes the TL model, the band curvature term generates two types of operators in general

$$\delta H = \frac{\sqrt{2\pi}}{6} \int dx \{ \eta_- [: (\partial_x \varphi_L)^3 : - : (\partial_x \varphi_R)^3 :]$$
$$+ \eta_+ [: (\partial_x \varphi_L)^2 \partial_x \varphi_R : - : (\partial_x \varphi_R)^2 \partial_x \varphi_L) :] \} , \tag{7}$$

where $\eta_\pm$ are coupling constants of order $1/m$ which can be calculated to lowest order by bosonization or fixed by phenomenological relations.[19,20] The important point is that these cubic terms spoil the solvability of the TL model since they introduce interactions between the bosonic modes. Nevertheless, we might hope they can be treated perturbatively. To illustrate the problem with perturbation theory, it suffices to consider the η_- term, which does not mix the two chiral components of the bosonic field. Calculating the first-order correction to the single fermion Green's function, one finds[21]

$$\frac{\delta G_r(x,t)}{G_r(x,t)} = -\frac{i\lambda^3 \eta_-}{v} \left[\frac{1}{x - rvt} - \frac{x + rvt}{(x - rvt)^2} \right]$$
$$- \frac{i\bar{\lambda}^3 \eta_-}{v} \left[\frac{1}{x + rvt} - \frac{x - rvt}{(x + rvt)^2} \right] . \tag{8}$$

For $x \to \infty$ or $t \to \infty$, the expression in Eq. (8) decays faster than the result for the TL model.[a] This is expected since simple power counting tells us that the η_- perturbation has scaling dimension three and is irrelevant at the LL fixed point. Hence the argument for the universality of LL theory:[1] If we are interested in low-energy thermodynamic properties, band curvature effects are harmless because they only give subleading corrections to correlation functions at large distances $x \gg \alpha, vt$.

The problem shows up when we are interested in *dynamical* response functions which depend on *both* momentum and frequency, such as the spectral function $A(k,\omega)$. Remarkably, the correction in Eq. (8) is more singular at the light cone $x = \pm vt$ than the unperturbed LL result in Eq. (5). For the spectral function, this means that the corresponding correction $\delta A(k,\omega)/A(k,\omega) \sim \eta_- \delta k^2/(\omega - v\delta k)$ has a singular frequency dependence and diverges more strongly at the lower threshold

---

[a]Actually, the correction due to η_- vanishes for $t = 0$. Equivalently, the correction for equal-time correlation functions calculated around Eq. (5.4) of Ref. 1 vanishes if the "angle" of the Bogoliubov transformation that mixes right and left movers is set to zero.

$\omega = v\delta k$ than the LL result. The singularity actually gets worse at higher-orders of perturbation theory in η_-, as one obtains more powers of $1/(x \pm vt)$ in $\delta G_r(x,t)$.

The same problem is present in the calculation of the density–density correlation function $\chi(x,t) = \langle n(x,t)n(0,0)\rangle$. In the boson description, the fluctuation of the fermion density operator is represented by

$$n(x) =: \Psi^\dagger(x)\Psi(x) :\approx \sqrt{K/\pi}\partial_x\phi(x) - (1/2\pi\alpha)\cos[\sqrt{4\pi K}\phi(x) - 2k_F x]\,, \qquad (9)$$

where $\phi(x) = [\varphi_L(x) - \varphi_R(x)]/\sqrt{2}$. The correlation function for the long wavelength part of $n(x)$ is equivalent to the boson propagator

$$\chi(x,t) \sim \sum_{r,r'}\langle: \psi_r^\dagger\psi_r : (x,t) : \psi_{r'}^\dagger\psi_{r'} : (0,0)\rangle = \frac{K}{\pi}\langle\partial_x\phi(x,t)\partial_x\phi(0,0)\rangle\,. \qquad (10)$$

The Fourier transform of $\chi(x,t)$ for small momentum $q \ll k_F$ yields the dynamical structure factor $S(q,\omega) = -2\text{Im}\chi_{\text{ret}}(q,\omega) \approx K|q|\delta(\omega - v|q|)$. The delta function peak obtained as an approximation for $S(q \ll k_F,\omega)$ within the TL model corresponds to the spectral function of the coherent bosonic mode with well-defined energy and momentum. Following the analogy with Landau's Fermi liquid theory, one would expect that going beyond the TL model and introducing band curvature effects would lead to a finite boson lifetime. However, the attempt to calculate a boson self-energy using perturbation theory fails.[22] Similar to the fermion Green's function, perturbation theory in η_- for $\chi(x,t)$ generates corrections which are increasingly more singular at the light cone $x = \pm vt$. To understand this singularity, we note that in momentum and frequency domain the three-legged vertex η_- in Eq. (7) allows the single boson with momentum q to decay into two bosons with momenta q_1 and $q_2 = q - q_1$ [see Fig. 1(a)]; however, the energy of the intermediate state when the boson lines are put on shell is always $\omega = vq_1 + vq_2 = vq$, independent of the internal momenta. This huge degeneracy is present at any finite order of perturbation theory. The result is that to any finite order the boson decay rate due to the η_- perturbation diverges on shell as $\sim \delta(\omega - vq)$.

The degeneracy of the many-boson states stems from the linear dispersion approximation of the TL model. In the model of free fermions with nonlinear dispersion, particle-hole pairs with the same total momentum — a linear combination of which defines the bosonic excitations — are not degenerate. Thus the cause of the breakdown of LL theory is the impossibility of starting from the TL model and breaking Lorentz invariance within finite order perturbation theory. The picture of a "quasi-boson" with well behaved self-energy due to band curvature simply does not work. It is frustrating that the bosonization method, so helpful in resumming the divergences of Fermi liquid theory in one dimension, cannot handle the innocent looking perturbation in Eq. (6), which is quadratic in fermions.

The question then is whether there is an alternative representation which captures the essential physics and resums the divergences of band curvature effects in

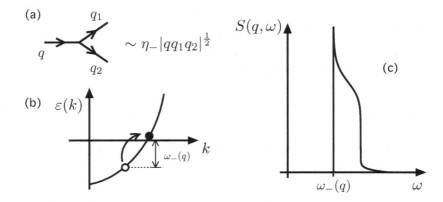

Fig. 1.   (a) In the boson description, band curvature is represented by an interaction vertex that scales with the momenta of the bosons. But due to Lorentz invariance of the unperturbed TL model, the two boson state after the decay is degenerate with the initial single boson state, regardless of the value of q_1 (or $q_2 = q - q_1$). (b) The divergences of perturbation theory in band curvature operators can be resummed by refermionizing to new fermions with renormalized nonlinear dispersion. Density excitations are now represented by nondegenerate particle-hole pairs. (c) Broadened boson peak in the density structure factor $S(q, \omega)$ of a nonlinear LL for $q \ll k_F$.

LL theory. Fortunately, the answer is yes. The trick is to refermionize the Hamiltonian with the cubic perturbation η_- to obtain a model of noninteracting fermions with nonlinear dispersion[11,23]

$$H + \delta H \approx \sum_{r=R,L} \int dx : \tilde{\psi}_r^\dagger \left[v(-ir\partial_x) + \frac{\eta_-}{2}(-i\partial_x)^2 \right] \tilde{\psi}_r : \qquad (11)$$

The new fermions are defined such that $: \tilde{\psi}_r^\dagger \tilde{\psi}_r := -r\partial_x \varphi_r / \sqrt{2\pi}$ and differ from the original fermions by string operators.[11] We note that the fermion interactions g_2, g_4 in Eq. (2) are absorbed into the renormalization of the velocity v and effective mass η_-. Including the η_- operator as the parabolic term in the dispersion of the new fermions lifts the degeneracy of the many-boson intermediate states. The approximation in Eq. (11) is to neglect the η_+ operator defined in Eq. (7) as well as more irrelevant operators (with dimension four and higher). Within this approximation, the free boson peak in $S(q, \omega)$ broadens into a two-fermion continuum with rectangular line shape and width $\delta\omega_q \sim \eta_- q^2$.[19,24] This result is exactly what one expects from summing the infinite series of diagrams in the η_- perturbation.[20]

   The renormalized fermion band with nonlinear dispersion relation offers a convenient starting point to study dynamical response functions. With Lorentz invariance broken at the outset, the thresholds of the exact spectrum are not the same as in LL theory. While the η_- band curvature term sets the width of the two-fermion continuum for $q \ll k_F$, additional irrelevant operators generate interactions between the new fermions and give rise to power law singularities at the edges of the spectrum. Rather than governed by light cone effects, the new singularities of nonlinear LLs are in analogy to the X-ray edge problem of optical absorption in metals.[24] The lower threshold $\omega_-(q)$ below which $S(q, \omega)$ vanishes is given by the minimum energy of a particle-hole excitation with momentum q created in the renormalized fermion band. For positive band curvature $\eta_- > 0$ as in Fig. 1(b), this corresponds to a particle at the Fermi point and a hole as deep as possible with energy $\omega_-(q) = vq - \eta_- q^2/2$. Like in the X-ray edge problem, this "deep hole" can be described as an effective quantum impurity which propagates with different velocity than the low energy modes. The power law singularity at the edge $\omega_-(q)$ has an exponent proportional to q in the case of short-range interactions.[24] Moreover, interactions between right and left movers make the spectral weight extend above the upper threshold of the two-fermion continuum $\omega_+(q) = vq + \eta_- q^2/2$ predicted by the approximation in Eq. (11) and $S(q, \omega)$ acquires a tail that decays as $\sim \eta_+^2 q^4/\omega^2$ at high frequencies $\omega - vq \gg \eta_- q^2$.[19,24]

The picture that emerges for the broadening of the peak in $S(q \ll k_F, \omega)$ due to irrelevant operators is illustrated in Fig. 1(c). Unlike the Lorentzian quasiparticle peak in the spectral function of Fermi liquids, the "quasi-boson" peak of nonlinear LLs is asymmetric and has an X-ray edge type singularity above the lower threshold $\omega_-(q)$.

Accounting for band curvature effects, the singularities of the single fermion spectral function of LLs are also modified.[25] In LL theory particle and hole Green's functions coincide because the TL model is particle-hole symmetric, but in the presence of band curvature this is no longer the case. For positive band curvature, the excitation that creates a single deep hole defines the lower threshold of the support of the hole spectral function for a given momentum k. As a result of kinematics, the power law singularity at the deep hole threshold cannot be broadened by any interactions. In contrast, the support of the particle spectral function does extend below the energy of the single-particle excitation. In this case, three-body scattering processes in generic (i.e., nonintegrable) models allow the single-particle to decay into the continuum and a Lorentzian peak with decay rate $\gamma_k \propto (k - k_F)^8$ is obtained.[25] The only surviving power law in the particle spectral function has a positive exponent and is found at the absolute lower threshold of the support, located at the energy of the deep hole excitation.

Going back to real space and time, we should expect the long time decay of correlation functions of LLs to be strongly affected by the X-ray edge type singularities at the thresholds of the nonlinear spectrum. This will be the subject of the next section.

3. High Energy Contributions to Time-Dependent Correlation Functions

Once Lorentz invariance is broken, the exponents for the long time decay of correlation functions are not constrained to be the same as the ones for large distance decay. In fact, contrary to conventional wisdom, the long time behavior at zero temperature is not even dominated by low energy modes, but by high energy saddle point contributions which take advantage of the dispersion nonlinearity.

To see how this comes about, consider the simple case of free fermions with parabolic dispersion relation $\varepsilon(k) = k^2/2m$. The hole Green's function is given exactly by

$$G_h(x,t) = \langle \Psi^\dagger(x,t)\Psi(0,0)\rangle = \int_{-k_F}^{k_F} \frac{dk}{2\pi} \, e^{ikx - i(k^2 - k_F^2)t/2m} . \tag{12}$$

The real part of Eq. (12) is plotted in Fig. 2(a). There are clearly two distinct regions in the (x,t) plane. Outside the light cone, $|x| > v_F t$, the Green's function oscillates with distance, but not with time. This is consistent with the usual contribution from the low energy modes which come with factors of $e^{\pm ik_F x}$. For large distances, $|x| \gg k_F^{-1}, v_F t$, the power law decay of $G_h(x,t)$ is well-described by Eq. (5) with $K = 1$. In contrast, outside the light cone, $|x| < v_F t$, the Green's function exhibits time oscillations in addition to spatial oscillations. Such time oscillations can only come from modes with finite frequency. Indeed, a moment's reflection shows that the integral in Eq. (12) picks up significant contributions from the saddle point away from the Fermi surface where $d/dk(kx - k^2 t/2m)|_{k=k^*} = 0$, which implies $k^* = mx/t$. In the long time limit $v_F t/|x| \to \infty$, the saddle point moves to the bottom of the band, $k^*/k_F \to 0$. The corresponding contribution to the hole Green's function oscillates with frequency $k_F^2/2m$ and, since the dispersion is parabolic about $k = 0$, it decays as $1/\sqrt{t}$. The decay is slower than that of the low energy contributions, which due to the linear dispersion about $\pm k_F$ decay as $1/t$. This dominant role of the saddle point contribution is peculiar to one dimension. In general, in d dimensions, the saddle point contribution with $k \approx 0$ would decay as $t^{-d/2}$, while the Fermi surface contribution always decays as $1/t$ in the noninteracting case.[b]

In the case of parabolic dispersion there is no saddle point above the Fermi level for $v_F t/|x| \to \infty$, thus the long time behavior of the particle Green's function $G_p(x,t) = \langle \Psi(x,t)\Psi^\dagger(0,0)\rangle$ is well-described by LL theory.[c] On the other hand, for a tight-binding model with nearest neighbour hopping J the dispersion relation is $\varepsilon(k) = -2J \cos k$ (setting the lattice parameter to 1). In this case the single-particle spectrum is bounded from above as well as from below and there is another saddle point at the top of the fermion band. Thus the particle Green's function for a lattice model also oscillates in time inside the light cone.

Numerical methods such as the time-dependent density matrix renormalization group (tDMRG)[26] and the time-evolving block decimation (TEBD) algorithm[27] provide direct information about the real time evolution of one-dimensional systems. These methods reveal that interacting models also exhibit high frequency oscillations

---

[b]Turning on weak interactions in $d \geq 2$, power law decay is replaced by exponential decay due to the finite lifetime of quasiparticles in Fermi liquids.
[c]The problem would be the short time limit, due to the lack of a natural high energy cutoff for particle states.

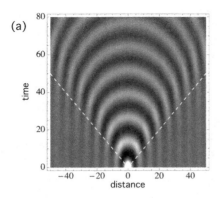 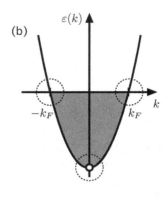

Fig. 2. (a) Real part of hole Green's function for free fermions with parabolic dispersion as a function of distance x (in units of k_F^{-1}) and time t (in units of m/k_F^2). The dashed line indicates the light cone $|x| = v_F t$. (b) The long time behavior inside the light cone is governed by high energy states at the bottom of the fermion band. The effective model for the interacting case keeps states with $k \approx 0$ and $k \approx \pm k_F$.

in equilibrium correlation functions inside the light cone $vt > |x|$, with renormalized velocity v.[28] Although there are no exact analytical results for time-dependent Green's functions of interacting models, not even for integrable ones, the exact solution for the noninteracting case suggests an approximation to describe the long time decay in the interacting case. Besides the low energy chiral components, it is important to consider modes with parabolic dispersion at the bottom of a renormalized fermion band. This can be achieved by pushing the X-ray edge methods for nonlinear LLs[24] beyond the low energy regime.[29] Starting from the noninteracting band, one expands the fermion field in the form $\Psi(x) \sim e^{ik_F x}\psi_R(x) + e^{-ik_F x}\psi_L(x) + d^\dagger(x)$, where the high energy field $d^\dagger(x)$ is defined from hole states with momentum $k \approx 0$ [see Fig. 2(b)]. Using this mode expansion for the kinetic energy and interaction terms in the Hamiltonian leads to the familiar model of a mobile impurity in a LL.[30,31] The impurity, defined in momentum space in this case, is a single deep hole at the bottom of the band. The long time limit of the hole Green's function in the interacting case is controlled by decay of the deep hole due not only to the parabolic dispersion, but also to scattering by low energy particle-hole pairs. The coupling between the deep hole and the low energy modes can be treated exactly within the effective impurity model using a canonical transformation that shifts the bosonic fields.[32] The result is that in addition to the (subleading) CFT terms in Eq. (5) the hole Green's function for $vt \gg |x| \gg k_F^{-1}$ has a time-oscillating term

$$G_h(x,t) \sim \frac{e^{-iWt+iMx^2/2t}}{\sqrt{t}(v^2t^2 - x^2)^{\nu/2}}. \tag{13}$$

Here W and M are the energy and effective mass of the deep hole, respectively. The exponent for large t in $G_h(vt/|x| \to \infty) \sim e^{-iWt}/t^{\frac{1}{2}+\nu}$ differs from the noninteracting result by an orthogonality catastrophe correction

$$\nu = \frac{1}{2K}\left(\frac{\delta}{\pi}\right)^2 , \tag{14}$$

where δ is interpreted as the phase shift of the Fermi surface states due to the creation of the deep hole. For a weak short-range density–density interaction $H_{int} = (1/2)\int dx dx' V(x - x')n(x)n(x')$, one finds to lowest-order[25] $\delta \approx (\tilde{V}_0 - \tilde{V}_{k_F})/v_F$, where $\tilde{V}_k$ is the Fourier transform of the interaction potential $V(x)$. For integrable models, it is possible to extract phase shifts from Bethe ansatz equations and then compute the exact exponents for strong interactions.[29,33] More generally, phase shifts can be determined from information about the exact high energy spectrum using phenomenological relations.[34]

The long time behavior of the density–density correlation function $\chi(x,t) = \langle n(x,t)n(0,0)\rangle$ also involves high energy modes. For free fermions, $\chi(x,t)$ factorizes into particle and hole Green's functions. In the case of a parabolic dispersion relation, the longest lived particle-hole excitation has total momentum $\pm k_F$ and corresponds to a hole at the bottom of band and a particle at either one of the Fermi points. With $1/\sqrt{t}$ decay for the hole Green's function and $1/t$ decay for the particle Green's function, the density–density correlation function for free fermions oscillates with frequency $k_F^2/2m$ and decays as $1/t^{3/2}$. Again, this should be compared with the decay predicted by LL theory. According to Eq. (9), there are low energy contributions with momentum $q \approx 0$ (particle-hole pair around a single Fermi point) and $q \approx \pm 2k_F$ (particle-hole excitation between the two Fermi points). In the noninteracting case, both contributions decay as $1/t^2$, more rapidly than the high energy contribution. Turning on interactions between the fermions, the parameters of the dispersion are renormalized and the exponent of the high energy term in $\chi(x,t)$ is modified by X-ray edge type effects. The general decay is of the form

$$\chi(x,t) \sim \frac{e^{\pm i k_F x - iWt + iMx^2/2t}}{\sqrt{t}(vt \mp x)^{(\lambda - \sqrt{\nu/2})^2}(vt \pm x)^{(\bar{\lambda} - \sqrt{\nu/2})^2}} , \tag{15}$$

with λ, $\bar{\lambda}$ defined in Eq. (4). In the long time limit, $\chi(t \gg |x|/v) \sim e^{-iWt}/t^\eta$ with exponent[29]

$$\eta = \frac{1+K}{2} + \frac{1}{2K}\left(1 - \frac{\delta}{\pi}\right)^2 . \tag{16}$$

In this case, the correction to the free fermion exponent is of first-order in the interaction: $\eta \approx 3/2 - \delta/\pi$ for $\tilde{V}_0 \ll v_F$. The exponent η is related to the lower edge singularity of the dynamical structure factor $S(q,\omega)$ for $q = k_F$. The fact that η decreases with an increasing repulsive interaction is manifested in $S(k_F,\omega)$ as a divergence at the lower edge, similar to the effect observed in the quasi-boson peak at low energies [see again Fig. 1(c)].

For a noninteracting lattice model, the longest lived particle-hole excitation is the one obtained from the saddle point contribution for both particle and hole Green's functions. This is the excitation with total momentum $q = \pi$ that has the maximum energy allowed for a single particle-hole pair. For simplicity, let us restrict ourselves to the particle-hole symmetric case of a half-filled lattice, $k_F = \pi/2$. The density–density correlation function picks up two factors of $e^{-iWt}/\sqrt{t}$, where $W = 2J$ is half the bandwidth, from the decay of hole and particle with parabolic dispersion. As a result, at large times we obtain $\chi(t \gg |x|/v_F) \sim e^{-i2Wt}/t$. However, it turns out that this contribution is strongly suppressed by repulsive interactions. The reason is that the problem of two high energy particles (or a particle and hole) at the threshold of a continuum where there is an inverse square-root divergence in the joint density of states — related to the slow $1/t$ decay of $\chi(x, t)$ in real space and time — is analogous to the exciton problem in one dimension. For arbitrarily weak interactions, resonant scattering between the two particles removes the divergence at the threshold of the density of states.[35] Consequently, the exponent must change discontinuously when the fermion interaction is switched on. For the integrable model with nearest neighbor repulsion, $\tilde{V}(q) = 4J\Delta \cos q$ with $0 < \Delta < 1$ (which is equivalent to XXZ spin chain with anisotropy parameter Δ), it is verified[29] that the $1/t$ decay turns into a $1/t^2$ decay for times $t \gg 1/J\Delta^2$. In nonintegrable models the suppression must be even stronger because this contribution is connected with the upper threshold of the particle-hole continuum in frequency domain and the power law at this threshold is broadened by coupling to the continuum of multiple particle-hole pairs.[25] In real time, this implies an exponential decay for times larger than the corresponding decay rate. The conclusion is that also for lattice models with repulsive fermion interactions the long time behavior of $\chi(x, t)$ is governed by the excitation with one single high energy particle (or hole) and described by Eq. (15).

Summarizing this section, at zero temperature correlation functions of critical one-dimensional systems with nonlinear dispersion oscillate at large times and decay as power laws with nonuniversal exponents. The exponents depend not only on the Luttinger parameter but also on phase shifts associated with high energy modes with parabolic dispersion. Calculating the exact exponents requires information about the exact spectrum. There are, however, exceptions where the exponents do assume universal values because they are constrained by the high symmetry of the model.[34] For example, in spin chains with SU(2) symmetry the phase shift δ is fixed to $\delta = \pi/2$ and the oscillating term in the time-dependent spin correlation function decays as e^{-iWt}/t independently of details of the interactions (this includes nonintegrable models with finite range spin exchange interactions and even the Haldane–Shastry model[36] with $1/r^2$ long-range interactions). On the other hand, at the SU(2) symmetric point the Luttinger parameter becomes[37] $K = 1/2$ and the staggered ($q = 2k_F = \pi$) low energy contribution decays as $1/t^{2K} = 1/t$ at large times, i.e., with the same exponent as the high energy contribution. Moreover, for SU(2) symmetric models one should also expect logarithmic corrections in the long time decay due to marginally irrelevant couplings between high energy modes and low energy SU(2) currents.[38]

4. Long Time Decay at Finite Temperatures

Conformal invariance implies that in the TL model the correlation function for a field $\Phi(x)$ with conformal dimensions (Δ_+, Δ_-) decays at finite temperatures as[5]

$$\langle \Phi(x,t)\Phi^\dagger(0,0)\rangle \sim \left[\frac{\pi T/v}{\sinh \pi T(x/v - t)} \right]^{2\Delta_+} \left[\frac{\pi T/v}{\sinh \pi T(x/v + t)} \right]^{2\Delta_-}, \qquad (17)$$

where T is the temperature. For values of x and vt which are small compared to the inverse temperature $|x \pm vt| \ll v/T$, one observes the power law decay characteristic of zero temperature correlations. For $|x \pm vt| \gg v/T$, the correlation function decays exponentially $\sim e^{-2\pi\Delta_+ T(x/v-t)}e^{-2\pi\Delta_- T(x/v+t)}$ with a thermal correlation length $\xi \sim v/T$.

As discussed in the Sec. 3, for models with nonlinear dispersion the slowest decaying term in correlation functions involves the propagator of a deep hole coupled to low energy modes. In Eqs. (13) and (15), for example, the long time decay is determined by the factor $e^{-iWt}/\sqrt{t}$ from the Green's function of the free hole with energy W and parabolic dispersion together with the factors from low energy chiral fields. The conformal dimensions of the latter are determined after the canonical transformation that decouples the deep hole. For $T \ll W$, the main effect of thermal fluctuations is to replace the low energy factors in the result from the effective impurity model by the corresponding finite temperature expressions according to Eq. (17). As a result, the oscillating term in time-dependent correlation functions also decays exponentially within a time scale $\sim 1/T$. For instance, for the density–density correlation function one gets

$$\chi(t \gg T^{-1}, x/v) \sim \frac{e^{\pm ik_F x - iWt}}{\sqrt{t}} e^{-\pi T(\eta - 1/2)t}. \qquad (18)$$

The exponential decay in real time is connected with the thermal broadening of X-ray edge singularities in frequency domain. Beyond the expression in Eq. (18), one should also include the decay of the high energy hole due to three-body scattering processes, which lead to relaxation time $\tau_h \sim W/T^2$.[39]

However, the long time behavior of low energy correlators can also be affected by irrelevant operators. Since irrelevant operators introduce interactions between bosonic modes, the interesting possibility is that inelastic collisions lead to diffusive behavior in LLs. The word diffusion is used here in the sense of phenomenological theories[40] for many-body systems at high temperatures which predict that the autocorrelation for the density of a globally conserved quantity decay as $1/t^{d/2}$ in d dimensions due to scattering-dominated random walk of the excitations. In one dimension, this means a $1/\sqrt{t}$ decay, which is clearly much slower than the exponential decay predicted by scaling form in Eq. (17).

A finite boson lifetime that gives rise to diffusion is indeed found for lattice models at half-filling.[41,42] At half-filling, particle-hole symmetry rules out cubic band curvature terms. In this case the leading perturbations to the TL model are quartic band curvature operators, which have scaling dimension four, and the nonoscillating umklapp term, which has scaling dimension $4K$. For the integrable model with nearest neighbor interactions only, the coupling constants for these perturbations are known exactly.[43] While the calculation of the boson self-energy at zero temperature is again plagued by on-shell singularities, at finite temperature the calculation is well-behaved in the regime $|\omega \pm vq| \ll T$. The result[41] is that band curvature operators only contribute to the real part of the self-energy. The finite temperature decay rate is due entirely to umklapp scattering $\delta H_u = \lambda \int dx \cos(4\sqrt{\pi K}\phi)$. Computing the self-energy to first-order in band curvature and second-order in umklapp, the long wavelength part of the density–density correlation function (i.e., the boson propagator) becomes[d]

$$\chi_{\text{ret}}(q, \omega) = \frac{A(T)q^2/\pi}{\omega^2 - v^2(T)q^2 + i2\gamma(T)\omega}. \tag{19}$$

Here $A(T) \approx vK/[1 + b(T)]$ and $v(T) \approx v[1 + c(T) - b(T)]^{1/2}$ with $b(T), c(T) \sim \mathcal{O}(T^2, T^{8K-4})$ determined by the real part of the boson self-energy. The decay rate is given by[41,42]

$$\gamma(T) \approx \lambda^2 \left(\frac{2\pi}{v}\right)^{8K-2} \frac{K}{2^{8K-2}} \cos^2(2\pi K)\Gamma^2(2K)\Gamma^2\left(\frac{1}{2} - 2K\right) T^{8K-3}. \tag{20}$$

The imaginary part of Eq. (19) yields a Lorentzian peak with width $\gamma(T) \ll T$ for all $K > 1/2$. Particularly at the value $K = 1/2$, for which the umklapp operator becomes marginal, the decay rate picks up logarithmic corrections and we obtain $\gamma(T) \sim T/\ln^2(W/T)$. We stress that this Lorentzian approximation is only valid in the regime where $\gamma(T)$ dominates over the $T = 0$ broadening due to band curvature for finite q. Using Eq. (19) to calculate the correlation function in real space and time, one finds that inside the light cone $t > |x|/v$ there is an additional contribution to $\chi(x, t)$ besides the standard CFT terms. For times $t \gg 1/\gamma(T) \gg |x|/v$, this contribution can be calculated analytically and reads[41]

$$\chi(x, t) \approx \frac{KT}{v^2}\sqrt{\frac{2\gamma(T)}{\pi t}}\, e^{-\gamma x^2/2v^2 t}. \tag{21}$$

This is precisely the $1/\sqrt{t}$ decay expected for classical diffusion in one dimension, even though the assumptions of phenomenological theories definitely do not hold for

---

[d]This expression assumes that the model is far from an integrable point and there is no ballistic channel contribution to $\chi(q, \omega)$.[42]

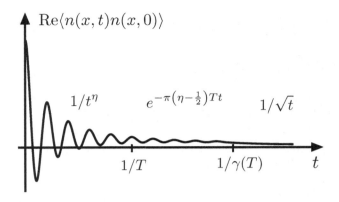

Fig. 3. Schematic time decay of density autocorrelation function $\chi(t) = \langle n(x,t)n(x,0)\rangle$ for weakly inter-acting lattice model at half-filling and low temperature T. For $t \ll 1/T$, $\chi(t)$ oscillates and decays as a power law. For $t \gtrsim 1/T$, the high frequency contribution decays exponentially. For $t \gg 1/\gamma(T)$, the low energy contribution becomes diffusive, $\chi(t) \sim T\sqrt{\gamma(T)/t}$.

LLs. As a result, the long time behavior of the density–density correlation function at finite temperatures is governed by low energy modes, since the high energy terms die out exponentially for $t \gtrsim 1/T$. Notice that the important effect here is the correction to scaling due to the irrelevant umklapp operator, not to band curvature operators that break Lorentz invariance. The real time decay of the autocorrelation function $\chi(x = 0, t)$ for $T \neq 0$ is illustrated in Fig. 3.

The mechanism of diffusion in the density correlator is important to explain the spin–lattice relaxation rate observed in spin-1/2 chains.[44] It is also connected with the question of ballistic versus diffusive transport in integrable one-dimensional systems.[42]

5. Conclusion and Outlook

We have revisited time-dependent correlation functions of critical one-dimensional systems in light of recent theoretical advances that have taken us beyond the paradigms of LL theory. Although very successful in describing thermodynamic properties, LL theory breaks down in the calculation of dynamical properties in the presence of band curvature. When treated as perturbations to the Lorentz invariant TL model, formally irrelevant band curvature operators generate divergent correc-tions to the boson propagator. However, it is possible to resum the divergences in perturbation theory by refermionizing the bosonic excitations and considering particle-hole pairs in a band with renormalized nonlinear dispersion. This proce-dure yields a line shape for the "quasi-boson" peak in the dynamic density–density response which is remarkably different than the quasiparticle peak in the spectral function of a Fermi liquid.

Also due to band curvature effects, LL theory misses the leading terms in the asymptotic long time behavior of correlation functions at zero temperature. In one dimension, high energy modes with parabolic dispersion located at the band edges give contributions to time-dependent correlation functions which oscillate in time and decay as power laws with smaller exponents than the standard contributions from low energy modes.

At finite temperatures, low energy and high energy contributions in general decay exponentially. However, for lattice models at half-filling inelastic umklapp scattering can give rise to diffusive behavior of the low energy modes. When this happens, the density–density correlation function decays in time as a power law with universal exponent $\sim 1/\sqrt{t}$ for times much larger than the relaxation time.

It should be made clear that experimental probes which are only sensitive to low frequencies, such as measurements of the local density of states close to the Fermi level, $\rho(\omega) \sim \omega^{(K+K^{-1}-2)/2}$, are not affected by the high energy contributions discussed here. However, the effects in real time evolution could be observed in experiments with ultra cold atoms in optical lattices. Coherent equilibrium dynamics can be investigated by preparing the system in the ground state and then creating a local perturbation, which is possible with the development of techniques to address individual atoms.[45] Time oscillations similar to the ones we discussed are also seen in numerical simulations of nonequilibrium dynamics of one-dimensional systems.[46] The oscillatory behavior is beyond the light cone effect predicted by CFT theory methods for quantum quenches,[47] but is presumably interpreted in terms of effective band edges for highly excited states. Experiments suggest that questions about the decay of correlations in equilibrium and nonequilibrium dynamics require a better understanding of the role played by the integrability of the model.[48]

The prediction of a diffusive long time tail at finite temperature is consistent with nuclear magnetic resonance experiments which probe the dynamics of spin-1/2 chains.[44] The decay rate $\gamma(T)$ has been confirmed numerically through the decay of the current–current correlation function,[41] but not directly in the density–density correlation function in the low temperature regime. It would be interesting to investigate the effects of spin diffusion in other correlation functions, such as the single-particle Green's function for spin-1/2 fermions. The study of finite temperature dynamics of LLs should benefit from the recent progress in numerical methods.[49,50]

Acknowledgments

I am grateful to my collaborators on this topic, in special I. Affleck, J.-S. Caux, J. Sirker and S. R. White. This work is supported by CNPq grant 309234/2011-5.

References

1. F. D. M. Haldane, *J. Phys. C: Solid State Phys.* **14**, 2585 (1981).
2. S. Tomonaga, *Prog. Theor. Phys.* **5**, 544 (1950).
3. J. M. Luttinger, *J. Math. Phys.* **4**, 1154 (1963).

4. T. Giamarchi, *Quantum Physics in One Dimension* (Claredon Press, Oxford, 2004).
5. J. Cardy, *Scaling and Renormalization in Statistical Physics* (Cambridge University Press, Cambridge, UK, 1996).
6. A. G. Izergin, V. E. Korepin and N. Y. Reshetikhin, *J. Phys. A: Math. Gen.* **22**, 2615 (1989).
7. H. Frahm and V. E. Korepin, *Phys. Rev. B* **42**, 10553 (1990).
8. V. E. Korepin, N. M. Bogoliubov and A. G. Izergin, *Quantum Inverse Scattering Method and Correlation Functions* (Cambridge University Press, Cambridge, UK, 1993).
9. V. V. Cheianov and M. B. Zvonarev, *Phys. Rev. Lett.* **92**, 176401 (2004).
10. M. B. Zvonarev, V. V. Cheianov and T. Giamarchi, *Phys. Rev. Lett.* **99**, 240404 (2007).
11. A. Imambekov and L. I. Glazman, *Science* **323**, 228 (2009).
12. N. Kitanine *et al.*, arXiv:1206.2630.
13. A. Imambekov, T. L. Schmidt and L. I. Glazman, arXiv:1110.1374.
14. J. Voit, *Rep. Prog. Phys.* **58**, 977 (1995).
15. A. Luther and I. Peschel, *Phys. Rev. B* **9**, 2911 (1974).
16. V. Meden and K. Schonhammer, *Phys. Rev. B* **46**, 15753 (1992).
17. J. Voit, *J. Phys. Condens. Matter* **5**, 8305 (1993).
18. I. E. Dzyaloshinskii and A. I. Larkin, *Sov. Phys. JETP* **38**, 202 (1974).
19. R. G. Pereira *et al.*, *Phys. Rev. Lett.* **96**, 257202 (2006).
20. R. G. Pereira *et al.*, *J. Stat. Mech. Theory Exp.* P08022 (2007).
21. H. Karimi and I. Affleck, *Phys. Rev. B* **84**, 174420 (2011).
22. K. V. Samokhin, *J. Phys.: Condens. Mat.* **10**, L533 (1998).
23. A. V. Rozhkov, *Phys. Rev. B* **74**, 245123 (2006).
24. M. Pustilnik *et al.*, *Phys. Rev. Lett.* **96**, 196405 (2006).
25. M. Khodas *et al.*, *Phys. Rev. B* **76**, 155402 (2007).
26. S. R. White and A. E. Feiguin, *Phys. Rev. Lett.* **93**, 076401 (2004).
27. A. J. Daley *et al.*, *J. Stat. Mech. Theory Exp.* P04005 (2004).
28. R. G. Pereira, S. R. White and I. Affleck, *Phys. Rev. B* **79**, 165113 (2009).
29. R. G. Pereira, S. R. White and I. Affleck, *Phys. Rev. Lett.* **100**, 027206 (2008).
30. A. H. C. Neto and M. P. A. Fisher, *Phys. Rev. B* **53**, 9713 (1996).
31. L. Balents, *Phys. Rev. B* **61**, 4429 (2000).
32. K. D. Schotte and U. Schotte, *Phys. Rev.* **182**, 479 (1969).
33. V. V. Cheianov and M. Pustilnik, *Phys. Rev. Lett.* **100**, 126403 (2008).
34. A. Imambekov and L. I. Glazman, *Phys. Rev. Lett.* **102**, 126405 (2009).
35. G. D. Mahan, *Many-Particle Physics* (Plenum, New York, 1981).
36. F. D. M. Haldane and M. R. Zirnbauer, *Phys. Rev. Lett.* **71**, 4055 (1993).
37. I. Affleck, in *Fields, Strings and Critical Phenomena*, eds. E. Brézin and J. Zinn-Justin (North-Holland, Amsterdam, 1990), p. 563.
38. R. G. Pereira, K. Penc, S. R. White, P. D. Sacramento, and J. M. P. Carmelo, arXiv:1111.2009.
39. T. Karzig, L. I. Glazman and F. von Oppen, *Phys. Rev. Lett.* **105**, 226407 (2010).
40. L. P. Kadanoff and P. C. Martin, *Ann. Phys. (NY)* **24**, 419 (1963).
41. J. Sirker, R. G. Pereira and I. Affleck, *Phys. Rev. Lett.* **103**, 216602 (2009).
42. J. Sirker, R. G. Pereira and I. Affleck, *Phys. Rev. B* **83**, 035115 (2011).
43. S. Lukyanov, *Nucl. Phys. B* **522**, 533 (1998).
44. K. R. Thurber *et al.*, *Phys. Rev. Lett.* **87**, 247202 (2001).
45. C. Weitenberg *et al.*, *Nature* **471**, 319 (2011).
46. P. Barmettler *et al.*, *New J. Phys.* **12**, 055017 (2010).
47. P. Calabrese and J. Cardy, *J. Stat. Mech.: Theory Exp.* **0504**, P04010 (2005).
48. T. Kinoshita, T. R. Wenger and D. S. Weiss, *Nature* **440**, 900 (2006).
49. J. Sirker, *Phys. Rev. B* **73**, 224424 (2006).
50. T. Barthel, U. Schollwöck and S. R. White, *Phys. Rev. B* **79**, 245101 (2009).

AN EXPANDED LUTTINGER MODEL*

DANIEL C. MATTIS

Department of Physics, University of Utah,
Salt Lake City, UT 84112 USA
mattis@physics.utah.edu

This paper generalizes Luttinger's model by introducing curvature $(d^2\varepsilon(k)/dk^2 \neq 0)$ into the kinetic energy. An exact solution for arbitrary interactions is still possible in principle, but it now requires disentangling the eigenvalue spectrum of an *harmonic string* of interacting boson fields at *each* value of q. The additional boson fields, extracted from the excitation spectrum of the Fermi sea, are self-selected according to the nature and strength of the dispersion.

Keywords: Luttinger model; one dimension; fermions.

1. Introduction

After it was first solved exactly by "bosonization"[1] and its solution inverted later by "fermionization",[2,3] Luttinger's model[4] of interacting fermions continued to deliver a huge amount of information concerning two-body correlations and other properties — far more than one might have expected from the standard apparatus of mathematical physics.[5-8] No wonder! In this reductive two-branch model of SU(2) fermions in 1D, half move to the right at a constant velocity while the others move left, also at constant speed. A right-going fermion in a plane wave state k has kinetic energy $\varepsilon(k) = c(k - k_F)$, and a left-going fermion has kinetic energy $-c(k + k_F)$ over a common range $-\infty < k < +\infty$. For reasons related to the constant speed and to the separate number-conservation of the right- and of the left-going particles, even in the presence of arbitrary two-body forces $U(x - x')$ the Hamiltonian of this model is easily diagonalized.

The present work expands both the model and its solution to the many instances in which the speed of the fermions is *not* constant, barring only "backscatterings" from one branch to the other.[a] We show that under these circumstances it is still possible to transform the interacting fermions into a multitude of bosons — all culled out of a common Fermi sea. Once the dispersion $d^2\varepsilon(k)/dk^2$ ceases to be identically zero, it becomes necessary to consider not just a single harmonic oscillator at each value of the momentum transfer q but a full-fledged harmonic *string* at each q. Quadratic forms in bosons (or, for that matter, fermions) can *always*

---

*This article first appeared in International Journal of Modern Physics B, Vol. 26, No. 22 (2012).
[a]To keep the number of particles within each branch constant.

be diagonalized exactly and all their eigenstates determined although not always *analytically* in closed form. The extended model presented below remains — in principle — a quadratic form, hence exactly solvable. In examples such as the truncated two-site string version explicitly worked out below, *Mathematica* on a home PC proved sufficiently powerful to extract the roots in closed form and to plot them.

2.   Review of the (Original) Dispersionless Model

The motional (kinetic) energy of individual fermions in the Luttinger model is given by $\varepsilon_\tau(k) = c(\tau k - k_F)$, where $\tau = \pm 1$ labels the right- (+) and left- (−) going particles. The relevant operator is,

$$KE = \sum_k \sum_\tau \sum_\sigma \varepsilon_\tau(k) c^\dagger_{\sigma,\tau}(k) c_{\sigma,\tau}(k), \quad \text{where} \quad \varepsilon_\tau(k) = c(\tau k - k_F), \qquad (1)$$

where spin index $\sigma = \uparrow$ or $\downarrow$ (or $\pm 1/2$).

The interaction Hamiltonian H_2 involves $U(x - x')$, the potential that connects fermions at x and x', *via* its Fourier transform $V(q)/L$. Delta-function interactions, $V(q) = $ constant, are distinguished from longer-ranged interactions $V(q) \propto 1/|q|$ or $1/|q|^2$ by powers of $1/|q|$. The density operator for the right-hand-goers ($\tau = +1$) can be written in terms of density creation/annihilation bosonic-type operators, as follows:

$$\rho_+(q) = \sum_{k,\sigma} c^+_{\sigma,+}(k - q/2) c_{\sigma,+}(k + q/2)$$

$$= \begin{cases} \sqrt{\dfrac{qL}{2\pi}} \displaystyle\sum_\sigma a_\sigma(q) & \text{if } q > 0, \\[3ex] \sqrt{\dfrac{-qL}{2\pi}} \displaystyle\sum_\sigma a^\dagger_\sigma(-q) & \text{if } q < 0, \end{cases}$$

with complementary relations for the left-goers. The commutator bracket relations satisfied by the a's are the obvious ones:

$$[a_\sigma(q), a_{\sigma'}(q')] = 0 = [a^\dagger_{\sigma'}(q'), a^\dagger_\sigma(q)] \quad \text{and} \quad [a_\sigma(q), a^\dagger_{\sigma'}(q')] = \delta_{\sigma,\sigma'} \delta_{q,q'}.$$

When rewritten in such operators, the interaction Hamiltonian is:

$$H_2 = \frac{1}{2} \times \frac{1}{2\pi} \sum_{q=-\infty}^{+\infty} \sum_\sigma \sum_{\sigma'} V(q)|q|((a^\dagger_\sigma(q) + a_{-\sigma}(-q))a_{\sigma'}(q) + \text{H.c.}). \qquad (2)$$

3. Neglect of Exchange Terms

In order to remain simple and solvable, the above formulation omits *exchange* interactions. To date, most papers dealing with the Luttinger model have ignored the effects both of exchange terms in the interactions and of dispersion in the kinetic energy. In this paper we study the effects of dispersion, but the reader may ask, are not exchange terms equally vital to our understanding? In the case of delta function interactions, the exchange corrections completely cancel all self-interactions. The Hamiltonian in that case can be decoupled by a linear transformation that splits it into two parts: a spinless portion that governs density fluctuations and a spinful part that maps onto the nonlinear sigma model. Even when dispersion is introduced into the kinetic energy the density fluctuations remain easily handled, using the methods outlined in the present paper; however, the solution of the concomitant sigma model requires special considerations that go beyond the elementary techniques that are introduced in the present work, albeit only slightly.

If however we wish to go beyond the local delta-function two-body potentials to consider nonlocal potentials (e.g., the Coulomb two-body interaction) in the presence of dispersion, the exchange terms in our generalized Luttinger model become nonlocal in the spin string variables. Diagonalization of such Hamiltonians requires a form of scattering theory that takes us well beyond the introductory aspects of the present work. While the problem still lies in the domain of linear algebra its solution is vastly more complicated than anything treated here. It is the subject of a paper presently under preparation and to be published elsewhere. Interested readers may request a preprint.

In summary, the topic of "exchange corrections" to the two-body interactions is avoided in the remainder of the present paper. Henceforth we deal here only with the so-called "direct interactions".

4. Important Features of the Original Luttinger Model

The total Hamiltonian $H = KE + H_2$ decouples into nonoverlapping sectors labeled q. Each sector involves only four boson operators: $a_\sigma(q)$, $a^\dagger_{-\sigma}(-q)$, $a_{-\sigma}(-q)$, $a^\dagger_\sigma(q)$. Assuming $cq > 0$, the commutator bracket equations of motion $[a_\sigma(q), KE] = cqa_\sigma(q)$ and $[a^\dagger_\sigma(q), KE] = -cqa^\dagger_\sigma(q)$ show these to be lowering and raising operator of KE in Eq. (1), respectively. For $cq < 0$ there are similar results. Thus the commutators of each member of this quartet with H, and with each other, do not generate any *new* operators and constitute a small Lie algebra within each sector. Once the sectors are individually diagonalized and their energies summed, the original model can be said to be solved in closed form.[1]

5. Dispersion: The New Feature

In our expanded version of the model, the kinetic energy of a right-going particle takes the form $\varepsilon(k) = c \times (k - k_F) + h(k - k_F)$ subject only to some mild conditions

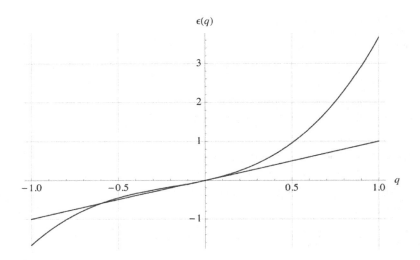

Fig. 1. Kinetic energy $\varepsilon(q)$ of right-hand going fermions (curve) at small values of $q = k - k_F$ about the Fermi surface, compared to the original (straight-line dispersion) cq.

on the *function* $h(k)$: that both it and its derivative $h'(k)$ vanish at $k = k_F$, to ensure that the Fermi level and the Fermi velocity are unaffected. (Similarly for left-goers, with $c \to -c$ and $k_F \to -k_F$.)[b] We note that once $h(k)$ ceases to be identically zero, no matter how large or small it might be relative to the original kinetic energy $c|k|$, the operators $a_\sigma(q)$ and $a^\dagger_{-\sigma}(-q)$ cease to be exact lowering or raising operators of KE, although H_2 is unaffected in its appearance. We expand the function h in a Taylor series about k_F:

$$h(k - k_F) = h(0) + (k - k_F)h_1 + \frac{1}{2!}(k - k_F)^2 h_2 + \frac{1}{3!}(k - k_F)^3 h_3 + \cdots,$$

requiring only that $h(0) = 0$ and $h_1 = 0$ as well. A typical energy curve is shown in Fig. 1.

The presence of dispersion vitiates the algebra that solved the Luttinger model, but the one essential feature that does remain ultimately leads to a solution. It is following:

By conservation of momentum, the decomposition into sectors labeled q *remains completely valid*. We shall see that all results, including correlation functions, can be expanded in powers of q. Thus they must reduce to the corresponding Luttinger model formulas in the long wavelength "correspondence limit", at $q \to 0$.

In the hypothetical example of Fig. 1 we have set $c = 1$, the second-derivative model parameter $h_2 = +2$, the third-derivative model parameter $h_3 = +10$, and all other h_j's = 0. Note the asymmetry in quasiparticle energy $|\varepsilon(q)|$ due to the mixing

---

[b]If it were additionally required that in each branch the energy $\varepsilon(k)$ remain single-valued, we could specify that $h'(k) > -c$ for the right-hand goers (and similarly, $h'(k) < c$ for left-goers) at *all* values of k, but this optional condition was not imposed in the small q expansion considered here.

of even and odd terms. (The right-left symmetry is restored by the left-hand goers, but quasiparticle–quasihole symmetry about the Fermi level is broken.)

6. Notation

In the expanded model the old operators require a new notation:

$$a_{\sigma,0}(q) = \delta_{\frac{q}{|q|},\tau} \times \sqrt{\frac{2\pi}{L|q|}} \sum_{k=-\infty}^{+\infty} c_{\sigma,\tau}^{\dagger}\left(k - \tau\frac{q}{2}\right) c_{\sigma,\tau}\left(k + \tau\frac{q}{2}\right). \tag{3a}$$

The above are the lowering operators. Their Hermitean conjugates are,

$$a_{\sigma,0}^{\dagger}(q) = \delta_{\frac{q}{|q|},\tau} \times \sqrt{\frac{2\pi}{L|q|}} \sum_{k=-\infty}^{+\infty} c_{\sigma,\tau}^{\dagger}\left(k + \tau\frac{q}{2}\right) c_{\sigma,\tau}\left(k - \tau\frac{q}{2}\right), \tag{3b}$$

where $\tau = \pm 1$ refers to right- or left-goers.

As first proved in Ref. 1 these boson operators satisfy $[a_{\sigma',0}(q'), a_{\sigma,0}^{\dagger}(q)] = \delta_{\sigma,\sigma'}\delta_{q,q'}$. With n or m standing for the composite label (σ, j), the fields in the *new* model also satisfy an extended commutator algebra, $[a_m(q), a_n^{\dagger}(q')] = \delta_{n,m}\delta_{q,q'}$ and $[a_m(q), a_n(q')] = 0$. The proof that they do (in the "thermodynamic limit" $L \to \infty$) is given in Appendix A.

Except for the extended subscript the direct interaction Hamiltonian remains that of Eq. (2),

$$H_2 = \frac{1}{2} \times \frac{1}{2\pi} \sum_{q=-\infty}^{+\infty} \sum_{\sigma} \sum_{\sigma'} V(q)|q|((a_{\sigma,0}^{\dagger}(q) + a_{-\sigma,0}(-q))a_{\sigma',0}(q) + \text{H.c.}). \tag{4}$$

That is, the *appearance* of the interactions is not affected by the nonlinear dispersion introduced into the model problem.

7. Equations of Motion in the Generalized Model

Once dispersion has been inserted into the kinetic energy, the corresponding kinetic energy Hamiltonian H_1,

$$H_1 = \sum_{k} \sum_{\tau} \sum_{\sigma} \varepsilon_{\tau}(k) c_{\sigma,\tau}^{\dagger}(k) c_{\sigma,\tau}(k), \quad \text{where} \quad \varepsilon_{\tau}(k) = c \times (\tau k - k_F) + h(\tau k - k_F) \tag{5}$$

does change its appearance and its spectrum can no longer be represented by a single boson field. However, insofar as scattering processes from one branch to the other are excluded (see footnote a) H_1 can still be decomposed into a sum of right- $(+)$ and left- $(-)$ goers' kinetic energy, as: $H_1 = \sum_{\tau=\pm 1} H_{1,\tau}$. Therefore we analyze just right-hand goers, $q > 0$, as the analysis is similar for left-hand-goers, mutatis mutandis.

The equation of motion of $a_{\sigma,0}(q)$ introduces a new boson field $a_{\sigma,1}(q)$ sharing the same values of σ and q, as follows:

$$[a_{\sigma,0}(q), H_{1,+}] = [a_{\sigma,0}(q), H_1] = A_0^{(0)}(q)a_{\sigma,0}(q) + A_1^{(0)}(q)a_{\sigma,1}(q). \tag{6}$$

The equation of motion of the new operator $a_{\sigma,1}(q)$ produces yet another, $a_{\sigma,2}(q)$,

$$[a_{\sigma,1}(q), H_{1,+}] = A_1^{(1)}(q)a_{\sigma,1}(q) + A_0^{(1)}(q)a_{\sigma,0}(q) + A_2^{(1)}(q)a_{\sigma,2}(q), \tag{7}$$

and so on, *ad infinitum*. The coefficients $A_p^{(m)}$ in this iterative procedure can be written as an infinite-dimensional, symmetric, tridiagonal array:

$$\mathbf{A}(q) = \begin{pmatrix} A_0^{(0)} & A_1^{(0)} & 0 & 0 & 0 & \cdots \\ A_1^{(0)} & A_1^{(1)} & A_2^{(1)} & 0 & \cdots & \\ 0 & A_2^{(1)} & A_2^{(2)} & A_3^{(2)} & \cdots & \\ 0 & 0 & A_3^{(2)} & \cdots & & \end{pmatrix}. \tag{8}$$

At fixed q this matrix is isomorphic to that of the equations of motion of a string of masses and springs. To simplify notation we define a row-vector creation operator: $\boldsymbol{\alpha}_\sigma^\dagger(q) = (a_{\sigma,0}^\dagger(q), a_{\sigma,1}^\dagger(q), a_{\sigma,2}^\dagger(q), \ldots)$ and its conjugate column-vector, $\boldsymbol{\alpha}_\sigma(q) = (\boldsymbol{\alpha}_\sigma^\dagger(q))^\dagger$. Each $a_{\sigma,j}(q)$ is a distinct quadratic form in the c operators. This allows for a simplified notation, in which the motional energy takes the form,

$$H_{1,+} = \sum_{q>0} \sum_\sigma \boldsymbol{\alpha}_\sigma^\dagger(q) \cdot \mathbf{A}(q) \cdot \boldsymbol{\alpha}_\sigma(q).$$

Similar manipulations at $q < 0$ take care of the left-goers. Finally we represent the kinetic energy H_1 as follows:

$$H_1 = \sum_{q=-\infty}^{+\infty} \sum_\sigma \boldsymbol{\alpha}_\sigma^\dagger(q) \cdot \mathbf{A}(q) \cdot \boldsymbol{\alpha}_\sigma(q). \tag{9}$$

This H_1 is functionally equivalent to the fermion expression in Eq. (5).

It remains only to identify the individual members of the hierarchy and, simultaneously, to calculate the matrix elements in Eq. (8). Unlike high-energy string theory, here they are not given but have to be extracted from the equations of motion.

For the right-goers, upon substitution of sums by integrals, $\sum_k \Rightarrow (L/2\pi) \int dk$, the first diagonal entry in (8) is calculated as follows:

$$A_0^{(0)}(q) = [a_{\sigma,0}^\dagger(q), [a_{\sigma,0}(q), H_1]]$$

$$= cq + \frac{1}{q} \int_{k_F - |q|/2}^{k_F + |q|/2} dk \{h(k - k_F + q/2) - h(k - k_F - q/2)\}$$

$$\approx cq + \frac{q^3}{12} \left.\frac{d^3 h(k - k_F)}{dk^3}\right|_{k_F} + O(q^5)$$

$$\equiv cq + \frac{q^3}{12} h_3 + O(q^5). \tag{10}$$

For left-goers, replace q by $-q$. Although $h(0) = h_1 = 0$ (by convention), the coefficients $h_2, h_3, \ldots$, are significant and are, so far, arbitrary. Accordingly, from (6),

$$A_1^{(0)}(q) a_{\sigma,1}(q) = [a_{\sigma,0}(q), H_1] - A_0^{(0)}(q) a_{\sigma,0}(q)$$

$$= \sqrt{\frac{2\pi}{Lq}} \sum_k \{cq + h(k + q/2) - h(k - q/2)$$

$$- A_0^{(0)}(q)\} c_{\sigma,+}^\dagger \left(k - \frac{q}{2}\right) c_{\sigma,+} \left(k + \frac{q}{2}\right). \tag{11}$$

Given that $a_{\sigma,1}(q)$ is normalized, it follows that:

$$[A_1^{(0)}(q) a_{\sigma,1}(q), A_1^{(0)}(q) a_{\sigma,1}^\dagger(q)]$$

$$= (A_1^{(0)}(q))^2$$

$$= \frac{2\pi}{Lq} \sum_k \{cq + h(k - k_F + q/2) - h(k - k_F - q/2) - A_0^{(0)}(q)\}^2$$

$$\times \left\{\tilde{n}_{\sigma,+} \left(k - k_F - \frac{q}{2}\right) - \tilde{n}_{\sigma,+} \left(k - k_F + \frac{q}{2}\right)\right\}$$

$$= \frac{1}{q} \int_{k-q/2}^{k+q/2} dk \sum_k \{cq + h(k + q/2) - h(k - q/2) - A_0^{(0)}(q)\}^2$$

$$= \frac{1}{2} \int_{-1}^{1} dx \left\{\left(\frac{1}{2} q^2 h_2\right) x + \left(\frac{1}{12} q^3 h_3\right) \left(\frac{3}{2} x^2 - \frac{1}{2}\right) + O(q^4) \cdots\right\}^2$$

$$= \frac{(h_2)^2}{12} q^4 + \frac{(h_3)^2}{720} q^6 + \cdots \tag{12}$$

Table 1.   Matrix M in first iteration.

$cq + \frac{q^3}{12}h_3 + 2V(q)\lvert q\rvert/\pi$	$2V(q)\lvert q\rvert/\pi$	$\sqrt{\frac{(h_2)^2}{12}q^4 + \frac{(h_3)^2}{720}q^6}$	0
$-2V(q)\lvert q\rvert/\pi$	$-cq - \frac{q^3}{12}h_3 - 2V(q)\lvert q\rvert/\pi$	0	$-\sqrt{\frac{(h_2)^2}{12}q^4 + \frac{(h_3)^2}{720}q^6}$
$\sqrt{\frac{(h_2)^2}{12}q^4 + \frac{(h_3)^2}{720}q^6}$	0	$cq + q^3\left(\frac{11}{60}\right)h_3$	0
0	$-\sqrt{\frac{(h_2)^2}{12}q^4 + \frac{(h_3)^2}{720}q^6}$	0	$-cq - q^3\left(\frac{11}{60}\right)h_3$

as calculated[c] to leading orders in powers of q. Then,

$$A_1^{(1)}(q) = [[A_1^{(0)}(q)a_{\sigma,1}(q), H_1], A_1^{(0)}(q)a_{\sigma,1}^\dagger(q)]/(A_1^{(0)}(q))^2$$

$$= cq + q^3\left(\frac{11}{60}\right)h_3 + O(q^5). \tag{13}$$

This Lanczös-type procedure is iterated until it is deemed to have converged and all significant entries $A_m^{(n)}$ to matrix (8) are made explicit.

8.   Eigenvalue Equation at Long Wavelengths

If the interactions are not too strong and q is sufficiently small, the procedure of solving the equations of motion of the boson quadratic form $H = H_1 + H_2$ converges quickly. After a first iteration the boson frequencies are the two positive eigenvalues of the 4×4 matrix of coefficients, M, exhibited in Table 1. This matrix was calculated from the equations of motion at arbitrary h_2 and h_3 and small q. Its first four eigenvalues are $\pm\omega_0(q)$ and $\pm\omega_1(q)$. In second iteration (three-site strings) M has three branches and its six eigenvalues are $\pm\omega_0(q)$, $\pm\omega_1(q)$, $\pm\omega_2(q)$. After N iterations M has dimension $2N$. If $q \to 0$, $N = 1$ suffices and Luttinger's model is recovered precisely.

The resulting frequencies are plotted in Fig. 2 for each of three possible or plausible interaction potentials.

---

[c]Initial calculations are simplified somewhat upon observing that

$$\{cq + h(k - k_F + q/2) - h(k - k_F - q/2) - A_0^{(0)}(q)\}$$
$$= \{h(k - k_F + q/2) - h(k - k_F - q/2) - \langle h(k - k_F + q/2) - h(k - k_F - q/2)\rangle\},$$

where

$$\langle f(k \pm q/2)\rangle \equiv \frac{1}{q}\int_{-q/2}^{q/2} dk\, f(k \pm q/2)$$

for any function f.

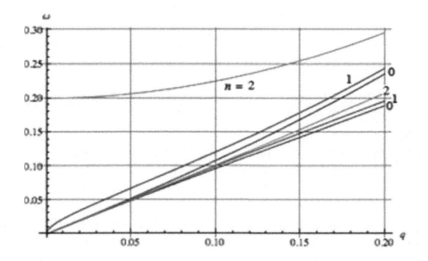

Fig. 2. $\omega_0(q)$ and $\omega_1(q)$, eigenvalues of M of Table 1, as functions of q, for two-fermion interactions of the types: $(2V(q)/\pi)|q| \equiv g|q|^{1-n}$, variously labeled $n = 0, 1, 2$. $n = 0$ corresponds to delta-function repulsion and $n = 2$ to Coulomb repulsion. Here the (arbitrarily chosen) kinetic parameters are $c = 1$, $h_2 = 2$, $h_3 = 10$. As coupling constant we chose the relatively small value, $g = 0.02$. The lower three curves are representative of the infinite set of "string modes" that are relatively unaffected by the interactions, regardless of the value of g or n. The upper three modes, consisting principally of density fluctuations, are strongly affected by interactions. Example at a finite $\omega = 0.20$ observe the plasma frequency that appears already at $q = 0$ for $n = 2$, also the strong concavity of the upper curve labeled $n = 0$.

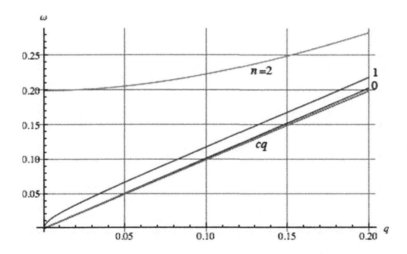

Fig. 3. Luttinger-model solutions. Same as in Fig. 2 but with $h_2 = h_3 = 0$. Individual curves are again labeled by their power-law interactions $n = 0, 1, 2$ (with the same coupling constant $g = +0.02$ as in Fig. 2). Of the 3 upper curves, the straight line marked "$n = 0$" (corresponding to repulsive delta function interactions) has slope *slightly* higher than c, as it should. The lower branches of Fig. 2 that do not appear in the original formulation[1] of the Luttinger model are here entirely decoupled from the interactions, and have each collapsed into one single line at exactly cq. Note the upper branches are similar but not identical to those in Fig. 2, as they display marginally less upward curvature in the absence of dispersion.

9. Attractive Versus Repulsive Forces

For inverse power laws $n = 1$ or 2, an attractive coupling constant ($g < 0$) leads to collapse of the system. One can still meaningfully examine the case of the delta function potential, $n = 0$, at either sign of g.

In the Luttinger model, an attractive delta function two-body interaction $-g\delta(x - x')$ lowers the speed of the collective normal mode below its original value c whereas a repulsive interaction raises it. This is the case in Fig. 3, although it is hard to see in this figure because of the small value of g. But regardless whether g is positive or negative, in the original Luttinger model with delta-function interactions, all *renormalized speeds* are *independent* of q.

Let us re-examine this feature in the expanded model with $c = 1$, $h_2 = 2$, $h_3 = 10$. Figure 4 deals with both attractive and repulsive delta-function interactions (what we denoted $n = 0$ in Figs. 2 and 3) but in strong-coupling, using a more noticeable value of the coupling constant ($g = \pm 0.5$). We plot not the frequencies, but frequencies divided by the unperturbed cq, related to what one might term the relative "phase velocity."

At $g = -0.5$, maximal attraction, the phase velocity in the strongly affected collective mode drops markedly to near zero at $q = 0$, and shows considerable dispersion at finite q. For $g = +0.5$, two-body repulsion, it rises (by 40%) because of the interaction, and exhibits similar dispersion at finite q.

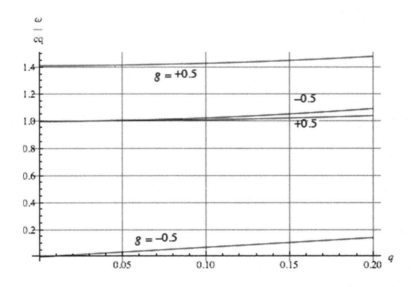

Fig. 4. Relative phase velocities, $\omega(q) \div |cq|$ versus q, calculated using Table 1 with $h_2 = 2$, $h_3 = 10$. (Delta-function repulsion or attraction ($n = 0$) in strong coupling, assuming $g = \pm 0.5$.) As $q \to 0$ the nontrivial modes (the one at $\omega(q) \div |cq| = 1.4$ for $g = +0.5$, the other at $\omega(q) \div |cq| = 0$ for $g = -0.5$) become identical to the limiting values found in the Luttinger model at the same values of the interactions. The extra mode shown, one for each value of g, are absent in Luttinger's model. They have $\omega(q) \div |cq| = 1$ at $q \to 0$ but, at finite q, both show some extra upward dispersion reflecting the choice of h_2 and h_3. Higher modes, if we were to calculate them, would also lie close to $\omega(q) \div |cq| = 1$ at small q.

In both instances the relative phase velocity of the *second* solution hovers around one, while displaying somewhat more upward curvature in the attractive than in the repulsive case.

10. Conclusion

The original Luttinger model[1] of interacting fermions in one dimension (and *only* that model) is, fortuitously, expressible as a quadratic form in a single field of boson operators, the a_0's.

Introducing nonlinear dispersion (e.g., mass) into the kinetic energy of fermions in the Luttinger model *hardly changes the results at long wavelengths* (for we have kept the speed at the Fermi surface the same as before), but it does cause a novel set of orthonormal modes, $a_1, \ldots, a_j \ldots$, to be injected into the problem. These extra modes are associated with an harmonic *string* at each q, a feature that enlarges and changes the Hilbert space qualitatively, not just quantitatively.

Such extra normal modes are (obviously) present in Tomonaga's original model[9] (although they have not been explicitly discussed before), but they were excluded in Luttinger's *by construction*.

Some of the novel modes may be detectable experimentally, optically or by injection phenomena. To calculate them theoretically we study their equations of motion. In the present work we used a long-wavelength approximation to expand the properties of each string as powers of q. We expressed the results in terms of the curvature parameters h_2 and h_3 and powers of q. At small q a truncated two-site string was sufficient for present purposes. More generally, numerical procedures are required.

Once the frequencies $\omega_j(q)$ for $j = 0, 1, 2, \ldots$ are known, one easily constructs the Hamiltonian in a diagonal form:

$$H = \sum_{q=-\infty}^{+\infty} \sum_{\sigma} \boldsymbol{\alpha}_\sigma^\dagger(q) \cdot \mathbf{H}(q) \cdot \boldsymbol{\alpha}_\sigma(q) + \Delta E_0 \tag{14}$$

and calculates the renormalized vacuum energy ΔE_0. Here the Hamiltonian matrix is,

$$\mathbf{H}(q) = \begin{pmatrix} |\omega_0(q)| & 0 & 0 & 0 & \cdots \\ 0 & |\omega_1(q)| & 0 & 0 & \cdots \\ 0 & 0 & |\omega_2(q)| & 0 & \cdots \\ 0 & 0 & \cdots & & \end{pmatrix}. \tag{15}$$

In the present examples as illustrated in the figures, finite dispersion affected the normal modes only at finite or large q. That is a consequence of the restrictions put on the functions $h(k-k_F \pm q)$ near the Femi surface. For this reason in $\lim |\omega_j(q)| \xrightarrow[q \to 0]{}$ cq the nontrivial solution $\omega_0(0)$ takes on a value which is the same as in the Luttinger model (e.g., the plasma frequency ω_{pl} for $n = 2$.) Similar decoupling would occur at short wavelengths, $q \to \infty$, if we enforced an extra requirement, $h(\pm\infty) \to 0$ (requiring that the dispersion becomes linear again asymptotically at $k \to \pm\infty$.)

Once we know the normal mode frequencies, the vacuum energy is, quite generally,

$$\Delta E_0 = \frac{1}{2} \sum_{\text{all } q} \sum_{j=0}^{\infty} (|\omega_j(q)| - c|q|). \tag{16}$$

As for the inverse process, fermion operators $\Psi(x)$ are still defined in the same way as before;[2] however, the exponents are no longer simple raising/lowering operators of H but, in the dispersive medium, are appropriate linear combinations of string operators.

Appendix A

Here we show that boson commutation relations are, at the very least, plausible for the various a's.[d]

Consider a Fermi sea of noninteracting right-going fermions, their Fermi level set at some k_F. An arbitrary bilinear form in fermion operators that decreases total momentum in an amount q, can be written in the form: $a(q) \equiv \sqrt{(2\pi/L|q|)} \sum_k \Phi(k,q) c^{\dagger}_{k-q/2} c_{k+q/2}$. If the a's are to be bosons, their amplitudes Φ need to be normalized such that $[a(q), a^{\dagger}(q)] = 1$. That is, $(1/q) \int_{-q/2}^{+q/2} dk |\Phi(k,q)|^2 = \int_{-1/2}^{+1/2} dx |\Phi(xq,q)|^2 \equiv \langle|\Phi(k,q)|^2\rangle = 1$ is required, after the substitution of sums by integrals $(\sum_k \Rightarrow (L/2\pi) \int dk)$. An infinite number of acceptable linearly independent functions exist on the unit interval $(-1/2 < x < 1/2)$ at every given q. To distinguish them let us index the quantities and operators by a subscript: $a_j(q) \equiv \sqrt{(2\pi/L|q|)} \sum_k \Phi_j(k,q) c^{\dagger}_{k-q/2} c_{k+q/2}$ and require these functions to form an orthonormal set on the unit interval: $\int_{-1/2}^{+1/2} dx \Phi_j^*(xq,q) \Phi_m(xq,q) = \delta_{j,m}$.

Operators constructed at a differing value of the momentum transfer, say at $q' \neq q$, are similarly designated $a_j(q')$. Generalizing the quantum statistics to arbitrary q we must also have $[a_n(q), a_m^{\dagger}(q')] = \delta_{n,m} \delta_{q,q'}$, as well as $[a_n(q), a_m(q')] = [a_m^{\dagger}(q'), a_n^{\dagger}(q)] = 0$ at all n, m, q, q'. Let us now prove what should be obvious.

$$[a_n(q), a_m(q')] = \frac{2\pi}{L|qq'|} \sum_k \{\Phi_n(k - q'/2, q)\Phi_m(k + q/2, q')$$

$$- \Phi_n(k + q'/2, q)\Phi_m(k - q/2, q')\} c^{\dagger}_{k - \frac{q+q'}{2}} c_{k + \frac{q+q'}{2}}. \tag{A.1}$$

---

[d]For notational simplicity we leave out all unnecessary subscripts in this Appendix.

The right-hand side of Eq. (A.1) defines a new operator $b_{n,m}$. When normalized, such that $[b_{n,m}(q+q'), b_{n,m}^\dagger(q+q')] = 1$, it is:

$$b_{n,m}(q+q') \equiv \frac{1}{|D_{n,m}(q+q')|} \sqrt{\frac{2\pi}{L|q+q'|}} \sum_k \{\Phi_n(k-q'/2, q)\Phi_m(k+q/2, q')$$

$$- \Phi_n(k+q'/2, q)\Phi_m(k-q/2, q')\} c^\dagger_{k-\frac{q+q'}{2}} c_{k+\frac{q+q'}{2}}. \qquad (A.2)$$

The square of the normalization parameter $D_{n,m}$, is a dimensionless quantity:

$$D_{n,m}^2(q+q') \equiv \frac{1}{|q+q'|} \int_{-\frac{q+q'}{2}}^{+\frac{q+q'}{2}} dk |\Phi_n(k-q'/2, q)\Phi_m(k+q/2, q')$$

$$- \Phi_n(k+q'/2, q)\Phi_m(k-q/2, q')|^2$$

Now, compare (A.1) and (A.2) to verify that the commutator in (A.1) is,

$$[a_n(q), a_m(q')] = \frac{1}{\sqrt{L}} \left(|D_{n,m}(q+q')| \sqrt{\frac{2\pi|q+q'|}{|qq'|}}\right) b_{n,m}(q+q'). \qquad (A.3)$$

Let us test that the right-hand side of (A.3) vanishes in the thermodynamic limit $L \to \infty$. Consider,

$$[[a_n(q), a_m(q')], [a_m^\dagger(q'), a_n^\dagger(q)]] = \frac{1}{L} \left(D_{n,m}^2(q+q') \frac{2\pi|q+q'|}{|qq'|}\right), \qquad (A.4)$$

where $D_{n,m}^2(q+q')$ is a finite, dimensionless number defined by the integral just below (A.2) and it is bounded. Hence the parenthetical expression in (A.4) is also bounded. Denote it: $l_{n,m}(q, q')$, a length independent of L. (Note: $l_{0,0}(q, q') = 0$). It follows that (A.3) and (A.4) vanish in limit $L \to \infty$, the first as $\sqrt{l_{n,m}(q, q')}/L$ and the second as $l_{n,m}q, q')/L$.

Various other commutation relations, expected for continuous boson fields, are all equally satisfied in the thermodynamic limit $L \to \infty$, as can be verified by similar calculations.

References

1. D. C. Mattis and E. H. Lieb, *J. Math. Phys.* **6**, 304 (1965).
2. D. C. Mattis, *J. Math. Phys.* **15**, 609 (1974).
3. S. Mandelstam, *Phys. Rev. D* **11**, 3026 (1975).
4. J. M. Luttinger, *J. Math. Phys.* **4**, 1154 (1963).
5. F. D. M. Haldane, *J. Phys. C* **14**, 2585 (1981).
6. F. D. M. Haldane, *Phys. Lett. A* **81**, 153 (1981).
7. M. Stone (ed.), *Bosonization* (World Scientific Publishing, Singapore, 1994).
8. V. E. Korepin and F. H. L. Essler, *Exactly Solvable Models of Strongly Correlated Electrons* (World Scientific Publishing, Singapore, 1994).
9. S. Tomonaga, *Progr. Theoret. Phys. (Kyoto)* **5**, 544 (1950).

Chapter III
Applications and Experimental Test

QUANTUM HALL EDGE PHYSICS AND ITS
ONE-DIMENSIONAL LUTTINGER LIQUID DESCRIPTION[*]

ORION CIFTJA

Department of Physics, Prairie View A&M University,
Prairie View, Texas 77446, USA

We describe the relationship between quantum Hall edge states and the one-dimensional Luttinger liquid model. The Luttinger liquid model originated from studies of one-dimensional Fermi systems, however, it results that many ideas inspired by such a model can find applications to phenomena occurring even in higher dimensions. Quantum Hall systems which essentially are correlated two-dimensional electronic systems in a strong perpendicular magnetic field have an edge. It turns out that the quantum Hall edge states can be described by a one-dimensional Luttinger model. In this work, we give a general background of the quantum Hall and Luttinger liquid physics and then point out the relationship between the quantum Hall edge states and its one-dimensional Luttinger liquid representation. Such a description is very useful given that the Luttinger liquid model has the property that it can be bosonized and solved. The fact that we can introduce a simpler model of noninteracting bosons, even if the quantum Hall edge states of electrons are interacting, allows one to calculate exactly various quantities of interest. One such quantity is the correlation function which, in the asymptotic limit, is predicted to have a power law form. The Luttinger liquid model also suggests that such a power law exponent should have a universal value. A large number of experiments have found the quantum Hall edge states to show behavior consistent with a Luttinger liquid description. However, while a power law dependence of the correlation function has been observed, the experimental values of the exponent appear not to be universal. This discrepancy might be due to various correlation effects between electrons that sometimes are not easy to incorporate within a standard Luttinger liquid model.

Keywords: Luttinger liquid; one-dimensional electronic system; Fermi liquid; quantum Hall effect.

1. Introduction

The Luttinger liquid model[1] proposed by Luttinger in 1963 is a theoretical model that describes the behavior of interacting electrons (or more generally, fermions) in one dimension (1D). While the picture of interacting Fermi liquids in two dimensions (2D) and three dimensions (3D) is fairly well understood, a Fermi liquid model breaks down in 1D. A precursor to the Luttinger liquid model was the Tomonaga model first proposed by Tomonaga in 1950.[2] The original Tomonaga model describes a 1D electron gas with a linear dispersion relation. The Tomonaga procedure is to examine such a Hamiltonian and then solve it approximately. The important conclusion is that, for such a model, the excitations of the 1D electron gas are approximate bosons, although the elementary particles of the system are fermions.

The Luttinger liquid model is, in some sense, a reformulated version of the Tomonaga model. The basic feature of the Luttinger liquid model is the intro-

---

[*]This article first appeared in International Journal of Modern Physics B, Vol. 26, No. 22 (2012).

duction of two types of fermions (left moving and right moving ones). One type of fermion has an energy spectrum given by $\epsilon_k = +\hbar v_F k$, while the other type has an energy spectrum given by $\epsilon_k = -\hbar v_F k$. In the above expressions, v_F is the Fermi velocity, k is the wave vector and $\hbar$ is the reduced Planck's constant. Differently from the energy dispersion of the Tomonaga model (where $\epsilon_k = \hbar v_F |k|$), the energy dispersion of the Luttinger liquid model allows an infinite number of each kind of particle since the occupied energy states extend to negative infinity (thus, some extra care is needed while dealing with such subtlety). As it turns out, the Luttinger liquid model has the advantage of being exactly solvable. It was solved correctly in 1965 by Mattis and Lieb.[3]

Strictly speaking, the Luttinger liquid model is a generic 1D many-particle state. The simplest way to understand how we can arrive to such a model is to consider the ground state of a noninteracting 1D system of fermions. For simplicity, we assume that the fermions are spinless. In the noninteracting ground state, each fermion is represented by a plane wave state with wave vector, k such that:

$$-k_F \leq k \leq +k_F \,, \tag{1}$$

where k_F is the Fermi wave vector. The uniform number density of a 1D system of noninteracting spinless fermions can be written as:

$$\rho_0 = \frac{k_F}{\pi} \,, \tag{2}$$

while the single-fermion dispersion energies are:

$$\epsilon_k = \frac{\hbar^2 k^2}{2m} \tag{3}$$

where m is the mass of the fermions. In absence of interactions, the fermions occupy all the plane wave states with $|k| \leq k_F$. All other states with $|k| > k_F$ are empty. In presence of interactions, the most affected states are the one that lie close to the Fermi energy, $\epsilon_F = \hbar^2 k_F^2/(2m)$. Thus, at low energies, only excitations close to the Fermi points are possible. To a good degree of approximation, one can therefore expand the dispersion energy at the Fermi points and keep only the lowest (i.e., first) order contribution. Thus, the dispersion energy can be easily linearized at the two Fermi points, $\pm k_F$ and one ends up with:

$$\epsilon_{k,R} = \epsilon_F + \hbar v_F (k - k_F) \,; \quad k \approx +k_F \,, \tag{4}$$

and

$$\epsilon_{k,L} = \epsilon_F - \hbar v_F (k + k_F) \,; \quad k \approx -k_F \,, \tag{5}$$

where $v_F = \hbar k_F/m$ is the Fermi velocity. Relative to the Fermi energy we have:

$$E_{k,R,L} = \epsilon_{k,R,L} - \epsilon_F = \pm \hbar v_F (k \mp k_F) \,. \tag{6}$$

Apart from insignificant constant terms, the two linear dispersion expressions $(\pm \hbar v_F k)$ for respectively, R and L moving fermions are an accurate description of the single-particle energies close to the 1D Fermi surface, which in this case consists of only two points. States labeled "R" have positive velocity, thus move to the "right". Those labeled "L" on the contrary move to the "left".

We now imagine that the "vacuum" state consists of an infinite Fermi sea in which all R movers with $k < k_F$ are occupied. Similarly, all L movers with $k > -k_F$ are occupied. Other states are empty. With this assumption, the momentum occupation numbers of the "vacuum" state can be written as:

$$n_{\chi,k} = \Theta(k_F - \chi k), \tag{7}$$

where $\chi = \pm 1$ is the so-called chirality, $+1$ for R movers and -1 for L movers. The function $\Theta(x)$ is the usual step function.

By adopting a second-quantization formalism, we can introduce the creation and destruction operators for R and L movers, $\hat{c}^\dagger_{\chi,k}$ and $\hat{c}_{\chi,k}$. Such operators satisfy the usual fermionic anticommutation relations. At this point, we are interested only in the excitations above the vacuum. Thus, if we measure the energy relative to the ground state, $\sum_{k,\chi} E_{k,\chi} \Theta(k_F - \chi k)$, the Hamiltonian can be written as:

$$\hat{H}_0 = \hbar v_F \sum_{k,\chi} \Theta(\chi k)(\chi k - k_F)[\hat{c}^\dagger_{\chi,k}\hat{c}_{\chi,k} - \Theta(k_F - \chi k)]. \tag{8}$$

Relative to the new ground state, $\hat{c}_{\chi,k}|0\rangle = 0$ if $\chi k > k_F$ and $\hat{c}^\dagger_{\chi,k}|0\rangle = 0$ if $\chi k < k_F$. If we introduce a "normal ordered" number operator which counts the number of particles relative to the vacuum, one easily recognizes that:

$$\hat{N}_{\chi,k} =: \hat{c}^\dagger_{\chi,k}\hat{c}_{\chi,k} := \hat{c}^\dagger_{\chi,k}\hat{c}_{\chi,k} - \Theta(k_F - \chi k). \tag{9}$$

The eigenvalues of $\hat{N}_{\chi,k}$ are the relative (with respect to the vacuum) occupation numbers which are $+1, 0$ for particles and $-1, 0$ for holes. At this point, one notes that the low energy noninteracting Hamiltonian can be compactly written as:

$$\hat{H}_0 = \hbar v_F \sum_{k,\chi} (\chi k - k_F)\hat{N}_{\chi,k}. \tag{10}$$

Note that the single-particle energies have been defined so as to vanish at $k = \chi k_F$. The eigenvalues of $\hat{H}_0$ are so-defined that they are all nonnegative.

Inclusion of interaction between particles in the Luttinger liquid model is a little bit more elaborate. One starts the usual way of expressing the fundamental interaction Hamiltonian as:

$$\hat{H}_{\text{int}} = \frac{1}{2L} \sum_{q \neq 0} v(q)\hat{\rho}_{-q}\hat{\rho}_q, \tag{11}$$

where $\hat{\rho}_q$ is the (total) fermion density operator, $v(q)$ is the Fourier transform of a "well-behaved" interaction potential and L is the length of the system. As usually, the $q = 0$ term is excluded from the sum. The expression for $\hat{\rho}_q$ in terms of the separate density fluctuation operators for R and L movers is more complicated than simply adding together the two of them. For further details we refer the reader to more specialized treatments, for example p. 506 of Ref. 4. The basic idea is that, by following these technical procedures, one eventually obtains:

$$\hat{H}_{\text{int}} = \frac{1}{2L} \sum_{q \neq 0} V_1(q)[\hat{\rho}_{R,-q}\hat{\rho}_{R,q} + \hat{\rho}_{L,-q}\hat{\rho}_{L,q}]$$

$$+ \frac{1}{2L} \sum_{q \neq 0} V_2(q)[\hat{\rho}_{R,-q}\hat{\rho}_{L,q} + \hat{\rho}_{L,-q}\hat{\rho}_{R,q}], \tag{12}$$

where

$$V_1(q) = v(q), \tag{13}$$

and

$$V_2(q) = v(q) - v(2k_F). \tag{14}$$

By combining together the noninteracting and interacting terms, one writes the complete Luttinger liquid Hamiltonian as: $\hat{H} = \hat{H}_0 + \hat{H}_{\text{int}}$. As showed by Mattis and Lieb,[3] the Luttinger liquid model can be exactly solved by using bosonization techniques.[5,6] [a] The repercussions of this result have had a far reaching influence and have stimulated great interest in many scientific disciplines. Besides the fact that it can be solved exactly, the Luttinger liquid model is interesting to a wide range of areas for many reasons: (i) The behavior of Luttinger liquids is drastically different from the higher-dimensional counterparts, which can be described fairly well as Fermi liquids; (ii) The Luttinger liquid model can be applied to any generic 1D Fermi system and, thus, solved in terms of a few parameters; and (iii) The Luttinger liquid model provides a good example to understand quantum phases and quantum phase transitions in 1D systems. The Luttinger liquid model is also able to explain that the response of charge (or particle) density to some external perturbation is represented by waves ("plasmons" or charge density waves) propagating at a velocity that is determined by the strength of the interaction and the average density. For a noninteracting system, this wave velocity is equal to the Fermi velocity. Such spin density waves (whose velocity, to lowest approximation, is equal to the unperturbed Fermi velocity) propagate independently from the charge density waves. Thus, unlike the quasiparticles of a Fermi liquid which carry both spin and charge, the elementary excitations of a Luttinger liquid are separate charge and spin waves. The mathematical description becomes very simple in terms of these waves (solving

---

[a]See, for instance Refs. 5 and 6.

the one-dimensional wave equation), and most of the work consists in transforming back to obtain the properties of the particles themselves. This very important fact is known as spin-charge separation.

Even though the Luttinger liquid model has historically been treated as a toy model for theorists, experiments are getting close to realizing such a system. Plausible examples of physical systems believed to be described by the Luttinger liquid model include artificial "quantum wires" (1D chains of electrons) created by applying gate voltages to a 2D electron gas, electrons in carbon nanotubes, electrons hopping along 1D chains of molecules (e.g., certain organic molecular crystals), fermionic atoms in quasi-1D atomic traps, and last, but not least electrons moving along the edge states of a quantum Hall system, the topic of the current review paper.

2. Luttinger Liquid and Bosonization

The bosonization technique is widely used in many areas of theoretical physics. When applied to a Luttinger liquid model, this technique makes it possible to study the low energy limit of a 1D system. Let us first recall the definition of the normal order of (fermionic) operators as the procedure in which all destruction operators are placed on the right of all creation operators. For example,

$$: \hat{A}_1^\dagger \hat{A}_2 \hat{A}_3^\dagger := -\hat{A}_1^\dagger \hat{A}_3^\dagger \hat{A}_2 \,. \tag{15}$$

The most important property of a normal ordered product of creation and destruction operators is to have a zero expectation value with respect to the vacuum ground state. From now on normal ordering of operators is implicitly assumed.

After having defined the chiral creation and annihilation operators, $\hat{c}_{\chi,k}^\dagger$ and $\hat{c}_{\chi,k}$ we then define the respective field operators:

$$\hat{\psi}_\chi(x) = \frac{1}{\sqrt{L}} \sum_k e^{ikx} \hat{c}_{\chi,k} \,, \tag{16}$$

and its conjugate, $\hat{\psi}_\chi^\dagger(x)$. Here L is some length. The chiral density operator is defined as:

$$\hat{\rho}_\chi(x) =: \hat{\psi}_\chi^\dagger(x)\hat{\psi}_\chi(x) := \frac{1}{L} \sum_q e^{iqx} \hat{\rho}_{\chi,q} \,, \tag{17}$$

where (for $q \neq 0$) we have:

$$\hat{\rho}_{\chi,q} = \sum_k \hat{c}_{\chi,k}^\dagger \hat{c}_{\chi,k+q} \,. \tag{18}$$

When $q = 0$ we simply obtain $\hat{\rho}_{\chi,q=0} = \hat{N}_\chi = \sum_k : \hat{c}_{\chi,k}^\dagger \hat{c}_{\chi,k}$: where $\hat{N}_\chi$ is the number operator for the χ type fermions (relative to the "vacuum"). Since density operators play a crucial role in the bosonization technique one first needs to understand the commutation relations between density operators and chiral field operators and between density operators themselves, as well. While the commutation relations between a density operator and a chiral field operator is easy to obtain:

$$[\hat{\rho}_{\chi,q}, \hat{\psi}_\chi(x)] = e^{-iqx}\hat{\psi}_{\chi'}(x)\delta_{\chi,\chi'}, \tag{19}$$

the commutation relation between density operators is much more cumbersome to derive. Here, we skip all the technical details and simply give the final result:

$$[\hat{\rho}_{\chi,q}, \hat{\rho}_{\chi',-q'}] = \chi\delta_{\chi,\chi'}\delta_{q,q'}\frac{qL}{2\pi}. \tag{20}$$

This result can be interpreted as representing a form of the so-called $U(1)$ Kac–Moody algebra. The commutator between the density operator and the noninteracting Hamiltonian is much easier to calculate. For $q \neq 0$ one ends up with:

$$[\hat{H}_0, \hat{\rho}_{\chi,q}] = -\hbar v_F \chi q \hat{\rho}_{\chi,q}. \tag{21}$$

For the $q = 0$ case, we simply obtain $[\hat{H}_0, \hat{N}_\chi] = 0$. The commutation relationship between density operators turns out to be bosonic-like. Thus, one can introduce bosonic creation and annihilation operators, $\hat{b}_q^\dagger$ and $\hat{b}_q$ of the form:

$$\hat{b}_q^\dagger = \sqrt{\frac{2\pi}{L|q|}}\sum_\chi \Theta(\chi q)\hat{\rho}_{\chi,-q}, \tag{22}$$

and

$$\hat{b}_q = \sqrt{\frac{2\pi}{L|q|}}\sum_\chi \Theta(\chi q)\hat{\rho}_{\chi,q}. \tag{23}$$

It is also easy to show that the $\hat{b} - s$ are proper boson operators:

$$[\hat{b}_q, \hat{b}_{q'}^\dagger] = \delta_{q,q'}; \quad [\hat{b}_q, \hat{b}_{q'}] = [\hat{b}_q^\dagger, \hat{b}_{q'}^\dagger] = 0. \tag{24}$$

The commutator of the noninteracting Hamiltonian, $\hat{H}_0$ and these boson operators gives:

$$[\hat{H}_0, \hat{b}_q^\dagger] = \hbar v_F |q|\hat{b}_q^\dagger; \quad [\hat{H}_0, \hat{b}_q] = -\hbar v_F |q|\hat{b}_q. \tag{25}$$

This clearly indicates that $\hat{b}_q$ and $\hat{b}_q^\dagger$ act as lowering and raising operators for the energy. From this point on, one can use the standard approach to build the whole family of eigenstates of $\hat{H}_0$ by repeated applications of the raising operator. Thus, one can follow all the routine steps of setting up the usual bosonic states and write a bosonized noninteracting Hamiltonian of the form:

$$\hat{H}_0 = \sum_{q \neq 0} \hbar v_F |q| \hat{b}_q^\dagger \hat{b}_q + \hbar v_F \frac{\pi}{L} \sum_\chi N_\chi^2 \,, \tag{26}$$

where the second term in the expression above can safely be omitted.

One can write the complete Luttinger liquid Hamiltonian, $\hat{H} = \hat{H}_0 + \hat{H}_{\text{int}}$ (including the interaction part) in terms of $\hat{b}_q^\dagger$ and $\hat{b}_q$ operators with help from Eqs. (26) and (12). The resulting Hamiltonian is quadratic in boson operators, and can therefore be solved exactly by means of a Bogoliubov transformation (with $\hat{\beta}_q^\dagger$ and $\hat{\beta}_{-q}$ operators).

By casting the total Hamiltonian in a noninteracting form, one concludes that the Luttinger liquid model is equivalent to a system of independent massless bosons:

$$\hat{H} = \hat{H}_0 + \hat{H}_{\text{int}} = \sum_{q \neq 0} \hbar \omega_q \hat{\beta}_q^\dagger \hat{\beta}_q \,. \tag{27}$$

The dispersion energy of the collective modes is:

$$\epsilon_q = \hbar \omega_q \,; \quad \omega_q = c(q)|q| \,, \tag{28}$$

where ω_q is the frequency and $c(q)$ is a parameter that depends in a nontrivial way on v_F, $V_1(q)$ and $V_2(q)$. Among several quantities that can be calculated exactly for a Luttinger liquid model, an important one is the correlation function:

$$\langle \hat{\psi}_\chi^\dagger(x) \hat{\psi}_\chi(x') \rangle \,. \tag{29}$$

Its calculation is very important to understand transport and tunneling behavior in the Luttinger liquid regime. It can be shown that in the large $(x - x')$ limit (and large L limit), the correlation function takes the asymptotic form:

$$\langle \hat{\psi}_\chi^\dagger(x) \hat{\psi}_\chi(x') \rangle \sim \frac{1}{|x - x'|^\gamma} \,, \tag{30}$$

where the exponent γ has a universal value within the framework of a Luttinger liquid model. In absence of interactions, $\gamma = 1$, but with interactions the value of the exponent γ is renormalized.

3. Quantum Hall Effect and the Laughlin States

At very low temperatures, a two-dimensional electron system (2DES) in a perpendicular magnetic field exhibits remarkable quantum phenomena where the integer quantum Hall effect (IQHE)[7] and the fractional quantum Hall effect (FQHE)[8] stand out as some of the most important discoveries in condensed matter physics for the last decades. The FQHE represents a novel strongly correlated electronic quantum phase arising from the interplay between low-dimensionality, near absolute zero temperature and strong magnetic field.[9-13] In particular, FQHE is a unique example of a novel collective quantum liquid state of matter that originates from strong electronic correlations only. Consistent with the rules of quantum mechanics, a perpendicular magnetic field leads to the creation of massively degenerate discrete quantum states known as Landau levels (LLs). The energy quantization associated with such states is the foundation of many new experimental and theoretical advances in condensed matter physics. The filling factor, defined as the ratio of the number of electrons to the degeneracy (number of available states) of each LL, represents an important characteristic parameter of the system. At low enough temperature and low enough amount of disorder, the FQHE represents the condensation of electrons into an incompressible quantum fluid state formed at specific filling factors, $\nu = N/N_s$ where N is the number of electrons in the system and N_s is the LL degeneracy.

Typical liquid FQHE states are more pronounced in the extreme quantum limit of a very high perpendicular magnetic field when the lowest Landau level (LLL) is fractionally filled with electrons. Stabilization of these novel electronic liquid phases happens at filling factors that generally have odd denominators. The most striking experimental signature of this new collective state of matter is the observation of quantized plateaus of the Hall resistance,

$$R_H = \frac{h}{\nu e^2} \approx \frac{25812.807}{\nu} \Omega \,, \tag{31}$$

where the filling factor ν is a fractional number, h is the Planck constant and e is the magnitude of electron charge. The strong magnetic field quantizes the electrons' motion on the plane and quenches the kinetic energy of each electron to a discrete set of LLs separated by the relatively large cyclotron energy, $\hbar\omega_c = \hbar eB/m_e$, where $-e$ ($e > 0$) is the electron charge and m_e is the (effective) mass of electrons. For simplicity, we assume that the spin of electrons is fully polarized by the strong magnetic field rendering them *effectively spinless*. The uniform electron density of the system can be written as:

$$\rho_0 = \frac{\nu}{2\pi l_0^2} \,, \tag{32}$$

where $l_0 = \sqrt{\hbar/(eB)}$ is the electronic magnetic length. The most robust FQHE states occur in the LLL and correspond to filling factors $\nu = 1/3$ and $\nu = 1/5$. Laughlin's theory[14] describes very well such states in terms of trial wavefunctions that belong to the Hilbert space of the LLL.[15] Following Laughlin's discovery, a great deal of subsequent theoretical work has shed light on the incompressible nature of such electronic liquid phases[16-20] as well as the nature of other states at different filling fractions.[21-24]

Since a 2DES in the FQHE regime consists of electrons with negative charge, it is customary to assume that the direction of the perpendicular magnetic field is $\mathbf{B} = (0, 0, -B)$. The choice of the negative sign of $\mathbf{B}$ is a matter of convenience, allowing us to express the LLL wavefunctions for electrons in terms of the complex variable $z = x + iy$ rather than its complex conjugate. For filling factors of the form $\nu = 1/m$ ($m = 3, 5$) the trial Laughlin[14] wavefunction for N electrons can be written as:

$$\Psi_m(z_1, \ldots, z_N) = \prod_{i<j}^{N} (z_i - z_j)^m \prod_{j=1}^{N} \exp\left(-\frac{|z_j|^2}{4l_0^2}\right), \tag{33}$$

where $z_j = x_j + iy_j$ is the position of the jth electron in complex coordinates. This wavefunction[25] gives an excellent description of the true ground state of the electrons for $m = 3$ and 5. For $m \geq 7$ the electrons tend to form a Wigner crystal state[26] consistent with the experimental observation[27] that the FQHE does not occur for filling factors $\nu \leq 1/7$.

Given the form of the Laughlin's wavefunction in Eq.(33), it is clear that in this case we have adopted a symmetric gauge for the magnetic vector potential and, thus, we are employing a disk geometry. In a disk geometry model[28] we consider N electrons of charge $-e$ ($e > 0$) immersed in a uniform positively charged finite disk of area $\Omega_N = \pi R_N^2$ where R_N is the radius of the disk. The density of the system (number of electrons per unit area) or otherwise the uniform density of the background, $\rho_0 = N/\Omega_N$, is constant. For a given filling factor, the density of the system can also be written as $\rho_0 = \nu/(2\pi l_0^2)$, therefore the radius of disk varies with N as:

$$\frac{R_N}{l_0} = \sqrt{\frac{2N}{\nu}}; \quad N \geq 2. \tag{34}$$

The quantum Hamiltonian of the system,

$$\hat{H} = \hat{K} + \hat{V}, \tag{35}$$

consists of kinetic and potential energy operators where

$$\hat{K} = \frac{1}{2m_e} \sum_{i=1}^{N} [\hat{\mathbf{p}}_i + e\mathbf{A}_i]^2, \tag{36}$$

is the kinetic energy operator written in a symmetric gauge. The potential energy operator:

$$\hat{V} = \hat{V}_{ee} + \hat{V}_{eb} + \hat{V}_{bb}, \tag{37}$$

consists of electron-electron (ee), electron-background (eb) and background-background (bb) interaction potentials written as:

$$\hat{V}_{ee} = \sum_{i<j}^{N} v(\mathbf{r}_i - \mathbf{r}_j), \tag{38}$$

$$\hat{V}_{eb} = -\rho_0 \sum_{i=1}^{N} \int_{\Omega_N} d^2 r v(\mathbf{r}_i - \mathbf{r}), \tag{39}$$

and

$$\hat{V}_{bb} = \frac{\rho_0^2}{2} \int_{\Omega_N} d^2 r \int_{\Omega_N} d^2 r' v(\mathbf{r} - \mathbf{r}'), \tag{40}$$

where $v(\mathbf{r}_i - \mathbf{r}_j) = e^2/|\mathbf{r}_i - \mathbf{r}_j|$ is the Coulomb interaction potential. In all expressions above, $\mathbf{r}_i$ (or $\mathbf{r}_j$) denote electronic 2D position vectors, while $\mathbf{r}$ and $\mathbf{r}'$ are background coordinates. Each of the N position variables of electrons, $\{\mathbf{r}_i\}$ extends all over space ($-\infty$ to $+\infty$). However, background coordinates, $\mathbf{r}$ (or $\mathbf{r}'$) are confined within the finite disk, namely $0 \le |\mathbf{r}| \le R_N$ (or $0 \le |\mathbf{r}'| \le R_N$).

Since the Laughlin wavefunction lies entirely in the LLL, the expectation value of the kinetic energy per electron:

$$\frac{\langle \hat{K} \rangle}{N} = \frac{1}{2} \hbar \omega_c, \tag{41}$$

is an irrelevant constant. Therefore, the only important contribution in the quantum mechanical Hamiltonian originates from the total potential energy operator.

The ground state interaction energy per particle can be written as:

$$\epsilon = \epsilon_{ee} + \epsilon_{eb} + \epsilon_{bb}, \tag{42}$$

where $\epsilon = \langle \hat{V} \rangle/N$, $\epsilon_{ee} = \langle \hat{V}_{ee} \rangle/N$, $\epsilon_{eb} = \langle \hat{V}_{eb} \rangle/N$ and $\epsilon_{bb} = \langle \hat{V}_{bb} \rangle/N$ are, respectively, the total, ee, eb and bb interaction energies per particle (namely, the standard quantum expectation values for the respective operators). As usual, a charge neutralizing background is required to guarantee the stability of electronic systems with repulsive Coulomb interactions. Since our interest is to study states at a given (fixed) magnetic field, this implies that the system should be kept at a fixed density. Stated otherwise, this means that when we increase the number N of electrons in our finite system, the area, Ω_N of the background disk also increases in a way that the density, N/Ω_N remains constant. Note that the $\langle \hat{V}_{bb} \rangle$ term does not depend in the form of the wavefunction and can be calculated analytically. For a disk with uniform density, $\rho_0 = \nu/(2\pi l_0^2)$ and radius R_N we calculate that the bb interaction potential is:

$$\epsilon_{bb} = \frac{\langle \hat{V}_{bb} \rangle}{N} = \frac{8}{3\pi} \sqrt{\frac{\nu N}{2}} \frac{e^2}{l_0}. \tag{43}$$

We can write the quantity $\langle \hat{V}_{eb} \rangle$ as:

$$\langle \hat{V}_{eb} \rangle = -\rho_0 \int d^2 r_1 \rho(\mathbf{r}_1) \int_{\Omega_N} d^2 r v(\mathbf{r}_1 - \mathbf{r}), \tag{44}$$

where $\rho(\mathbf{r}_1)$ is the single-particle density function:

$$\rho(\mathbf{r}_1) = N \frac{\int d^2 r_2 \cdots d^2 r_N |\Psi_m(\mathbf{r}_1 \ldots, \mathbf{r}_N)|^2}{\int d^2 r_1 \cdots d^2 r_N |\Psi_m(\mathbf{r}_1, \ldots, \mathbf{r}_N)|^2} . \tag{45}$$

Clearly, considering the strongly correlated nature of the Laughlin states, the energy of the system and related quantities can be calculated exactly only for small systems of electrons.[29–31] In all other circumstances, one must resort to numerical and computational approaches.

Numerous techniques have been employed to calculate the above quantities including various Monte Carlo (MC) schemes which are essentially exact in the thermodynamic limit.[32] In the usual Metropolis MC method,[33] the expectation value of an operator can be computed by averaging the value of the operator over numerous configurations $\{\mathbf{r}_1, \ldots, \mathbf{r}_N\}$ of the many-body system that obey detailed balance, that is, the probability ratios between pairs of discrete configurations are related by the ratios of the probability distribution for the system. Usually several million configurations are used for each N and the results are extrapolated to the thermodynamic limit by considering a sequence of increasing system sizes, N.

A MC step (MCS) consists of attempts to move one by one all the electrons of the system by a small distance of order Δ in a random direction. After each attempt (to move the ith electron from $\mathbf{r}_i^{\text{old}}$ to $\mathbf{r}_i^{\text{new}}$, the probability ratio between the "new" state and the "old" state is then computed. In the usual Metropolis scheme,[33] if this ratio is bigger than a uniformly distributed number in the $[0, 1]$ range the attempt is accepted, otherwise it is rejected. The parameter Δ is adjusted so that the acceptance ratio is close to 50%. After attempting to move all electrons (one MCS), the electron configurations are then used to calculate the operator under consideration. Averaging over numerous MCSs converges gradually (as $1/\sqrt{\text{number of MCS}}$) to the desired expectation value. Normally it is convenient to disregard numerous (several thousands) initial configurations to reach a good "thermalization" before the averaging begins. This approach significantly reduces the spurious effects of the somewhat arbitrary initial configurations. All the results reported here were obtained after discarding 100,000 "thermalization" MCSs and using 2×10^6 MCSs for averaging purposes. In Table 1 we show the correlation energy per particle for finite systems of N electrons in the Laughlin states $\nu = 1/3$ and $1/5$ obtained using the standard MC method described above. The results are rounded in the last digit. Energies are in units of e^2/l_0.

To obtain the bulk (thermodynamic estimate) of the correlation energy per particle one needs to perform a careful extrapolation of the finite-N results. We fitted the energies of Table 1 for $N = 4, 16, 36, 64, 100, 144, 196$ and 400 electrons to a

Table 1. Total interaction energy per particle for the Laughlin state at filling factors $\nu = 1/3$ and 1/5. These results were obtained after a standard MC simulation in a disk geometry. Energies are in units of e^2/l_0.

N	$\nu = 1/3$	$\nu = 1/5$
4	-0.38884	-0.32159
16	-0.39766	-0.32328
36	-0.40129	-0.32446
64	-0.40323	-0.32510
100	-0.40445	-0.32550
144	-0.40521	-0.32577
196	-0.40579	-0.32594
400	-0.40675	-0.32624

polynomial function[32] and obtained:

$$\frac{\langle \hat{V} \rangle_{1/3}}{N} = \left(-0.4094 + \frac{0.0524}{\sqrt{N}} - \frac{0.0225}{N} \right) \frac{e^2}{l_0}, \tag{46}$$

and

$$\frac{\langle \hat{V} \rangle_{1/5}}{N} = \left(-0.3273 + \frac{0.0200}{\sqrt{N}} - \frac{0.0172}{N} \right) \frac{e^2}{l_0}. \tag{47}$$

Our energy results for thermodynamic limit are similar to those found in the literature.[16] In Fig. 1 we display a snapshot of the positions of $N = 64$ electrons at the end of a MC simulation run for the Laughlin state at $\nu = 1/3$. While electron configurations in real space give a good visual perspective indicating a system with more or less uniform density, many more details (that otherwise may be missed) can be extracted by calculating other quantities such as the single-particle density function, $\rho(r)$ or the pair distribution function, $g(r)$.

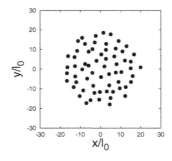

Fig. 1. A typical electron configuration at the end of a MC simulation run for a system of $N = 64$ electrons in a disk geometry. The system represents a Laughlin state at filling factor, $\nu = 1/3$. Distances are measured in terms of the magnetic length.

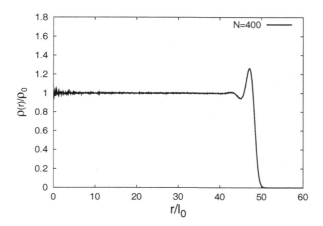

Fig. 2. One-body density function, $\rho(r)/\rho_0$, for the Laughlin state at filling factor $\nu = 1/3$ as a function of the distance r/l_0 from the center of the disk for a system of $N = 400$ electrons. Note the persistence of an "edge region" of finite width and the development of a "bulk region".

Incidentally, the calculation of the single-particle density function defined in Eq. (45) will tell us something rather interesting about the behavior of the Laughlin states close to the edge of the disk. Given that the Laughlin wavefunction describes an isotropic liquid state and is rotationally invariant, the single-particle density depends only on the radial distance from the center of the disk. We may compute the single-particle density by counting the number of electrons $N_l(\Delta r)$ found in several 2D shells of width Δr centered around a discrete set of distances to the center $r_l = (l + (1/2))\Delta r$ ($l = 0, 1, \ldots$):

$$\rho(r_l) \equiv \left\langle \frac{N_l(\Delta r)}{\Omega_l(\Delta r)} \right\rangle, \tag{48}$$

where $\Omega_l(\Delta r) = \pi(\Delta r)^2[(l+1)^2 - l^2]$ is the area of each 2D shell. In the $\Delta r \to 0$ limit the computed quantity corresponds unequivocally to the electron density:

$$\rho(r) = \left\langle \sum_{i=1}^{N} \delta(r - r_i) \right\rangle. \tag{49}$$

In Fig. 2 we show the one-body density function, $\rho(r)/\rho_0$, for the Laughlin state $\nu = 1/3$ as a function of the distance r/l_0 from the center of the disk for a large system of $N = 400$ electrons. One can clearly see that the density function, $\rho(r)$ is uniform inside the disk within a large "bulk region". However, it is very interesting to note a density spike at the "edge region". It will turn out that this behavior is not casual. In fact, the persistence of an edge density spike at the edge mimics some very rich physics, the physics of the quantum Hall edge states. Obviously, one can perform any number of simulations for systems with a varying number of electrons. In all cases, the computation of the single-particle density for the

Laughlin state, indicates a significant nonuniformity near the edge of the system. As the number of electrons increases, a significant portion of the system becomes uniform as expected. However, the nonuniformity near the edge always persists. As mentioned earlier, such a nonuniformity may be interpreted as signature of some interesting phenomenon occurring at the edge. This will eventually bring us to the idea of the quantum Hall edge states. As it will turn out, the edge states of a quantum Hall system have a very distinct nature from their bulk counterparts.

4. Luttinger Liquid Description of Quantum Hall Edge States

Strongly correlated 2DES subject to a high perpendicular magnetic field have provided various fascinating phenomena where among them we already mentioned the IQHE and FQHE. Studies on the quantum Hall regime have greatly contributed to the development of many novel ideas like the existence of fractionally charged quasiparticles,[14] topological quantum numbers,[34] composite fermion particles,[35–39] novel anisotropic liquid crystalline phases,[40–44] etc. At a basic level, the quantum Hall states are incompressible electronic liquid phases that arise due to strong correlations. It is also a known fact that the most robust quantum Hall states, such as the Laughlin states consist of gapped excitations.

The low energy structure at the edge of fractional quantum Hall liquids provides a fertile ground for studies of strongly correlated 2DES. Along these lines, the Luttinger liquid model provides a powerful framework for describing these edge excitations. Here we briefly review recent theoretical progress and then focus our attention on some important edge state transport experiments. We first introduce the Luttinger model for the IQHE state at filling factor $\nu = 1$ and then the FQHE Laughlin states at filling factor, $\nu = 1/m$ ($m = 1, 3$), where there is only a single edge mode. For the FQHE, this edge mode is a strongly correlated Luttinger liquid, which leads to a number of theoretical predictions for the transport/tunneling behavior at the edge. This is very important from an experimental perspective since the current carrying FQHE edge states are very robust and insensitive to disorder or external parameters. For the Laughlin sequence of filling factors, $\nu = 1/m$, the FQHE edge is predicted to consist of a single branch chiral Luttinger liquid. Thus, such edge states provide a unique laboratory to test the predictions of the 1D Luttinger liquid theories.

At this point, one might be puzzled by the question of how a 2D system with gapped excitations (a quantum Hall system) can be described by a 1D model with gapless bosonic excitations (a 1D Luttinger liquid model). As it turns out, the *bulk* quantum Hall states are very different from the quantum Hall *edge* states. Differently from the bulk, the quantum Hall edge states are the one that exhibit gapless modes and can be effectively described by a 1D chiral Luttinger liquid model.[45,46] This is easy to understand if one recalls that generally the bulk quantum Hall states are incompressible gapful liquids. As a result, one may argue that if there are gapless excitations they should reside only at the edge. Thus, the nontrivial transport properties of quantum Hall states come from the gapless edge excitations.

As emphasized by Wen,[45,46] the FQHE edge excitations cannot be described in terms of noninteracting electrons. For the case of Laughlin states, one can view the edge as a gas of fractionally charged quasiparticles free to move along the sample edge. In this sense, one can point out the close analogy between these edge excitations and the low energy excitations in interacting 1D electron liquid models. As shown by the theoretical work of Luttinger and others, the electron interactions have a much more profound effect in 1D electron gas systems than in 2D or 3D systems. As exemplified by the physics of the Luttinger liquid model, even weak interactions are able to invalidate a Fermi liquid description of the 1D electron systems. In a Luttinger liquid, the low energy excitations are not weakly dressed quasiparticles, but are collective density waves that move clockwise or anti-clockwise along the edge without scattering. In this framework, the edge excitations in a FQHE state at $\nu = 1/m$, which move only in one direction (right/left), are formally equivalent to the right/left moving half of a Luttinger liquid.

In the following, we will try to give a simple picture of how the Luttinger liquid model can describe the edge states of a quantum Hall system. Having said that, we also point out an excellent presentation of the Luttinger liquid model ideas from the perspective of the quantum Hall edge states given by Haldane.[47]

The simplest way to understand how the Luttinger liquid picture works for the quantum Hall edge states, is to consider the IQHE edge states for filling factor, $\nu = 1$. A simple picture of the $\nu = 1$ edge states[48] that applies to noninteracting electrons in a filled LLL has found that edge excitations can be represented by a 1D chiral Fermi liquid. To understand the basic physics of this description, we assume a Landau gauge and a vector potential of the form:

$$\mathbf{A} = (-By, 0, 0) \,. \tag{50}$$

Obviously, this vector potential gives rise to a magnetic field in the z-direction (perpendicular to the plane of the 2D electron gas). In addition, we imagine applying an electric field in the y-direction:

$$\mathbf{E} = (0, E_y, 0) \,. \tag{51}$$

Because of the electric field, the bulk Landau levels (LLs) are bent upward near the edge of the sample by the confining electric potential, $V(y) = -E_y y$. For a macroscopically large sample, there are many states near the Fermi energy of electrons even when there are no states near the Fermi energy in the bulk. We know that in the absence of an electric field, the LLL eigenfunctions at $\nu = 1$ (apart from normalization factors) are written as:

$$\psi_{k_x}(x, y) = e^{ik_x x} e^{-(y - k_x l_0^2)^2 / 2l_0^2} \,, \tag{52}$$

where $l_0^2 = \hbar/(eB)$ is electron magnetic length. In presence of the electric field in the y-direction the degeneracy of such states is lifted[49] and the energy of a state, $k_x = k$ becomes k_x-dependent through a term of the form $\epsilon_k = \hbar k v_D$ where

$$v_D = \frac{E_y}{B}, \tag{53}$$

is a drift velocity. Note that we are not specifically writing the other energy terms which do not depend on k_x. Clearly, the effect of the electric field in the wavefunction of Eq. (52) is to displace the center of the Gaussian wavefunction in a way that now such a state carries a finite Hall current. The physical surface of the system is at $y = 0$, with the states to the left being full and those at the right being empty. Since, $y = k_x l_0^2 - v_D/\omega_c$ (where ω_c is the cyclotron frequency) is the center of the Gaussian wavefunction, this physical edge of the droplet in y space can be identified with a "Fermi surface" at $k = 0$ and an effective "Fermi velocity" corresponding to the drift velocity, $v_F = v_D$. Since all states with energy, $\epsilon_k = \hbar k v_D$ move in the same direction of the droplet one can identify such a dispersion spectrum with a Dirac sea of chiral fermions and, thus, see it as a Luttinger liquid system with a linear dispersion energy like the one given by Eq. (6). In a Hall bar geometry (appropriate for the Landau gauge), the modes on the top and bottom edges move in opposite directions. For the $\nu = 1$ state there is a single R moving mode on the top edge and a L mover on the bottom edge. Because the two modes are spatially separated on two opposite sides of the Hall bar there is no intermode edge tunneling.

The theory for other quantum Hall states is a little bit more complicated. Thus, for the sake of simplicity, we consider here only the FQHE Laughlin states with uniform density: $\rho_0 = \nu/(2\pi l_0^2)$ where $\nu = 1/3, 1/5, \ldots$ is the LLL filling factor. As we made clear earlier, we are interested only on the edge states of the 2D quantum Hall system. Unfortunately, the theory of Laughlin quantum Hall edge states cannot be constructed from a noninteracting single electron picture since the incompressibility of a FQHE liquid phases originates from strong correlations between electrons. Therefore, the simplest way to understand the dynamics of FQHE edge excitations is to use the hydrodynamic picture introduced by Wen[50] in which the neutral edge excitation is a deformation of the edge. In this theory, it was argued that the FQHE states should also have gapless current-carrying edge modes. In particular, the theory predicted a single chiral Luttinger liquid mode for the Laughlin states. For a Hall bar geometry, as in the IQHE case with $\nu = 1$, we would have one R moving mode on the top edge and one L moving mode on the bottom edge. Thus, we focus our attention on a Laughlin state where the edge is described in terms of a single hydrodynamic field. For more general FQHE edge states see Refs. 50 and 51. In this effective theory, the classical theory of a surface wave on the quantum Hall droplet is quantized to obtain a description of the edge excitations. As before, we use a Landau gauge which eventually results in a Hamiltonian with parabolic confinement along the y-direction and translationally invariant (thus, plane wave states) along the x-direction. The edge profile is some function of x. Let $h(x)$ denote the local displacement of the electron fluid at that point. To describe the edge wave, we can define a 1D density along the edge:

$$\rho(x) = \rho_0 h(x) \,, \tag{54}$$

where ρ_0 is the uniform density in the bulk. The most straightforward signature of a quantum Hall Laughlin edge state is the shape of the one-particle density function which generally has a spike (peak) close to the edge of the sample.[52] The Lagrangian for a particle of charge q moving with velocity $\mathbf{v}$ in a magnetic field, $\mathbf{B} = \boldsymbol{\nabla} \times \mathbf{A}$ and electric potential, Φ is:

$$L = \frac{m}{2}|\mathbf{v}|^2 + q\mathbf{v}\mathbf{A} - q\Phi \,. \tag{55}$$

Since we are considering only the LLL Laughlin edge states, the kinetic energy is frozen to a constant value. Therefore, we can safely ignore the first term (kinetic energy) in the above expression. The Lagrangian $L_A = q\mathbf{v}\mathbf{A} = qv_x A_x$ where ($A_x = -By$) can be written as:

$$L_A = -e\rho_0 B \int_0^L dx \int_0^{h(x)} dy v_x y \,, \tag{56}$$

where $q = e$ is electron charge and L is the length of the system along the x-direction. The expression above is simplified if we assume that the velocity v_x does not depend on y. The current density, $j(x) = \rho(x)v_x$ can be calculated by integrating the continuity equation:

$$\frac{\partial j(x)}{\partial x} = \dot{\rho} \,, \tag{57}$$

where the "dot" denotes a time derivative. From there, we can obtain an expression for the velocity. Since we assume periodic boundary conditions along x, one can always write:

$$\rho(x) = \frac{1}{L}\sum_q e^{iqx}\rho_q \,. \tag{58}$$

After straightforward calculations one ends up with:

$$L_A = \frac{\hbar\pi}{\nu L}\sum_{q\neq 0}\frac{1}{iq}\dot{\rho}_q\rho_{-q} - \frac{\hbar\pi}{\nu L}N_e\sum_{q\neq 0}\frac{1}{iq}\dot{\rho}_q \,, \tag{59}$$

where N_e represents the total number of electrons in the edge region. Note that the above expression does not incorporate any of the constant terms, namely, we have dropped all constant terms. The Lagrangian $L_\phi = -q\Phi$ can be written as:

$$L_\phi = -e\rho_0 \int_0^L dx \int_0^{h(x)} dy E_y y \,, \tag{60}$$

where we wrote the scalar potential as $\Phi = -E_y y$ and $q = e$ is electron charge. From here it is easy to obtain:

$$L_\phi = -\frac{\hbar\pi v}{\nu L} \sum_{q\neq 0} \rho_q \rho_{-q} \,, \tag{61}$$

where $v = v_D = E_y/B$ is the drift velocity (as before, all constant terms have been dropped). After having dropped all constant terms appearing throughout the derivation, one writes the Lagrangian as:

$$L = \frac{\hbar\pi}{\nu L} \sum_{q\neq 0} \frac{1}{iq} \dot\rho_q (\rho_{-q} - N_e) - \frac{\hbar\pi v}{\nu L} \sum_{q\neq 0} \rho_q \rho_{-q} \,. \tag{62}$$

The canonical momentum is obtained from the Lagrangian:

$$\Pi_q = \frac{\partial L}{\partial \rho_q} \,. \tag{63}$$

Similarly, one uses the usual standard steps to derive the Hamiltonian from the Lagrangian and the canonical momenta:

$$H = \frac{\hbar\pi v}{\nu L} \sum_{q\neq 0} \rho_q \rho_{-q} \,. \tag{64}$$

While the form of the Hamiltonian above is very simple, just an electrostatic contribution, the quantum counterpart of the classical Hamiltonian is not easy to obtain. We skip the details here and simply remark that, by following well-described procedures,[53,54] one eventually gets to the quantum formulation of the problem. The most interesting results that we can derive are the various commutator relationships between density operators, field operators and the quantum Hamiltonian. Specifically, it can be shown that:

$$[\hat\rho_q, \hat\rho_{-q'}] = \nu \frac{qL}{2\pi} \delta_{q,q'} \,. \tag{65}$$

Likewise, the commutator between the Hamiltonian and the density operator turns out to be:

$$[\hat H, \hat\rho_q] = \hbar v q \hat\rho_q \,, \tag{66}$$

while the commutator between the density operator and the field operator reads:

$$[\hat\rho_q, \hat\Psi^\dagger(x)] = e^{-iqx} \hat\Psi^\dagger(x) \,. \tag{67}$$

One then compares the commutators in Eqs. (65)–(67) to the corresponding expressions for the Luttinger liquid model found in Eqs. (19)–(21). One finds that the results are identical except the negative chirality sign [compare Eqs. (66) and (21)] and the presence of the additional filling factor, ν [compare Eqs. (65) and (20)].

The key conclusion is that one can pursue a straightforward Luttinger liquid approach and rely on well-known results from such a theory. Except for the presence of the extra factor ν, one can implement the bosonization method in the same way as before and pretty much repeat automatically all the required calculations. An important ramification of the presence of the factor ν in the $U(1)$ Kac–Moody algebra is that at the end one obtains a correlation function for the Laughlin quantum Hall edge states of the form:

$$\langle \hat{\Psi}^\dagger(x)\hat{\Psi}(x')\rangle \sim \frac{1}{|x-x'|^{1/\nu}}\,. \tag{68}$$

This result implies that, for the Laughlin states, the value of the universal power law exponent should be $\gamma = 1/\nu$. Remarkably, the universal value of the parameter is solely and completely determined by the bulk FQHE state, thus, independent of details. This shows that the low energy Hilbert space of Laughlin's FQHE edge states can be represented by a suitable set of chiral Luttinger liquids[49,50] at least in the absence of edge reconstruction. This identification brought considerable interest to the study of quantum Hall edge states through experimental studies of 1D quantum transport properties.

At this juncture, it is important to note that so far we have constrained ourselves to Laughlin filling factors of the form $\nu = 1/m$ with m odd. Having said that, we point out that the edge theory of quantum Hall states can be generalized, by a Chern–Simons description, to account for other fractions as well.[55] This procedure is more complicated and gives rise to different power exponents for the correlation function.

5. Summary and Conclusions

Theoretical results clearly indicate that the quantum Hall edge states should manifest Luttinger liquid behavior. In this respect, the correlation function should have a power-law dependence consistent with Eq. (30). Additionally, the Luttinger liquid treatment also suggests that such a power law exponent should have a universal value of $\gamma = 1/\nu$, at least in the case of quantum Hall Laughlin edge states. Such a correlation function can be related to the current–voltage (I–V) characteristics for electron tunneling into the edge of quantum Hall samples:

$$I \propto V^\gamma\,. \tag{69}$$

Tunneling experiments are not very difficult to implement and many such experiments have been performed over the last years.[56–59] The basic idea is to tunnel electrons (quasiparticles) into or between quantum Hall states since the electrons (quasiparticles) naturally tunnel into the edge states due to the incompressibility of the bulk state. Tunneling between edge states serve as a unique tool to explore 1D transport and study the excitation properties of quantum Hall edge states.

Unlike other 1D systems (e.g., semiconductor-based quantum wires or carbon nanotubes) the quantum Hall edge states are chiral. Chirality constrains the edge current to propagate without back scattering by simply going around any impurity. The topological aspects of edge excitations, namely the fact that they are essentially insensitive to any sample detail or disorder, allows one to compare directly the experimental 1D transport results to the Luttinger liquid theory predictions.[45,46,62] Much effort has also been devoted to studies of various QH tunnel junctions using patterned gate electrodes who allow detailed point contact tunneling experiments.[61] Alternatively, when interpreting the experimental results on quantum Hall tunnel junctions, a careful consideration on the role of contacts can result in a better understanding[62] of experiments.[63]

Indeed, experiments do confirm a Luttinger liquid behavior as indicated by the characteristic power law I–V dependence. However, the experimentally measured exponent, γ does not agree with the universal prediction of being $1/\nu$ for the case of Laughlin states. So far no widely accepted theoretical explanation for the experimental results exists, but several numerical works suggest that the discrepancy between theory and experiment can be ascribed to the effects of the electron interaction.[64–66] Obviously, tunneling experiments are very sensitive to many factors that might lead to some sort of renormalization of the correlation function exponent. Another possibility is that the interaction between the edges, namely, between the different chiralities, may lead to substantial renormalization of the exponent.

Overall, a lot of progress has been made in understanding tunneling in quantum Hall tunnel junctions. However, there are still many open questions especially in relation to the interpretation of recent experiments. With better experimental probes one might be able to fabricate samples with very little edge reconstruction. In this ideal scenario, one can compare the experimental properties directly to theoretical predictions. Such electron tunneling experiments[63] have confirmed the existence of both the scaling regime[60] and the crossover behavior[62] predicted by the chiral Luttinger liquid picture, thus, providing more evidence to the validity of a Luttinger liquid description for the quantum Hall edge states. While it is clear that a 1D Luttinger liquid model describes well the physics of quantum Hall edge states, open questions still remain about the actual observed behavior of the tunneling exponent and its consistency with the physics of bulk quantum Hall states.[67,68]

To summarize, in this review article we discussed the general features of a 1D Luttinger liquid model and its relation to the quantum Hall edge states. In Sec. 1 we give an introduction to the Luttinger liquid model and describe its main features in the context of the physics of 1D Fermi systems. Then, we give a broader perspective of how many 1D Luttinger liquid ideas have found applications to phenomena occurring in higher dimensions.

In Sec. 2 we show how the bosonization technique makes it possible to solve exactly the Luttinger liquid model. This property, namely, the fact that we can introduce a simpler model of noninteracting bosons, even if the electrons in the model are interacting allows one to calculate exactly various quantities of interest such as the correlation function, a quantity that can be measured by means of transport/tunneling experiments.

In Sec. 3 we describe the basic features of the quantum Hall effect and the Laughlin states. The FQHE states represent a unique example of a novel collective quantum liquid states of matter that originates from strong electronic correlations only. The most robust FQHE states occur at filling factors $\nu = 1/3$ and $1/5$. Such states are well described by Laughlin's theory and his trial wavefunction. Given the importance of such states, we briefly describe their properties and speculate how the behavior of the one-particle density function anticipates the occurrence of interesting phenomena at the edge, namely, the appearance of the quantum Hall edge states.

In Sec. 4 we give an introduction to the physics of quantum Hall edge states. Quantum Hall systems, which consist of 2D electronic systems subject to a strong perpendicular magnetic field have an edge which turns out it can be described by a 1D Luttinger liquid model. We specifically restrict our attention to the Laughlin states in the LLL and explain how the quantum Hall edge states can be described by an effective 1D Luttinger liquid representation. In particular, we show how a general hydrodynamical approach for such edge states is able to predict a Luttinger liquid behavior. Such a representation turns out to be very useful given that the Luttinger liquid can be solved exactly. This mapping suggests that the correlation function for quantum Hall edge states, in the asymptotic limit, has a power law form. In addition, the Luttinger liquid mapping also suggests that such a power law exponent should have a universal value.

In Sec. 5 we give a summary of important experimental and theoretical results regarding the Luttinger liquid behavior of quantum Hall edge states. Many experiments have found behavior consistent with a Luttinger liquid description of the edge states. However, even though a large number of transport and tunneling experiments are consistent with a Luttinger liquid power law dependence of the correlation function, the value of the exponent appears not to be universal. This numerical discrepancy might be due to correlation effects between electrons that are not considered in various standard Luttinger liquid models. There are also suggestions that this discrepancy might be due to other unaccounted effects that do occur in tunneling experiments of this nature. Overall, our work tries to give a general overview of the relationship between the quantum Hall edge states and their 1D Luttinger liquid representation.

Acknowledgment

This research was supported in part by the National Science Foundation under NSF Grant No. DMR-1104795.

References

1. J. M. Luttinger, *J. Math. Phys.* **4**, 1154 (1963).
2. S. Tomonaga, *Prog. Theor. Phys.* **5**, 544 (1950).
3. D. C. Mattis and E. H. Lieb, *J. Math. Phys.* **6**, 304 (1965).
4. G. F. Giuliani and G. Vignale, *Quantum Theory of the Electron Liquid* (Cambridge University Press, UK, 2005).
5. A. O. Gogolin, A. A. Nersesyan and A. M. Tsvelik, *Bosonization and Strongly Correlated Systems* (Cambridge University Press, Cambridge, 1998).
6. M. Stone, *Bosonization* (World Scientific, Singapore, 1994).
7. K. von Klitzing, G. Dorda and M. Pepper, *Phys. Rev. Lett.* **45**, 494 (1980).
8. D. C. Tsui, H. L. Stormer and A. C. Gossard, *Phys. Rev. Lett.* **48**, 1559 (1982).
9. R. E. Prange and S. M. Girvin (eds.), *The Quantum Hall Effect* (Springer-Verlag, New York, 1990).
10. T. Chakraborty and P. Pietiläinen, *The Fractional Quantum Hall Effect* (Springer-Verlag, New York, 1988).
11. O. Heinonen, ed., *Composite Fermions* (World Scientific, New York, 1998).
12. G. F. Giuliani and G. Vignale, *Quantum Theory of the Electron Liquid* (Cambridge University Press, Cambridge, 2005).
13. J. K. Jain, *Composite Fermions* (Cambridge University Press, New York, 2007).
14. R. B. Laughlin, *Phys. Rev. Lett.* **50**, 1395 (1983).
15. O. Ciftja and S. Fantoni, *Europhys. Lett.* **36**, 663 (1996).
16. D. Levesque, J. J. Weis and A. H. MacDonald, *Phys. Rev. B* **30**, R1056 (1984).
17. S. M. Girvin and T. Jach, *Phys. Rev. B* **29**, 5617 (1984).
18. S. M. Girvin, A. H. MacDonald and P. M. Platzman, *Phys. Rev. B* **33**, 2481 (1986).
19. G. Fano and F. Ortolani, *Phys. Rev. B* **37**, 8179 (1988).
20. R. Morf and N. d'Ambrumenil, *Phys. Rev. Lett.* **74**, 5116 (1995).
21. R. K. Kamilla, J. K. Jain and S. M. Girvin, *Phys. Rev. B* **56**, 12411 (1997).
22. O. Ciftja, *Eur. Phys. J. B* **13**, 671 (2000).
23. B. I. Halperin, P. A. Lee and N. Read, *Phys. Rev. B* **47**, 7312 (1993).
24. O. Ciftja, *Phys. Rev. B* **59**, 10194 (1999).
25. R. B. Laughlin, *Phys. Rev. B* **27**, 3383 (1983).
26. P. K. Lam and S. M. Girvin, *Phys. Rev. B* **30**, R473 (1984).
27. E. Mendez *et al.*, *Phys. Rev. B* **28**, 4886 (1983).
28. O. Ciftja and C. Wexler, *Phys. Rev. B* **67**, 075304 (2003).
29. O. Ciftja, *Physica B* **404**, 2244 (2009).
30. O. Ciftja, *Physica B* **404**, 227 (2009).
31. O. Ciftja, *Physica B* **406**, 2054 (2011).
32. R. Morf and B. I. Halperin, *Phys. Rev. B* **33**, 2221 (1986).
33. N. Metropolis *et al.*, *J. Chem. Phys.* **21**, 1087 (1953).
34. D. J. Thouless, *Topological Quantum Numbers in Nonrelativistic Physics* (World Scientific, Singapore, 1998).
35. J. K. Jain, *Phys. Rev. Lett.* **63**, 199 (1989).
36. O. Ciftja, *Phys. Rev. B* **59**(12), 8132 (1999).
37. J. K. Jain and R. K. Kamilla, *Int. J. Mod. Phys. B* **11**, 2621 (1997).
38. O. Ciftja and C. Wexler, *Solid State Commun.* **122**, 401 (2002).

39. R. K. Kamilla and J. K. Jain, *Phys. Rev. B* **55**, 9824 (1997).
40. O. Ciftja *et al.*, *Phys. Rev. B* **83**, 193101 (2011).
41. H. Y. Kee, *Phys. Rev. B* **67**, 073105 (2003).
42. C. Wexler and O. Ciftja, *J. Phys.: Condens. Matter* **14**, 3705 (2002).
43. M. M. Fogler, *Europhys. Lett.* **66**, 572 (2004).
44. O. Ciftja, C. M. Lapilli and C. Wexler, *Phys. Rev. B* **69**, 125320 (2004).
45. X. G. Wen, *Phys. Rev. B* **43**, 11025 (1991).
46. X. G. Wen, *Phys. Rev. Lett.* **64**, 2206 (1990).
47. F. D. M. Haldane, *J. Phys. C: Solid State Phys.* **14**, 2585 (1981).
48. B. I. Halperin, *Phys. Rev. B* **25**, 2185 (1982).
49. M. Stone, *Ann. Phys.* **207**, 38 (1991).
50. X. G. Wen, *Phys. Rev. B* **41**, 12838 (1990).
51. A. Lopez and E. Fradkin, *Phys. Rev. B* **59**, 15323 (1999).
52. O. Ciftja, N. Ockleberry and C. Okolo, *Mod. Phys. Lett. B* **25**, 1983 (2011).
53. P. A. M. Dirac, *Lectures on Quantum Mechanics*, (Belfer Graduate School of Science, Yeshiva University, New York, 1964).
54. M. Henneaux and C. Teitelboim, *Quantization of Gauge Systems* (Princeton University Press, Princeton, New Jersey, 1992).
55. X. G. Wen, *Int. J. Mod. Phys. B* **6**, 1711 (1992).
56. A. M. Chang, L. N. Pfeiffer and K. W. West, *Phys. Rev. Lett.* **77**, 2538 (1996).
57. M. Grayson *et al.*, *Phys. Rev. Lett.* **80**, 1062 (1998).
58. A. M. Chang *et al.*, *Phys. Rev. Lett.* **86**, 143 (2001).
59. M. Hilke *et al.*, *Phys. Rev. Lett.* **87**, 186806 (2001).
60. C. L. Kane and M. P. A. Fisher, *Phys. Rev. B* **46**, 15233 (1992).
61. F. P. Milliken, C. P. Umbach and R. A. Webb, *Solid State Commun.* **97**, 309 (1996).
62. C. de C. Chamon and E. Fradkin, *Phys. Rev. B* **56**, 2012 (1997).
63. A. M. Chang, L. N. Pfeiffer and K. W. West, *Phys. Rev. Lett.* **77**, 2538 (1996).
64. S. S. Mandal and J. K. Jain, *Solid State Commun.* **118**, 503 (2001).
65. V. J. Goldman and E. V. Tsiper, *Phys. Rev. Lett.* **86**, 5841 (2001).
66. S. S. Mandal and J. K. Jain, *Phys. Rev. Lett.* **89**, 096801 (2002).
67. A. V. Shytov, L. S. Levitov and B. I. Halperin, *Phys. Rev. Lett.* **80**, 141 (1998).
68. J. Moore and X.-G. Wen, *Phys. Rev. B* **57**, 10138 (1998).

A LUTTINGER LIQUID CORE INSIDE HELIUM-4 FILLED NANOPORES*

ADRIAN DEL MAESTRO

Department of Physics, University of Vermont,
Burlington, VT 05405, USA
Adrian.DelMaestro@uvm.edu

As helium-4 is cooled below 2.17 K it undergoes a phase transition to a fundamentally quantum mechanical state of matter known as a superfluid which supports flow without viscosity. This type of dissipationless transport can be observed by forcing helium to travel through a narrow constriction that the normal liquid could not penetrate. Recent experiments have highlighted the feasibility of fabricating smooth pores with nanometer radii, that approach the truly one-dimensional limit where it is believed that a system of bosons (like helium-4) may have startlingly different behavior than in three dimensions. The one-dimensional system is predicted to have a linear hydrodynamic description known as Luttinger liquid (LL) theory, where no type of long range order can be sustained. In the limit where the pore radius is small, LL theory would predict that helium inside the channel behaves as a sort of quasi-supersolid with all correlations decaying as power-law functions of distance at zero temperature. We have performed large scale quantum Monte Carlo simulations of helium-4 inside nanopores of varying radii at low temperature with realistic helium–helium and helium-pore interactions. The results indicate that helium inside the nanopore forms concentric cylindrical shells surrounding a core that can be described via LL theory and provides insights into the exciting possibility of the experimental detection of this intriguing low-dimensional state of matter.

Keywords: Low-dimensional; Luttinger liquid; quantum Monte Carlo; helium-4.

1. Introduction

As the spatial dimension of an interacting many-body system is reduced, both thermal and quantum fluctuations are enhanced, providing a fascinating arena for the study of complex phenomena. It is well-known that in strictly one dimension, there is no broken continuous symmetry but instead the persistence of only quasi-long range order characterized by power-law decay of correlation functions. The simplest example of a model displaying these features is the one-dimensional (1D) nonrelativistic Galilean invariant Bose gas with delta function interactions. The ground state of this model has been known for almost fifty years and can be solved exactly using the Bethe Ansatz.[1] An alternative approach, based on the concept of a low-dimensional harmonic fluid[2-5] was first understood to be universal by Haldane[6,7] with the techniques formalized in these seminal works now generally known as bosonization and Luttinger liquid (LL) theory. The combination of these two approaches indicate that there is no well-defined quantum phase transition at zero

*This article first appeared in International Journal of Modern Physics B, Vol. 26, No. 22 (2012).

temperature as a function of the strength of the interactions in 1D. As interactions are increased, the system crosses over from a regime dominated by phase fluctuations, to one with tendencies towards density wave order. Likewise, there is no phase transition as a function of temperature.

Historically, the experimental study of low-dimensional physics has been confined to fermionic systems such as insulating spin chains and quantum wires[8] and a bosonic realization has been lacking. The identification of accessible bosonic systems is particularly relevant in light of the universal predictions of LL theory. Recently, weakly interacting ultracold bosonic gases consisting of laser trapped Alkali atoms confined to cigar shaped quasi-one-dimensional condensates[9] or rings[10] have been studied but it is difficult to obtain results at higher densities where the constituent atoms are more strongly interacting. The various theoretical approaches and considerations required to study such systems have been covered in detail in a recent review article.[11]

A considerably different approach involves the preparation of a quantum fluid of interacting bosons, such as neutral ^{4}He in a *physically* confining geometry at high density. In fact, the ability of helium-4 cooled below $T_\lambda = 2.17$ K (the bulk superfluid transition temperature in ^{4}He) to flow through a narrow constriction (superleak) is one of the original defining characteristics of the superfluid state of matter. Original experiments in this vein achieved physical confinement of helium through the quasi-1D cavernous networks of porous glasses such as Vycor[12] and more recently using folded sheets of mesoporous materials.[13] Advances in nanofabrication techniques have culminated in a more systematic approach that has been exploited by Savard *et al.*[14,15] in studying ^{4}He inside nanopores of variable radii by sculpting a pore through a Si_3N_4 membrane using a transmission electron beam. Although these experiments have thus far focused on flow properties of helium in the gas phase in nanopores[14] and the superfluid phase for wide nanoholes[15] they provide a tantalizing road map for the experimental detection of a bosonic LL. It is thus natural to ponder a simpler equilibrium system, that can be studied numerically as a function of pore radius and temperature below T_λ. We expect that when the pore radius is sufficiently small, the length sufficiently long and the temperature low enough with respect to T_λ the ^{4}He filled nanopore should begin to exhibit strictly 1D behavior where LL theory can be used to characterize the nature of the quasi-long-range order.

The remainder of this paper is concerned with the precise definition of *sufficiently* in the previous sentence and is organized as follows. We first identify the particulars of a theoretical model of confined helium-4 and introduce the stochastically exact numerical method employed to study its behavior. The numerical results are used to determine a phase diagram for helium inside the pore with special attention paid to the types of structures that are allowed by the Galilean invariant cylindrically symmetric confining potential. Density–density correlations inside the inner region of the pore are then analyzed in terms of LL theory which is introduced in an economical fashion with references provided to more complete and elaborate treatments elsewhere. The results highlight the applicability of the linear hydrodynamics of LL theory in describing helium inside nanopores and a discussion of the consequences and limitations of this finding is presented.

2. Path Integral Simulations of Confined Helium-4

The finite extent and translational invariance of the van der Waals interaction between ^{4}He atoms combined with their fundamental indistinguishably conspire to make the numerical study of quantum fluids a challenging task. Unbiased stochastically exact simulations of the quantum bosonic many-body problem in the continuum at low temperature have only recently become more feasible through the introduction of the continuous space Worm Algorithm (WA).[16,17] This method builds upon the conventional Path Integral Monte Carlo approach of Ceperley[18] that exploits the Feynman path integral interpretation of quantum mechanics to perform Metropolis sampling of the *worldlines* of particles in $d + 1$ dimensions. The WA extends the configuration space to include open worldlines known as *worms* that are not periodic in imaginary time and directly contribute to the single particle Matsubara Green function allowing for simulations to be performed in the grand canonical ensemble. The simulation cell is kept in thermal equilibrium with a surrounding bath with which it can exchange particles and the chemical potential μ is an input parameter providing access to important physical observables such as the compressibility. In addition, the WA employs an efficient method via the swap operator[16] to directly sample the quantum statistics of identical bosons through only local exchanges of worldlines. Such exchanges are the crucial step en route to the superfluid state of matter, with the superfluid density being measured through a topological winding number; the standard technique in continuum simulations.[19]

We wish to perform WA simulations on a realistic physical model of the experimental single nanopore geometry discussed in the introduction with details in Refs. 14 and 15 and thus must consider interactions between helium atoms, as well as the interactions between helium and the atoms composing the material surrounding the nanopore. Our starting point is the general microscopic many-body Hamiltonian

$$\hat{H} = \sum_{i=1}^{N} \left[-\frac{1}{2m} \hat{\boldsymbol{\nabla}}_i^2 + \hat{V}(\mathbf{r}_i) \right] + \sum_{i<j} \hat{v}(\mathbf{r}_i - \mathbf{r}_j) \tag{1}$$

where $\mathbf{r}_i$ are the spatial positions of the N helium atoms and we have set $\hbar = k_B = 1$. The natural units of length in the nanopore systems are angstroms and in these reduced units, the mass of the helium atoms is $m = 0.0826$ Å^{-2}K^{-1}. The interaction energy between two helium atoms v, is given by the 1979 Aziz potential[20]

$$v(r) = \epsilon u \left(\frac{r}{r_{\min}} \right), \tag{2}$$

$$u(x) = Ae^{-\alpha x} - \left(\frac{C_6}{x^6} + \frac{C_8}{x^8} + \frac{C_{10}}{x^{10}} \right) \begin{cases} e^{-(D/x-1)^2}, & x < D, \\ 1, & x \geq D, \end{cases} \tag{3}$$

with a set of parameters inferred from both first principles calculations and experimental measurements to be $r_{\min} = 2.9673$ Å, $A = 5.449 \times 10^5$, $\epsilon/k_B = 10.8$ K, $\alpha = 13.353$, $D = 1.2413$, $C_6 = 1.3732$, $C_8 = 0.42538$ and $C_{10} = 0.1781$. This potential, shown as an inset in Fig. 1, contains both hardcore repulsion for distances less than $r_{hc} \simeq 2.64$ Å, a minimum at $r_{\min}$ and weak attraction at large separations. The external potential, V can be computed by modeling the nanopore as long cylindrical cavity carved inside a continuous medium. The potential energy of interaction of a single helium atom with the atoms of the medium can be obtained by integrating a Lennard–Jones pair potential as described in Ref. 21 yielding

$$V(r; R) = \frac{\pi \varepsilon \sigma^3 n}{3} \left[\frac{7}{32} \left(\frac{\sigma}{R} \right)^9 u_9 \left(\frac{r}{R} \right) - \left(\frac{\sigma}{R} \right)^3 u_3 \left(\frac{r}{R} \right) \right] \tag{4}$$

for a perfect cylindrical cavity of radius R where n is the number density and σ and ε are the Lennard-Jones parameters of the surrounding medium. The functional

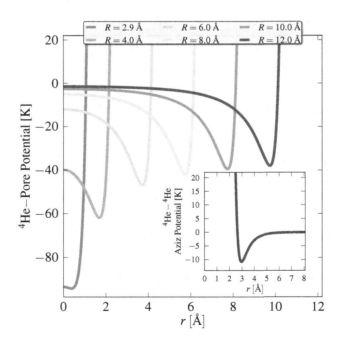

Fig. 1. The interaction potential V between helium and the walls of the nanopore constructed of amorphous Si_3N_4 for various radii ranging from $R = 2.9$ Å (left) to $R = 12.0$ Å (right) calculated via Eq. (4) where r is the radial distance of a helium atom from the axis of the cylinder. The inset shows the Aziz interaction potential v, described in Eq. (2) for two helium atoms separated by a distance r.

coefficients are given by

$$u_9(x) = \frac{1}{240(1-x^2)^9}[(1091 + 11516x^2 + 16434x^4 + 4052x^6 + 35x^8)E(x^2)$$

$$- 8(1-x^2)(1+7x^2)(97 + 134x^2 + 25x^4)K(x^2)], \tag{5}$$

$$u_3(x) = \frac{2}{(1-x^2)^3}[(7+x^2)E(x^2) - 4(1-x^2)K(x^2)], \tag{6}$$

where $K(x)$ and $E(x)$ are the complete elliptical integrals of the first and second kind respectively and r is the radial distance from the axis of the pore. Using the Lorentz–Berthelot mixing rules for the microscopic Lennard–Jones parameters of amorphous Si_3N_4[22–24] we set $\varepsilon = 10.22$ K and $\sigma = 2.628$ Å with $n = 0.078$ Å^{-3}. A plot of the pore potential for different radii is shown in Fig. 1. The use of parameters for different media such as silica glass[25] will alter the overall energy scale and may change the excluded volume experienced by atoms inside the pore but should not qualitatively affect the findings of this study.

3.   Helium-4 Inside Nanopores

Using this model for the interaction and confinement potentials, we have computed the low temperature thermodynamics of fluid ^{4}He inside nanopores with lengths between $L = 50$ and 200 Å and temperatures ranging from $T = 0.5$–2.0 K using the WA. We imagine the nanopore to be immersed in an essentially infinite bath of helium maintained at saturated vapor pressure (SVP) which sets the chemical potential in our grand canonical simulations to be $\mu = -7.2$ K. A wide range of cylindrical pores with radii R between 2.9 and 12.0 Å have been considered where we assume periodic boundary conditions along the axis of the cylinder.

We restrict the temperature in our simulations to be less than $T_\lambda \simeq 2.17$ K and wish to confirm that we are studying a *low energy* quantum regime. This is most straightforwardly determined by measuring the average kinetic energy per particle in our quantum Monte Carlo (QMC) simulations as shown in Fig. 2 and comparing with the temperature. We observe a general trend of decreasing kinetic energy, from more than 30 K per particle for $R = 2.9$ Å to 20 K per particle for $R = 12$ Å. The nonmonotonic behavior observed between $R = 2.9$ Å and $R = 4.0$ Å can be attributed to the complete transverse confinement of helium in the narrowest pore. The minimum value of the kinetic energy observed is an order of magnitude larger than all temperatures considered and we thus conclude our simulations are dominated by quantum effects. Unless otherwise specified, we fix the length of the nanopore to be $L = 100$ Å allowing three-dimensional number densities to be easily converted into particle numbers (usually near 1000 helium atoms). The errorbars in Fig. 2 are the result of a bootstrapping analysis of over 10^6 individual measurements

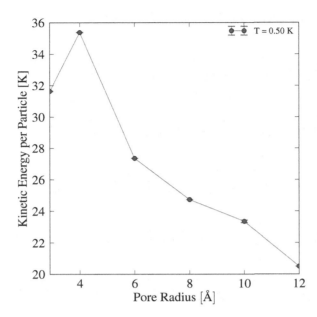

Fig. 2. The average kinetic energy per particle versus pore radius measured via a thermodynamic estimator at low temperature ($T = 0.5$ K) for a pore of length $L = 100$ Å held at $\mu = -7.2$ K.

of the kinetic energy using a thermodynamic estimator.[26] We have fixed the number of time slices (the discretization of the imaginary time interval corresponding to the inverse temperature) to be 125 per degree kelvin. Possible Trotter error introduced by using a finite imaginary time step $\Delta\tau$ has been analyzed and we find that it is well-described by a term proportional to $(\Delta\tau)^4$ consistent with the use of a fourth order path integral factorization of the density matrix.[26]

Assured we are in a low energy regime over the entire range of temperatures considered in this study, we now switch our attention to the structures formed by helium atoms inside the pore. There exists a large literature on the types of phases that are formed when confining a quantum fluid of helium in a cylindrical cavity and a complex phase diagram has been predicted for carbon nanotubes[27–32] and both smooth[33–35] and porous nanopores.[36,37] Advances owing to the development of the WA have allowed this and a previous work[35] to study considerably larger radius pores at high density and finite temperature in the grand canonical ensemble that approach those studied in recent experiments.[14,15]

For fixed chemical potential $\mu = -7.2$ K the average volume density $\rho_V = \langle N \rangle / \pi R^2 L$ inside the nanopore is shown in Fig. 3 where $\langle \cdots \rangle$ indicates a thermodynamic average performed in the QMC. The observed general trend is an increase in the volume density as a function of radius continuing to a limiting value for large R that is approaching the expected bulk density of 0.02198 Å^{-3}.[17] The resulting phases of helium inside the nanopore can be elucidated by measuring the radial density of particles, $\rho_R(r)$ defined to be the number of particles found a distance r

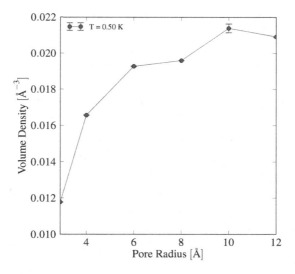

Fig. 3. The average volume density as a function of nanopore radius for helium at SVP and $T = 0.5$ K.

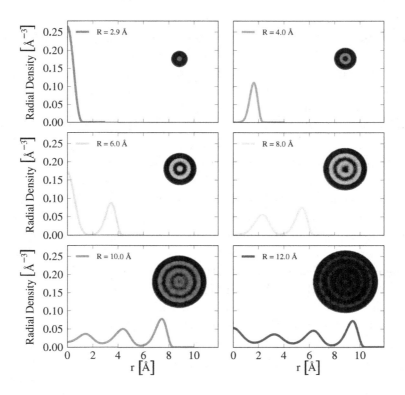

Fig. 4. The radial number density $\rho_R(r)$ of helium inside nanopores with radii $R = 2.9$, 4.0, 6.0, 8.0, 10.0, 12.0 Å with an inset showing a full instantaneous worldline configuration inside the pore measured in the QMC simulations projected on $z = 0$. All pores of $L = 100$ Å and are held at SVP corresponding to $\mu = -7.2$ K.

from the axis of a pore of radius R normalized such that the linear density ρ_L is

$$\rho_L \equiv \frac{\langle N \rangle}{L} = 2\pi \int dr r \rho_R(r), \tag{7}$$

where $\rho_R(r)$ includes an implicit average over the axial and angular coordinates. The results, shown in Fig. 4 display a progression of structures including a nearly one-dimensional chain of helium atoms for $R = 2.9$ Å to a series of three concentric shells surrounding a chain for $R = 12.0$ Å. The insets display a snapshot of the discretized helium worldline configuration (space and imaginary time coordinates) from the QMC simulations projected into the xy-plane at $z = 0$.

The observed oscillations in the radial density can be easily understood by contemplating the filling of an empty nanopore through a classical analysis of the potential interactions only (Fig. 1). Beginning with the widest pores, we observe an excluded volume effect due to the *hard* Si_3N_4 wall, as well as a deep minimum of the potential near the surface. This will naturally lead to an adsorption effect or wetting of the pore surface with helium atoms forming a shell due to the cylindrical symmetry. As helium atoms continue to enter the pore, the adsorption will cause the area density of the shell to increase until the average separation between atoms inside the shell begins to approach the hard core radius r_{hc} of the interaction potential. At this point, it will become energetically favorable to form a new layer inside the one adsorbed on the surface. As this process continues, a series of concentric cylindrical shells may be formed inside the pore analogous to the layering observed in quantum films of bosons.[38]

The results of the radial density for pores of different radii can be separated into two natural groups based on the presence of helium at high linear density near the center of the pore. The existence of such an *inner cylinder* (IC) which can be thought of as a quasi-1D chain of helium atoms for a given radius depends on the details of both the helium-pore and helium–helium interaction potentials. The pore wall (and thus the details of the wall material, Si_3N_4 here) produces a region of excluded volume, forcing the helium atoms to remain a distance greater than $r_e \simeq 1.32$ Å from the walls of the pore. This sets the location of the outermost shell. The separation between shells is then restricted to be near the minimum of the Aziz potential $r_m \sim 3$ Å with some weak dependence on the radius of the pore due to screening of the inner shells. We find that an IC exists whenever the pore radius is approximately divisible by three: $R = 2.9, 6.0$ and 12.0 Å.

The presence of an IC of helium is intriguing, as we expect that such a quasi-1D bosonic system should lack any long range order down to zero temperature and be described by LL theory. In Sec. 4 we test this prediction by introducing the universal theory for an effective harmonic fluid and use it to exhaustively study nanopores with radii $R = 2.9$ Å and $R = 12.0$ Å.

4. A Luttinger Liquid Core

In an attempt to understand the relevant low energy degrees of freedom for quasi-1D helium atoms inside the pore, we begin by studying a microscopic Hamiltonian like the one in Eq. (1) in second quantized form for a strictly one-dimensional system of bosons

$$H = \int_0^L dz \left[\frac{1}{2m} \partial_z \Psi^\dagger(z) \partial_z \Psi(z) + \frac{1}{2} \int_0^L dz' \rho(z) v_{1D}(z - z') \rho(z) \right], \qquad (8)$$

where $\Psi^\dagger(z) = \sqrt{\rho(z)} e^{-i\phi(z)}$ is a bosonic creation operator with $[\Psi(z), \Psi^\dagger(z')] = \delta(z - z')$ and the 1D density and phase operators satisfy

$$[\rho(z), e^{i\phi(z')}] = \delta(z - z') e^{i\phi(z)}. \qquad (9)$$

It is the ability to simulate such microscopic Hamiltonians over a range of energy scales that makes the WA so attractive. The stochastically exact equilibrium properties of a system of strongly interacting bosons at finite temperature can be determined using only fundamental input parameters such as the particle mass m and the details of the interaction potential.

4.1. *Luttinger liquid theory*

The manipulations (generally referred to as bosonization) that transform a microscopic one-dimensional interacting Hamiltonian like Eq. (8) into a universal description of the linear hydrodynamics of a quantum fluid (either bosonic or fermionic) are by now standard.[6,8,11,39] In particular, for bosonic systems, Ref. 39 provides a detailed and pedagogical derivation of the LL Hamiltonian defined by

$$H_{LL} - \mu N = \frac{1}{2\pi} \int_0^L dz [v_J (\partial_z \phi)^2 + v_N (\partial_z \theta)^2], \qquad (10)$$

where we have included only those terms which are formally relevant in the renormalization group sense. The velocities v_J and v_N are fixed by the microscopic details of the underlying high energy model and the angular field $\theta(z)$ appears through the redefinition of the density operator

$$\rho(z) \equiv \left[\rho_0 + \frac{1}{\pi} \partial_z \theta(z) \right] \sum_{m=-\infty}^{\infty} e^{i2m\theta(z)}, \qquad (11)$$

where ρ_0 is the number density at $T = 0$ in the thermodynamic limit. The underlying bosonic symmetry requires that Eq. (11) in conjunction with Eq. (9) produces the following commutation relation

$$[\partial_z \theta(z), \phi(z')] = i\pi \delta(z - z'). \qquad (12)$$

If the system described by Eq. (10) exhibits Galilean invariance we can identify

$$v_J = \frac{\pi \rho_0}{m}, \tag{13}$$

$$v_N = \frac{1}{\pi \rho_0^2 \kappa}, \tag{14}$$

where κ is the adiabatic compressibility in the limit $L \to \infty$, $T \to 0$.[6] The connection of these effective velocities to the underlying microscopic details of the original interacting Hamiltonian is now clear: a highly incompressible state with $v_N \gg 1$ should have a nearly constant density and thus $\partial_z \theta(z) \sim 0$ whereas a state displaying phase coherence has $\partial_z \phi(z) \sim 0$ and thus at $T = 0$ there should be a finite superfluid fraction with $v_J \gg 1$.

For a system with periodic boundary conditions in the z-direction, our original boson field must satisfy $\Psi^\dagger(z + L) = \Psi^\dagger(z)$ leading to a mode expansion of $\theta(z)$ and $\phi(z)$ indexed by wavevector $q = 2\pi n/L$ where n is an integer[6]

$$\theta(z) = \theta_0 + \frac{\pi z}{L}(N - N_0) - i \left(\frac{v_J}{v_N}\right)^{1/4} \sum_{q \neq 0} \left|\frac{\pi}{2qL}\right|^{1/2} e^{iqz}(b_q^\dagger + b_{-q})\text{sgn}(q), \tag{15}$$

$$\phi(z) = \phi_0 + \frac{\pi J z}{L} - i \left(\frac{v_J}{v_N}\right)^{-1/4} \sum_{q \neq 0} \left|\frac{\pi}{2qL}\right|^{1/2} e^{iqz}(b_q^\dagger - b_{-q}). \tag{16}$$

Here, $b_q^\dagger(b_q)$ is a bosonic creation (annihilation) operator for modes corresponding to long wavelength density fluctuations and the operator J has even integer eigenvalues indexing the topological winding number of the phase field $\phi(z)$, $N_0 = \rho_0 L$ and N is the boson number operator. Substituting Eqs. (15) and (16) into (10) yields the mode expanded Hamiltonian

$$H_{\text{LL}} - \mu N = \frac{\pi v}{2KL}J^2 + \frac{\pi v K}{2L}(N - N_0)^2 + \sum_{n \neq 0} v|q|b_q^\dagger b_q, \tag{17}$$

where we have dropped a nonuniversal constant and defined a new velocity $v = \sqrt{v_J v_N}$ measuring the famous linear dispersion of the low energy density modes. We have introduced the Luttinger parameter $K = \sqrt{v_N/v_J}$[a] and although it is well-known that no quantum phase transition can occur as a function of interactions in a truly 1D system described by Eq. (17), K can be tuned to initiate a $T = 0$ crossover between a state with algebraic density wave order at $K = \infty$ to one with quasi-long range superfluid correlations at $K = 0$.

The advantages of having the Hamiltonian in this form are unmistakable due to its quadratic nature and the ease with which we can compute averages in the

---

[a]Two predominant definitions of the Luttinger parameter K exist in the literature corresponding to K or $\widetilde{K} = 1/K$. We have chosen the present definition to be consistent with Ref. 35 but we caution the reader to be wary of this ambiguity.

oscillator basis. For example, one can exactly determine the grand partition function $\mathcal{Z} = \text{Tr} \exp[-(H_{\text{LL}} - \mu N)/T]$ in terms of known special functions[40] allowing for the straightforward (although possibly tedious) calculation of all two body correlation functions and thermodynamic observables in terms of the temperature T, system size L, Luttinger velocity v and Luttinger parameter K. We will focus on the derivation of a single observable, the density–density correlation function $\langle \rho(z)\rho(0) \rangle$. Such density correlations are of great interest, as they are easily measured in numerical simulations and are related via a Fourier transform to the experimentally measurable structure factor. In addition, their form provides an intuitive qualitative picture of the types of fluctuations (phase or density) which are dominant. This knowledge can help to pinpoint which region of the 1D crossover phase diagram a given system resides in. Starting from the definition of the density operator in Eq. (11) we have (keeping only the slowest decaying terms)

$$\langle \rho(z)\rho(0) \rangle \approx \rho_0^2 + \frac{1}{\pi^2} \langle \partial_z \theta(z) \partial_z \theta(0) \rangle + 2\rho_0^2 \langle e^{2i\theta(z)} e^{-2i\theta(0)} \rangle. \tag{18}$$

Next, using the mode expansion for $\theta(z)$ in Eq. (15) the expectation values can be computed to give[35]

$$\langle \rho(z)\rho(0) \rangle = \rho_0^2 + \frac{1}{2\pi^2 K} \frac{d^2}{dz^2} \ln \theta_1 \left[\frac{\pi z}{L}, e^{-\pi v/LT} \right]$$
$$+ \mathcal{A} \cos(2\pi\rho_0 z) \left\{ \frac{2\eta \left(\frac{iv}{LT} \right) e^{-\pi v/6LT}}{\theta_1 \left(\frac{\pi z}{L}, e^{-\pi v/LT} \right)} \right\}^{2/K}, \tag{19}$$

where $\theta_1(x, y)$ and $\eta(is)$ are the first Elliptical Theta function and Dedekind Eta function respectively. $\mathcal{A}$ is a nonuniversal constant dependent on the short distance properties of the system. Although Eq. (19) may appear daunting at first glance, in the thermodynamic limit $LT/v \to \infty$, it simplifies to[6]

$$\langle \rho(z)\rho(0) \rangle \to \rho_0^2 - \frac{1}{2\pi^2 K z^2} + \frac{\mathcal{A}}{z^{2/K}} \cos(2\pi\rho_0 z), \tag{20}$$

where it is now clear that $K \gg 1$ corresponds to a strong tendency towards density wave order.

We are now in a position to test the prediction that for some special radii, the core of a helium-4 filled nanopore can be fully described at large L and low T by LL theory. We can directly compare the pair correlation function measured in the QMC with the LL prediction of Eq. (19). Before doing so, let us make a brief digression to study the connection between the interactions in the truly one-dimensional Hamiltonian in Eq. (8) and those experienced by helium atoms inside the quasi-one-dimensional environment inside the nanopore.

4.2. *Effective interactions and linear density in the inner cylinder*

In order to determine the effective dimension that best describes the low energy properties of helium-4 atoms inside a nanopore of finite radius it is important to understand how interactions inside the core are affected by the possible presence of the cylindrical shells observed in the radial density (Fig. 4). Naively, we expect that such shells act to screen the confining potential and may mediate helium–helium interactions inside the central region of the pore through particle exchanges. To quantify this and make contact with the potential introduced in the strictly one-dimensional Eq. (8) we define

$$v_{1D}(z) = \frac{1}{\rho_L^2} \int d^2r \int d^2r' v(\mathbf{r} - \mathbf{r}') \rho_R(r) \rho_R(r') , \qquad (21)$$

where $\mathbf{r} = (r, \varphi, z)$ is a vector in cylindrical coordinates, $\rho_R(r)$ is the radial density defined in Eq. (7) and $v(\mathbf{r} - \mathbf{r}')$ is the Aziz interaction potential of Eq. (2) displayed as the inset in Fig. 7. Integrating over all transverse degrees of freedom in the pore using the measured radial density from our QMC simulations, the effective potential can be compared with the full Aziz interaction in 1D as shown in Fig. 5. We observe that the effective potential in the $R = 2.9$ Å pore very closely reproduces the interactions that would be experienced in a system of helium atoms interacting in strictly one dimension ($R = 0$). For the largest radius pore of $R = 12.0$ Å however, screening

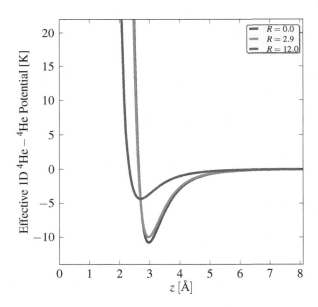

Fig. 5. The effective interaction energy between two helium atoms in strictly one dimension (lowest curve) as well as for the narrowest ($R = 2.9$ Å) and widest ($R = 12.0$ Å) pores considered in this study. The radial density used in the numerical evaluation of Eq. (21) was measured at $T = 0.5$ K, but the form of the potential is only weakly dependent on temperature when $T < 1.0$ K.

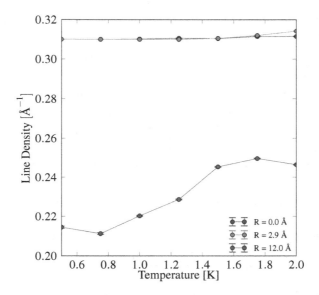

Fig. 6. The number of particles per angstrom for helium atoms in strictly one dimension (with $\mu = 85$ K), a pore with $R = 2.9$ Å and inside the IC with $r < r_{IC} = 1.75$ Å for $R = 12.0$ Å. All pores have length $L = 100$ Å and the finite radius pores are held at SVP with $\mu = -7.2$ K.

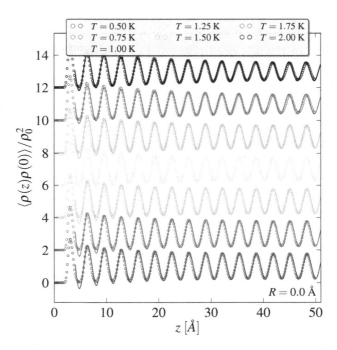

Fig. 7. Simulation data (circles) and a fit to Eq. (19) (lines) for the pair correlation function of strictly one-dimensional helium atoms ($R = 0$). The chemical potential ($\mu = 85$ K) has been set to approximate the energetically confining environment coming from the pore wall (see text). Error bars are smaller than the symbol size and data in the main panel has been given a vertical T-dependent shift for clarity with $T = 0.5$ K at the bottom and $T = 2.0$ K at the top.

from the concentric shells leads to a potential with a much weaker minimum (about half the depth of the unscreened bulk Aziz potential) with its location shifted to smaller separations. The effective 1D interactions in the axial direction experienced by atoms in the center of the pore influences the average separation between atoms which can be determined from the average linear density in Fig. 6. We observe almost no temperature dependence for small radii and the agreement of the linear density for $R = 0$ and $R = 2.9$ Å is by design, with the chemical potential having been tuned to $\mu_{\text{SVP}} - V(0; 2.9) \simeq 85$ K for the one-dimensional $(R = 0)$ system to reproduce the energetic confinement displayed in Fig. 1 at low temperature. The numerical value of ρ_0, defined as the zero temperature thermodynamic limit of the linear density in the core is measured to be $\rho_{0,0} \simeq \rho_{0,2.9} = 0.3100(1)$ Å^{-1} (with the number in brackets indicating the uncertainty in the final digit and a second subscript being the pore radius in angstroms). The fact that this value is slightly smaller than r_{min}^{-1} can be attributed to quantum effects. For $R = 12.0$ Å, the observed density of $\rho_{0,12} = 0.2144(5)$ Å^{-1} is only weakly related to the minimum of the effective 1D potential shown in Fig. 5 as there is a temperature dependent amount of exchange of helium atoms between the IC and surrounding shells.

4.3. *Density correlations*

Having studied the effective one-dimensional potential felt by helium atoms inside the pore, and determined the linear density in the IC we are now in a position to evaluate the efficacy of the LL prediction of Eq. (19) in describing density correlations in the nanopore. A first glance at this expression indicates that there are *four* fitting parameters, $(\rho_0, v, K, \mathcal{A})$ to be determined; a number that would seemingly allow large flexibility (and thus limited accuracy) of any least squares fitting procedure. However, one of the most attractive features of Eq. (19) is that it provides a prediction for the full analytical form of the finite size and temperature scaling behavior of $\langle \rho(z)\rho(0) \rangle$, a rare luxury indeed. We thus have a stringent procedure for confirming LL behavior in the nanopore: (1) Perform simulations for different pore radii, lengths and temperatures measuring thermodynamic estimators and correlation functions. (2) The average linear density in the central region of the pore can be determined as a function of temperature and extrapolated for each L and R to $T = 0$, this fixes $\rho_{0,R}$. (3) For each radius which displays a finite density of helium atoms at $r = 0$, perform a least squares fit of the density–density correlation function measured in the QMC to Eq. (19) at a single L and $T = 0.5$ K (the lowest temperature measured). This fixes the remaining three parameters, v_R, K_R and $\mathcal{A}_R$ where the subscript R will be used to distinguish results for different finite radius pores. These parameters are *intrinsic* to the zero temperature thermodynamic limit of Eq. (10) and provided that helium-4 inside the core is behaving as a LL, *cannot* exhibit any temperature dependence. (4) In other words, for a given radius, a single fit to Eq. (19) at fixed temperature and system size is enough to compute the LL prediction for the pair correlation function at all other temperatures and sizes and

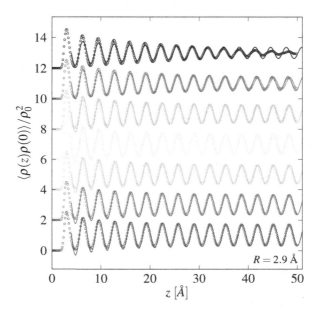

Fig. 8. Simulation data (symbols) and a fit to Eq. (19) (lines) for the pair correlation function along the axis of a nanopore with $R = 2.9$ Å and $L = 100$ Å. Different curves correspond to increasing temperature from $T = 0.5$ K (bottom) to $T = 2.0$ K (top) (vertical shifts have been added for clarity) with the legend displayed in Fig. 7.

we explore the resulting predictions for $R = 0, 2.9$ and 12.0 Å below. Obviously we are constrained by the low energy, long wavelength region of applicability of H_{LL} assumed throughout this study and it is easily confirmed that deviations appear at short lengths and high temperature.

To orient ourselves we begin by studying the strictly one-dimensional system, which we fully expect to be well-described by LL theory. In Fig. 7 we plot the results of QMC simulations for the density–density or pair correlation function (symbols) for a chain of helium atoms with $L = 100$ Å and $\mu = 85$ K. The full Galilean invariance of the one-dimensional system provides an additional constraint on our fitting procedure by fixing the value of $v_J \equiv v_0/K_0 = \pi \rho_{0,0}/m$ described in Eq. (13). Performing the aforementioned fits at $T = 0.5$ K, we find $v_0 = 75(1)$ Å K and $K_0 = 6.4(3)$ where stochastic errors in QMC data produce uncertainty in our regression procedure. The exact value of $\mathcal{A}_0$ is not relevant in the subsequent discussion and it will not be mentioned further. Fixing all parameters, Eq. (19) was used to produce the solid lines in Fig. (7) where a vertical shift has been included to allow for the visual differentiation of the effects of temperature. We observe spectacular agreement over the entire range of temperatures considered, noting that no additional fitting was required above $T = 0.5$ K.

We now perform an identical procedure for the narrowest finite radius pore, $R = 2.9$ Å, with the results for the density–density correlation function shown in Fig. 8. Similar to the case of 1D helium ($R = 0$), we observe persistent oscillations

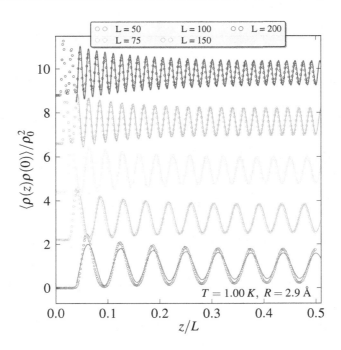

Fig. 9. Simulation data (symbols) and a fit to Eq. (19) (lines) for the pair correlation function along the axis of a nanopore with $R = 2.9$ Å for various pore lengths L (increasing from bottom to top). As in previous figures, a vertical shift has been added to distinguish the curves and accentuate the scaling behavior. Note the use of a dimensionless abscissa in order to plot spatial correlations for different pore lengths on the same scale.

out to the largest distances possible in the $L = 100$ Å pore with periodic boundary conditions indicating a strong tendency towards density wave order. As the helium-pore interaction is independent of the axial coordinate z and the radial density for $R = 2.9$ Å (Fig. 4) exhibits only a single central chain of atoms, we can again rely on Galilean invariance to restrict the ratio $v_{2.9}/K_{2.9} = \pi\rho_{0,2.9}/m$.[6] Extracting values in the presence of this constraint at $T = 0.5$ K yields $v_{2.9} = 70(3)$ Å K and $K_{2.9} = 6.0(2)$, in close agreement with the 1D chain of helium atoms as expected. For the finite radius pore however, we begin to observe deviations from LL predictions both at small distances and high temperature. Once the relevant thermal length scale, $\ell_T \sim v/T$ is on the order of the pore diameter, the system can no longer be thought of as quasi-1D and we expect significant corrections to arise from the thermal excitation of transverse modes. The values of the LL velocity and interaction strength determined in this way can also be used to test the predicted finite size scaling of Eq. (19) at fixed $T = 1.0$ K as seen in Fig. 9. We stress that the nearly perfect agreement seen between simulation data and LL theory in Fig. 9 for $L \geq 75$ does not require any additional fitting parameters and is a direct consequence of the predictive power of the universal hydrodynamics of H_{LL}.

Shifting attention to the $R = 12.0$ Å pore, in order to make an effective comparison we will focus the analysis on only those helium atoms which spontaneously find

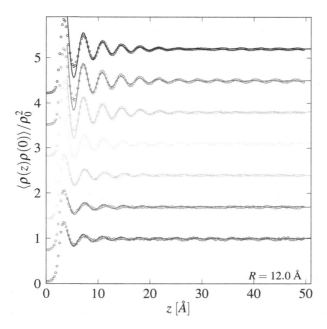

Fig. 10. Simulation data (symbols) and a fit to Eq. (19) (lines) for the inner cylinder pair correlation function along the axis of a nanopore with $L = 100$ Å and $R = 12.0$ Å for helium atoms with $r < r_{IC} = 1.75$ Å. The curves have been shifted for clarity and correspond to increasing temperature from $T = 0.5$ K (bottom) to top $T = 2.0$ K (top) with the legend displayed in Fig. 7.

themselves inside the IC. The precise definition of which helium atoms are inside the IC is somewhat arbitrary, but we define this to coincide with the location of the first minimum in the radial density in the presence of finite density at $r = 0$. For the pores considered here, this corresponds to $r_{IC} = 1.75$ Å. We expect that the presence of particle exchanges between the central chain of helium atoms and the surrounding cylindrical shells should begin to play an important role in the "melting" of the quasi-long range density wave order observed in narrow pores. The resulting axial pair correlation function measured in our QMC simulations is shown in Fig. 10. The data show density–density correlations in the IC that decay much faster than those seen for $R = 2.9$ Å. The lack of Galilean invariance due to atoms in the surrounding concentric shells adds some additional freedom to the now familiar fitting procedure (and thus more uncertainty) and it has been determined that $v_{12} = 42(2)$ Å K and $K_{12} = 1.3(2)$. This value of the Luttinger parameters is nearly a factor of five smaller than that found for the strictly one-dimensional system and indicates that this nanopore is in a region of phase space starting to be dominated by superfluid fluctuations.

The success of the LL prediction for $\langle \rho(z)\rho(0) \rangle$ in Eq. (19) in describing the density–density correlations measured in our QMC simulations is convincing evidence that a quantum fluid of helium-4 in pores with $R < 3$ Å can be described by LL theory and that the central region of wider pores may display emergent harmonic

Table 1. Values for the Luttinger velocity and parameter for helium atoms in 1D and in the quasi-1D core of nanopores by comparing the LL prediction with QMC measurements of the pair correlation function. The number in brackets indicates the uncertainty in the final digit.

Radius R [Å]	Luttinger Velocity v [Å K]	Luttinger Parameter K
0.0	75(1)	6.4(3)
2.9	70(3)	6.0(2)
12.0	42(2)	1.3(1)

fluid-like properties. In Sec. 5, we discuss the implications of these results and argue that the determination of K and v for a given system is not simply an academic exercise but has real consequences for the stability of the LL phase.

5. Discussion

Although it appears that LL theory works exceedingly well in describing pair or density–density correlations for the finite radius pores considered in this study, it is natural to ask whether there is something special about this particular observable. As mentioned in Sec. 4.1 the utility of the harmonic fluid description of 1D systems is that the resulting mode expanded Hamiltonian of Eq. (17) is quadratic (neglecting formally irrelevant operators) and thus the full grand partition function can be computed in closed form. We can therefore compare other two body correlation functions computed within LL theory with those measured in the QMC using the values of v_R and K_R determined in this study and displayed in Table 1. We have measured the axial one body matrix $n(z) = \langle \Psi^\dagger(z)\Psi(0) \rangle$ in the QMC[17] and find acceptable agreement with LL theory (with no new fitting parameters) at low T. The large value of $K \approx 6$ for $R = 0$ and $R = 2.9$ Å adds some complications when comparing numerical results for some scalar (noncorrelation function) observables with predictions coming from bosonization. For example, the strong tendency towards density wave order $\partial_z \theta(z) \sim 0$ displayed in Figs. 7 and 8 points to a compressibility $\kappa \sim \langle N^2 \rangle - \langle N \rangle^2 \approx 0$ and thus an analysis of the probability distribution function for particle number fluctuations like the one performed in Ref. 40 is not feasible. The conjugate relationship between density and phase variables in the LL theory [Eq. (12)] would predict that the boson phase $\phi(z)$ should be almost completely disordered resulting in the very small superfluid density observed in the numerical simulations performed here.

For the $R = 12$ Å pore, the analysis presented in the previous section is based on a fraction of the total number of ^{4}He atoms in the pore, those that dynamically have $r < r_{\rm IC}$ and find themselves in the core of the nanopore. The justification for this originates in the idea that we may be able to regard the pore as a coupled multi-component LL, with cylindrical shells replacing the legs of previously studied ladders.[41] Guided by these results we assume that only a single gapless degree of freedom may survive as a "center of mass mode" in the low energy effective field theory, due to tunneling between the shells. In the nanopore, this tunneling has two origins corresponding to the physical mobility of particles between shells as well as multi-particle quantum exchange cycles which may dynamically connect them at short time scales.

With this in mind, let us reanalyze the slight discrepancies between the QMC data in Fig. 10 and LL theory. At the lowest temperature considered, $T = 0.5$ K, the simulation data appears to show oscillations with a period that is slightly larger than that predicted by $1/(2\pi\rho_{0,12})$. This is most likely attributed to the physical exchange of particles between the IC and the surrounding shells producing a greater uncertainty in $\rho_{0,12}$ than is reflected in errorbars and making the extrapolation to zero temperature a difficult task. The amplitude of the oscillations in the pair correlation function is also slightly overestimated by LL theory at low temperature. Again, the finite radius of the pore is to blame, resulting in a finite superfluid fraction of helium (as measured via the usual winding number estimator[19] along the axis of the pore) of $\rho_s/\rho_0 \sim 0.2$ at $T = 1.0$ K. Transverse degrees of freedom that are not frozen out at this temperature lead to an increase in multi-particle exchanges that enhance superfluidity in the nanopore.

The opposite behavior appears to occur at high temperature with the simulation data showing weaker decay than predicted by LL theory with thermal fluctuations being unable to fully quench the proclivity towards density wave order. This is the opposite effect observed for $R = 2.9$ Å in Fig. 8 and it can possibly be explained through stabilization of the IC from attraction with surrounding atoms.

In a previous work,[40] finite size corrections to $\rho_{0,R}$ arising from the inclusion of higher order formally irrelevant terms in the effective Hamiltonian of Eq. (10) have been discussed at great length. However, due to the relatively large energy scales at play in the nanopore (Fig. 2) they are less important here and are not major players in any observed discrepancies between simulation data and the effective field theory.

We observe a general trend of K decreasing with increasing pore radius as seen in Table 1. However, the actual numerical values of the Luttinger parameter K for pores of varying radius can provide important information on the sensitivity of the LL to perturbations coming from commensuration effects or disorder; both of which are surely present in the real experiments of Savard et al.[14,15] For a strictly one-dimensional system, the introduction of a weak periodic substrate that is commensurate with the density will only lead to complete localization and the destruction of the harmonic fluid if $K > 1/2$.[6] Commensuration at other wavevectors is less relevant and would require a greater value of K to destabilize the LL. The introduction of a weak disorder potential, as might be present near the glassy walls of the pore, is known to be relevant only when $K > 2/3$.[42]

For the narrowest pores considered in this study $R < 3$ Å, we have found a value of $K \approx 6$ at SVP, indicating a strong tendency to form a solid, resulting from strong confinement and the deep minimum and accompanying hardcore found in the effective interaction potential $v_{1D}(z)$. The experimental confirmation of this result could be accomplished by noting that the formation of a quasi-solid should impede superfluid flow through a helium-4 filled nanopore at low temperature.

It may be useful to compare the values of K_R found here with other studies of low-dimensional helium. For example, in a WA study of helium-4 confined to flow in the channels formed by screw dislocations with $R \sim 3$ Å in solid helium, it was found that $K = 0.205(20)$,[43] nearly thirty times smaller than the comparable value of $K_{2.9}$ measured in this study. The sources of the discrepancy are rooted in the "softness" of the confining potential inside the screw dislocation as helium atoms are able to penetrate into the surrounding solid held at $\mu = 0.02$ K corresponding to the bulk melting point.

Much exciting work remains to be done in the nanopore system including a more systematic study of the superfluid density which can be measured in bundles of tubes or pores via torsional oscillator techniques.[44,45] The construction of more realistic models that contain both commensuration and disorder potentials would also enhance the applicability of numerical simulations. Further exploration of the available parameter space is also in order, including altering the chemical potential at fixed radius to simulate the effects of pressure which can be freely tuned in experiments.

In conclusion, we have studied a quantum fluid of bosonic helium-4 confined inside nanopores of varying radii via large scale continuum Worm Algorithm QMC Simulations at SVP below the bulk superfluid transition temperature. The results show a progression of phases inside the pore exhibiting a possible quasi-one-dimensional core surrounded by concentric shells of helium depending on the radius. When the core of the nanopore has a nonzero density of helium, the finite temperature and scaling properties of the density–density correlation function are fully described within harmonic LL theory. The description of the emergent low energy phase of confined helium in terms of the harmonic field theory allows for the extraction of the Luttinger parameter K which is found to be a decreasing function of radius. As the pore radius increases, the inner helium core is screened from the confining effects of the pore wall by the surrounding matter, resulting in a pronounced enhancement of quantum exchanges leading to superfluidity. The precise relationship between the material through which the pore has been sculpted and the resulting LL parameters could be further explored leading to new predictions and optimized experiments with the maximum likelihood of detecting a universal and stable one-dimensional quantum harmonic fluid of helium at low temperature.

Acknowledgments

The author would like to thank I. Affleck, M. Boninsegni and G. Gervais for many edifying discussions. This work was made possible through computational resources provided by the National Resource Allocation Committee of Compute Canada with all simulations taking place on Westgrid or SHARCNET.

References

1. E. H. Lieb and W. Liniger, *Phys. Rev.* **130**, 1605 (1963).
2. S.-I. Tomonaga, *Prog. Theor. Phys.* **5**, 544 (1951).
3. D. C. Mattis and E. H. Lieb, *J. Math. Phys.* **6**, 304 (1965).
4. A. Luther and I. Peschel, *Phys. Rev. B* **9**, 2911 (1974).
5. K. B. Efetov and A. I. Larkin, *Zh. Eksp. Teor. Fiz.* **69**, 764 (1975).
6. F. D. M. Haldane, *Phys. Rev. Lett.* **47**, 1840 (1981).
7. F. D. M. Haldane, *J. Phys. C: Solid State Phys.* **14**, 2585 (1981).
8. T. Giamarchi, *Quantum Physics in One Dimension* (Oxford University Press, Oxford, 2004).
9. M. Greiner *et al.*, *Phys. Rev. Lett.* **87**, 160405 (2001).
10. C. Ryu *et al.*, *Phys. Rev. Lett.* **99**, 260401 (2007).
11. M. Cazalilla *et al.*, *Rev. Mod. Phys.* **83**, 1405 (2011).
12. J. R. Beamish *et al.*, *Phys. Rev. Lett.* **50**, 425 (1983).
13. J. Taniguchi, Y. Aoki and M. Suzuki, *Phys. Rev. B* **82**, 104509 (2010).
14. M. Savard, C. Tremblay-Darveau and G. Gervais, *Phys. Rev. Lett.* **103**, 104502 (2009).
15. M. Savard, G. Dauphinais and G. Gervais, *Phys. Rev. Lett.* **107**, 254501 (2011).
16. M. Boninsegni, N. V. Prokof'ev and B. V. Svistunov, *Phys. Rev. Lett.* **96**, 070602 (2006).
17. M. Boninsegni, N. V. Prokof'ev and B. V. Svistunov, *Phys. Rev. E* **74**, 036701 (2006).
18. D. Ceperley, *Rev. Mod. Phys.* **67**, 279 (1995).
19. E. Pollock and D. Ceperley, *Phys. Rev. B* **36**, 8343 (1987).
20. R. A. Aziz *et al.*, *J. Chem. Phys.* **70**, 4330 (1979).
21. G. Tjatjopoulos, D. Feke and J. Mann Jr., *J. Phys. Chem.* **92**, 4006 (1988).
22. J. A. Wendel and W. A. Goddard, *J. Chem. Phys.* **97**, 5048 (1992).
23. C. Chakravarty, *J. Phys. Chem. B* **101**, 1878 (1997).
24. W.-Y. Ching *et al.*, *J. Am. Ceram. Soc.* **81**, 3189 (1998).
25. M. Boninsegni, *J. Low Temp. Phys.* **160**, 441 (2010).
26. S. Jang, S. Jang and G. A. Voth, *J. Chem. Phys.* **115**, 7832 (2001).
27. G. Stan and M. W. Cole, *Surf. Sci.* **395**, 280 (1998).
28. S. M. Gatica *et al.*, *J. Low Temp. Phys.* **120**, 337 (2000).
29. M. W. Cole *et al.*, *Phys. Rev. Lett.* **84**, 3883 (2000).
30. M. Gordillo, J. Boronat and J. Casulleras, *Phys. Rev. B* **61**, R878 (2000).
31. M. Gordillo, J. Boronat and J. Casulleras, *Phys. Rev. B* **76**, 193402 (2007).
32. M. C. Gordillo and J. Boronat, *J. Low Temp. Phys.* **157**, 296 (2009).
33. N. M. Urban and M. W. Cole, *Int. J. Mod. Phy. B* **20**, 5264 (2006).
34. E. S. Hernández, *J. Low Temp. Phys.* **162**, 583 (2010).
35. A. Del Maestro, M. Boninsegni and I. Affleck, *Phys. Rev. Lett.* **106**, 105303 (2011).
36. M. Rossi, D. Galli and L. Reatto, *Phys. Rev. B* **72**, 064516 (2005).
37. M. Rossi, D. E. Galli and L. Reatto, *J. Low Temp. Phys.* **146**, 95 (2006).
38. B. Clements *et al.*, *Phys. Rev. B* **48**, 7450 (1993).
39. M. A. Cazalilla, *J. Phys. B: At. Mol. Opt. Phys.* **37**, S1 (2004).
40. A. Del Maestro and I. Affleck, *Phys. Rev. B* **82**, 060515(R) (2010).
41. E. Orignac and T. Giamarchi, *Phys. Rev. B* **57**, 11713 (1998).
42. T. Giamarchi and H. J. Schulz, *Phys. Rev. B* **37**, 325 (1988).
43. M. Boninsegni *et al.*, *Phys. Rev. Lett.* **99**, 035301 (2007).
44. N. Wada *et al.*, *J. Low Temp. Phys.* **157**, 324 (2009).
45. N. Wada *et al.*, *J. Low Temp. Phys.* **162**, 549 (2011).

SOME EXPERIMENTAL TESTS OF TOMONAGA–LUTTINGER LIQUIDS*

T. GIAMARCHI

DPMC-MaNEP University of Geneva,
24 Quai Ernest-Ansermet CH-1211 Genève 4, Switzerland

The Tomonaga–Luttinger–Liquid (TLL) has been the cornerstone of our understanding of the properties of one dimensional systems. This universal set of properties plays in one dimension, the same role than Fermi liquid plays for the higher dimensional metals. I will give in these notes an overview of some of the experimental tests that were made to probe such TLL physics. In particular I will detail some of the recent experiments that were made in spin systems and which provided remarkable quantitative tests of the TLL physics.

Keywords: One dimensional systems; Luttinger liquids; strong correlations; localized magnetism; cold atomic gases.

1. Introduction

Interactions lead in one dimension to a physics radically different than for the higher dimensional counterparts.[1] Indeed if in high dimensions it is possible for excitations resembling individual particles (such as the famous Landau quasiparticle excitations)[2,3] to exist, this is not the case in one dimension where interactions will clearly turn any excitation into a collective one.

As a result a new type of physics exists, that can be described by the universal concepts of Tomonaga–Luttinger liquids (TLL).[1,4] This theory describing the physical properties of most of the one dimensional interacting systems (bosons, fermions, spins) has been first introduced as a perturbative expansion around idealized dirac dispersion. It then has been proven to be a universal concept, playing a similar key role in one dimension than the Fermi liquid played in higher dimensions.

Given the importance and universality of the TLL concept, and more generally of the unusual one dimensional physics, many experiments have sought to probe such a physics. I will describe in this brief review some "morceaux choisis" of the hunt for TLL. I will first briefly review the salient features of one dimensional quantum systems. I will then give a few examples of the experimental tests, and discuss in details some of the most recent achievements that allowed for the first time for a *quantitative* test of such TLL physics. Of course this review cannot pretend to be exhaustive and much more details on the vast body of experiments in the context of one dimensional physics can be found at other places in the literature (see e.g.,

---

*This article first appeared in International Journal of Modern Physics B, Vol. 26, No. 22 (2012).

Refs. 1, 5 and references therein). Nevertheless it should give an idea of the state of the art in the experimental search for TLL.

2. Some Basics of TLL

Let me recall in this section some of the properties of the TLL that will be useful in connection with the experiments. The goal is of course not to do a full review of the TLL. The reader wanting to get the full details on TLL is referred to Ref. 1. Two main properties are crucial in a TLL and we examine them separately.

2.1. *Power laws and universality*

One of the most visible properties of the TLL is the fact that all correlation functions behave as power laws with some nonuniversal exponents. For example, the density-density correlations for a system of quantum particles of density ρ_0 is given by:

$$\langle \delta_\rho(x,\tau)\delta\rho(0) \rangle = \frac{1}{r^2} + A_2 \cos(2\pi\rho_0 x)\left(\frac{1}{r}\right)^{2K} + A_4 \cos(4\pi\rho_0 x)\left(\frac{1}{r}\right)^{8K} + \cdots, \quad (1)$$

where x is the spatial position, τ the imaginary time, $\delta\rho(r) = \rho(r) - \rho_0$ and $r = \sqrt{x^2 + (u\tau)^2}$. The A_n are nonuniversal amplitudes that depend on the model, the value of the interactions etc. All the powerlaw behavior depends on two numbers, nicknamed the Luttinger liquid parameters: u which is the velocity of density excitations and a dimensionless parameter K. Both u and K depend on the model and the value of the interactions. There are various ways that these parameters can be computed.[1] The formula (1) presents several remarkable features.

The first one is of course the powerlaw behavior. Normally we associate this to a critical state, such as the one that occurs exactly at the critical temperature in a phase transition. One can physically understand why one dimensional quantum systems present such critical behavior by remarking that quantum fluctuations preclude, even at $T = 0$ the breaking of a continuous symmetry. The system thus cannot really order and is poised at the brink of order, thus behaving in a critical way. This powerlaw behavior is one of the hallmark of the TLL. It shows in all the correlation functions. It has many consequences that extend beyond the $T = 0$ results. Since the theory is conformally invariant one can deduce from the powerlaw (1) behaviors at finite temperature or for finite size systems.

The second, and very often overlooked remarkable fact, is the universality of the exponents. The various correlation functions may have different exponents, but they are *all* functions of the *sole* parameter K, which also controls the thermodynamics of the system. It means that although the exponents themselves are nonuniversal the relations between the exponents are. This is one of the strongest properties of a TLL, and rests on the fact that the fixed point Hamiltonian that describes the low energy properties of the system is:

$$H = \frac{1}{2\pi} \int dx (uK)(\pi\Pi(x))^2 + \frac{u}{K}(\nabla_x\phi(x))^2 \,, \tag{2}$$

where ϕ and Π are conjugate variables. This universality is of course what makes the TLL description so useful. It allows to go beyond the approximations that are usually inherent in models such as Hubbard, etc. for which one has to make a caricature of the interactions. With the TLL description one can simply take the value of K from one of the experimental value, and then be sure that one has a faithful description of the properties of the system. In that sense the TLL plays exactly the same role than the Fermi liquid plays in higher dimensions.

2.2. *Fractionalization of excitations*

Another crucial property of one dimensional interacting system is the possibility for the excitations to fractionalize. In the high dimensional world we are used to the fact that if we make an excitation with some quantum numbers that correspond to the *minimum* possible quantum numbers (e.g., one adds or removes one electron, such as in photoemission, or flips one spin $1/2$ thus creating a $S = 1$ magnon excitation) then these quantum numbers are stable. The excitation can scatter, acquire a lifetime etc. but cannot be viewed as a composite object of *smaller* quantum number excitations. The only exception (except for very bizarre models) are the Laughlin quasiparticles in the quantum hall effect, which can carry a charge smaller than one of the electron.

In one dimension fractionalization is the rule rather than the exception. The very reason is that because of the one dimensional nature of the system it is possible to make soliton excitations, something that would be much more difficult in higher dimensions. In Fig. 1 we show these two types of excitations. The spinon are the good excitations in one dimension. Therefore for spinons there is a good relation between the momentum of the excitation q and the energy of the excitation $E = J\cos(q)$. This is not the case for the magnon excitation, which is made of two spinon excitations. As a result instead of a well defined relation between the momentum q of the magnon and its energy E there is a continuum since

$$q = q_1 + q_2, \quad E = J\cos(q_1) + J\cos(q_2) \tag{3}$$

where $q_i \in [-\pi/2, \pi/2]$ since spinons move by two lattice spacings, and one can spread the total momentum q between the two spinon in any way compatible with the above constraints. The leads to the spectrum shown in Fig. 2.

This fractionalization of the excitations is even more spectacular in the case of an electron. In the presence of spin and charge, the good excitations are solitons of charge and solitons of spins. Hence the electron splits in two excitations, the holon carrying charge but no spin and the spinon which has spin but no charge.[1]

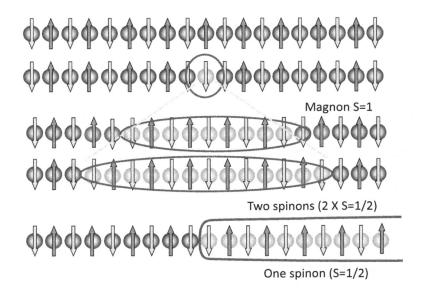

Fig. 1. In high dimensions the minimal spin excitation on an antiferromagnetic chain (top two lines) would be a magnon, in which a single spin is flipped. This is an $S = 1$ excitation (2nd line). In one dimension the magnon is not an individual excitation, and a magnon decomposes into two $S = 1/2$ excitations, the spinons (third and fourth line). Such excitations are soliton like objects with a string of spin flips on the right or the left of the position of the soliton (denoted by the blue circles), see bottom line for one spinon.

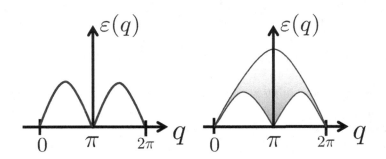

Fig. 2. (left) In high dimension a magnon excitation behaves as a particle, i.e., has a good relation between the momentum of the excitation q and its energy $\epsilon(q)$. (right) In one dimension the magnon decomposes into two spinons (see text). As a consequence, there is now a continuum of excitations, the des Cloiseaux–Pearson spectrum. Such continuum can be clearly distinguished in neutron scattering experiments.

Note that the notion of fractionalization is more general than the TLL in itself. It is more resting on the 1D nature and not confined to the low energy properties of the system. Of course for several systems such property is also appearing in the low energy properties since terms that couple the spin and charge excitations do in general exists (such as band curvature, impurities etc.). For more details on fractionalization we refer the reader to Ref. 1.

3. Power Laws and Such

Let us first look at experiments that are probing the powerlaw behavior of a TLL. Of course it is out of the question to even attempt to make an exhaustive review here of all the experiments. The reader interested in that will find many experiments discussed in Ref. 1 and for bosons and spins in the recent review.[5]

The first experimental observation of powerlaws in a TLL for an electronic system, was done in the organic conductors.[6] These materials are very anisotropic systems where stacks of organic molecules provide very different electronic overlaps for the band. Their bandstructure can be well represented by a tight-binding model[7-9]

$$H = - \sum_{\langle i,j \rangle, \sigma} t_{ij} c_{i\sigma}^{\dagger} c_{j\sigma} \tag{4}$$

with hopping along the three axis directions $t_a \sim 3000$ K, $t_b \sim 300$ K, $t_c \sim 20$ K, where a is the organic chain direction and b, c the two perpendicular directions. These systems have a commensurate filling and thus a tendency to become Mott insulators. Because of the commensurate potential provided by the scattering of the electrons on the lattice, the high frequency conductivity should behave as a power law of the form[10,11]:

$$\sigma(\omega) \propto \omega^{5-4n^2 K_\rho}, \tag{5}$$

where K_ρ is the TLL exponents for the charge excitations, and n the order of commensurability with the lattice ($n = 1$ for half filling, $n = 2$ for quarter filling etc.). Comparison of the above formula with measurements on the organic conductors is shown in Fig. 3.

The experiments clearly show over one decade in frequency the power law behavior. Additional experiments in such a class of materials, both on the temperature dependence of the conductivity,[7-9] or on the transverse transport[13-16] have been found in agreement with TLL behavior. Similar properties were observed in other compounds.[17] I refer the reader to the literature for more details on those points.

Shortly after the observation of the TLL power law in the organics conductors, transport in the nanotubes was investigated in a similar manner.[18] In that case the weak contact between leads and the nanotube ensured that one was performing a tunneling experiment. Such experiments showed power law behavior of the conductivity that was again in nice agreement with the expected behavior in a TLL. Variation of such a power law was observed depending on whether the tunneling occurred at the edge or in the middle of the tube, or if an impurity was producing additional backscattering within the nanotube.[19] I refer the reader to the literature for more details on such class of experiments.

Tunneling inside a TLL can also be probed in bulk material by doing a photoemission experiment. The interpretation of such experiments is often delicate given

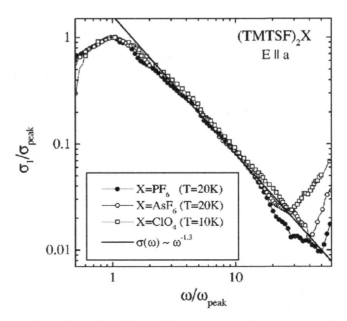

Fig. 3. Optical conductivity of various organic conductors-superconductors, normalized to the observed optical gap. From TLL theory a power-law behavior is expected above the optical gap in excellent agreement with the observed spectra. This constituted the first observation of the TLL behavior in an electronic system. From Ref. 12.

the possibility of surface artefacts. Several powerlaw behavior have nevertheless been observed.[20,21] Some powerlaw has been seen in photoemission in the organic conductors. However the range of energy over which such powerlaw was seen casts doubts on the fact that all of it is due to TLL physics, even if the extracted TLL parameter K_ρ would be in a good agreement with the other measurements on this compound.[22] Other one dimensional systems have also shown power law behavior under photoemission. One of the most interesting is provided by the purple bronze compound, where power laws in the density of states well consistent with the expected TLL behavior have been observed both in photoemission and STM.[23,24] The high energy behavior is well consistent with a TLL behavior and strong interactions, while there are still some puzzles in this compound to be explained for the low energy behavior.[25,26]

Another bread class of systems in which power laws were investigated was provided by realization of the TLL using edge states such as in the quantum hall effect.[27] Indeed the two chiral states existing at the edge of a Hall systems can be viewed as the right and left movers of a standard TLL. It can be shown that the TLL exponent K is related to the fractional Hall plateau of the conductance of the system.[1] Backscattering between the edge can thus be analyzed in terms of backscattering in a TLL and such experiments, although posing some questions on whether quasiparticles or electrons tunnel, have provided evidence of the powerlaw scaling expected from a TLL. More details on that class of experiments can be found in the literature.

Last but not least cold atoms have provided remarkable realizations of interacting quantum system,[28] in which the TLL behavior can be in principle quantitatively tested. Experiments on interacting bosons have successfully probed the so-called Tonks limit in which hard core bosons are found to behave as spinless fermions, showing the common behavior of the one dimensional quantum fluids that is covered by the TLL universality class.[5] However in experiments in optical lattices the ability to probe for the TLL power laws is impaired by the fact that such systems are inhomogeneous because of the trap, an r^2 varying chemical potential. Since the TLL depends on the density and interaction the inhomogeneity obscures the simple power law behavior. This is obvious in the Tonks limit for which the TLL behavior would have given a single particle correlation decaying as $1/\sqrt{x}$ (corresponding to $K = 1$ the limit for hard core bosons or free spinless fermions), and for which the averaging of many tubes lead to quite different and much less clear behavior. However experiments on atom chips do not suffer from such inhomogeneity and indeed interferences between two bosonic tubes have shown behavior well compatible with TLL.[29] However in these last experiments the interactions are quite weak for the moment, leading to extremely large TLL parameters ($K \sim 42$) making it very hard to unambiguously distinguish TLL theory for simpler ones such as a quasi-condensate or a time dependent Gross–Pitaevskii theory.[30] Fermionic systems are for the moment at too high a temperature for the TLL properties to be probed in them. It is clear that the field is rapidly progressing and that many of the objections of today will soon be overcome, in particular with the advent of local probes, that allow to avoid the inhomogeneity problems. In particular recently predicted[31] non-local order parameters, characteristic of 1D behavior have been observed.[32] There is thus little doubt that cold atoms will offer remarkable and controlled realizations of the TLL in the future.

4. Spin Charge Separation

Another crucial aspect of the 1D behavior, namely the fractionalization of excitations[1] has also been observed.

For such features the system of choice has proven to be quantum spin systems. Indeed in such systems the spectrum of excitations can be probed in a very clear way by neutrons scattering experiments. Although of course the neutrons scattering probes the full spin spins correlation functions one can view it as a gross determination of the spectrum if one uses the Lehmann representation of the spin–spin spectral function measured in a neutron scattering experiment

$$S(\omega, q) = \sum_{\nu} |\langle \nu | S_q^- | 0 \rangle|^2 \delta(\omega + E_0 - E_\nu), \tag{6}$$

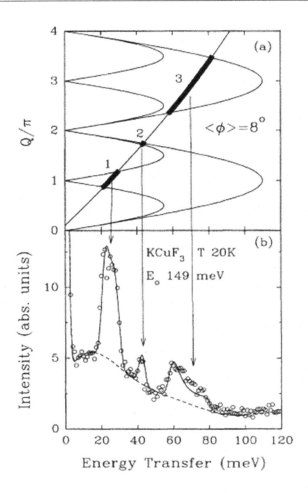

Fig. 4.   (a) The line represents the scan in energy E and momentum q which is done in a neutron scattering experiment. If one neglects the matrix elements specific to the spin–spin correlation function measured, one can expect intensity within the continuum described in Fig. 2. (b) The corresponding intensity in a neutron scattering experiment on the spin chain compound KCUF3. The intensity is in excellent agreement with the prediction of the fractionalization of the magnon excitation inherent to a one-dimensional situation. Note that to get the precise lineshape and not just the existence of the response, one would need the matrix elements in addition to the spectrum of excitations. From Ref. 33.

where $|0\rangle$ is the ground state of the system of energy E_0, $|\nu\rangle$ the exact eigenstates of energy E_n and the operator S_q^- flips a spin with momentum q. Ignoring as a first approximation the matrix elements $\langle\nu|S_q^-|0\rangle$ one can expect intensity only within the continuum of excitation. One can thus well distinguish between a quasi-particle like excitation, and a continuum. Some of the first experimental results for spin chains are shown in Fig. 4. The so-called Des Cloiseaux–Pierson spectrum is clearly identified, showing the fractionalization of the magnon excitations into two spinon ones in one dimension. Of course since then more refined experiments have been performed in a variety of systems and the continuum of excitations clearly confirmed. I refer the reader to Refs. 34–36 for more details on these more recent experiments.

Beyond spin chains, probing the fractionalization in spinfull systems, namely the spin charge separation is not an obvious task. One possible route is provided by photoemission, in which one would expect that because of the spin-charge separation the standard quasiparticle peak would split in two singularities, one at $\omega = u_\rho k$ and one at $\omega = u_\sigma k$ reflecting the existence of the holon and the spinon. Although some photoemission experiments claim to have observed the two excitations and their dispersion, the observation remains difficult and very often controversial. To the best of my knowledge the only experiment that unambiguously observed the spin-charge separation was the one involving the tunneling between two quantum wires.[37,38] Control provided in such experiments by a gate voltage (which controls the difference of energy between the two wires) and a magnetic field (allowing to control the difference of momentum) allows to tunnel from one of the Fermi points in one wire to one in another wire. The spin charge separation of the electron into two independent modes allows to see two different branches in the tunneling spectrum. Interference using the finite length of the wire allow to confirm that there are indeed two different modes with two different velocities.

Cold atoms, given the level of control should be the ultimate systems in which to test for the spin-charge separation. No experiment has been realized so far but theoretical proposals[39] have been put forward on how to test for it. The rapid progress in probing should give measurements of this feature in a not too distant future.

5. Quantitative Tests of TLL

As is obvious when looking at the previous selected sets of experiments for TLL physics, and of course the many other realizations (see Refs. 1 and 5 and the various chapters of the present book), the situation can be both considered as satisfactory and also quite frustrating. On the positive points:

(1) There is a growing number of one dimensional materials, both in the field of condensed matter and in the rapidly evolving field of cold atomic systems.
(2) Power laws have been observed in a variety of materials such as organics, nanotubes and cold atoms, and by a large variety of probes (optical conductivity, d.c. transport, photoemission, stm etc.).
(3) The effects of fractionalization has been evidenced, both for spinless (spectrum in spin chains) and spinful systems (quantum wires, photoemission).

The TLL physics is thus definitely on the map and applicable to a growing number of materials. On the other hand although some of the properties of TLL have been unambiguously seen, some problems still remain:

(1) In most of the tests for the electronic systems, the exponent is an adjustable parameter. This is due to the fact that for electronic systems it is extremely hard to determine from *ab initio* calculations the value of the TLL parameter K since the interaction is a screened long range coulombic interaction.

(2) In most of the systems the universality of the TLL description, i.e., the fact that all the exponents depend on a single value of the parameter K has not been tested. This comes from the fact that it is often difficult to do more than one type of measurement on a given compound. Some exception is provided by the organics. For more on that point we refer the reader to Refs. 8 and 9.

(3) In most of the experiments there is no control parameter, such as the band filling or the control of the interactions. In that respect it is of course possible to apply pressure to change the interactions, but the global effects are complicated to evaluate. In the cold atom context it is possible to vary the interactions or bandwidth in a controlled way but not on systems free of the harmonic trap, which clearly complicates the identification of Luttinger exponents.

It is thus highly desirable to have *quantitative* tests of the TLL and to be able to plug the above-mentioned holes. Cold atoms are of course candidate of choice in that direction. They can easily realize one dimensional bosonic or fermionic systems. The interactions are short range and very controlled. Unfortunately the presence of the harmonic trap and/or of many inequivalent tubes makes the identification of TLL powerlaw behaviors extremely difficult since the exponents depends on the interaction and density of particles. As discussed in the previous section, atom chips, which are very homogeneous, could provide a solution. For the moment however, interaction is these systems is yet to weak to unambiguously distinguish the TLL behavior from theories such as time dependent Gross–Pitaevskii physics.

It is thus suitable to have another type of systems in which such "quantum simulation" of TLL systems can be done. Quite remarkably such systems have been provided by quantum spin systems. I will thus introduce some aspects of these systems, and refer the reader to the published literature for more detailed references.

5.1. *Quantum ladders and spin systems as quantum simulators*

The drawback of being unable to precisely know the interactions in a condensed matter context can be circumvented in a Mott insulator. In that case the charges being localized, only spin superexchange remains. The localized quantum magnets thus offer the advantage to have short range, interactions. If one deals with the standard Heisenberg of XXZ hamiltonian one has:

$$H = \frac{J_{XY}}{2} \sum_i [S_i^+ S_{i+1}^- + S_{i+1}^+ S_i^-] + J_Z \sum_i S_i^z S_{i+1}^z - h \sum_i S_i^z . \tag{7}$$

Using the well known Holstein–Primakov representation $S_i^+ \to b_i^\dagger(-1)^i$ and $S^z = b_i^\dagger b_i - \frac{1}{2}$, where b represent hard core bosons one can map the problem (7) to an itinerant problem

$$H = -t \sum_i [b_i^\dagger b_{i+1} + \text{h.c.}] + V \sum_i \left(n_i - \frac{1}{2}\right)\left(n_{i+1} - \frac{1}{2}\right) - h \sum_i \left(n_i - \frac{1}{2}\right), \quad (8)$$

where $t = J_{XY}/2$ and $V = J_Z$. This is of course a very well known result.[40] One thus sees that a spin system could be used as a quantum simulator of hard core bosons with a nearest neighbor interaction. As mentioned one big advantage is that the interactions between the particles are short range and *a priori* well known if the exchange are known. Another important advantage of such a system is the use of the magnetic field as a chemical potential. One can control the "number" of particles by changing the magnetic field and measure the number by simply measuring the magnetization. That allows to realize the equivalent of a gate voltage for charged particles and thus to study the effects of interactions on the whole range of band filling. The empty band would be the fully polarized down chain, while a filled band would be a fully polarized up chain.

Although these results were well known for a long time, using them directly was difficult. First, the typical spin exchanges are usually in the range of several hundred Kelvin, which would mean impossible magnetic fields to polarize the chain and thus use the field as a control parameter. Second, since the two critical points, empty band and filled band correspond to a fully polarized system, there are usually parasitic couplings such as the dipolar forces that manifest themselves very strongly and thus spoil the nice mapping described above. So although the use of spin chains for this purpose is possible it is in general not easy. A system which is much more flexible is provided by dimers, either in a 3D, 2D or 1D (ladders) under magnetic field as shown in Fig. 5. I will discuss here mostly the case of the strong rung ladder, but similar arguments can be done for all cases. For strong rungs, in the absence of magnetic field, the ground state consists of singlets, while the magnetic field will favor triplets on the rung. One can use a similar mapping between the singlet and triplet states and the absence of presence of a boson on the rung. All commutation relations are obeyed. The ladders have the advantage that the two critical points are controlled by the two different energy scales $J_\perp$ and J. In addition the singlet state is very robust so parasitic interactions are usually less effective. The mapping one the bosons can thus be implemented, and these systems used to "simulate" itinerant quantum bosons. In particular they were very fruitful to realize Bose–Einstein condensation. I refer the reader to the review Ref. 41 for more discussion and references on that point.

In one dimension (i.e., for ladders) these systems can thus be used to simulate TLL. For one dimension the hard core bosons can be further mapped onto spinless fermions. The phase diagram of these systems is indicated in Fig. 6.

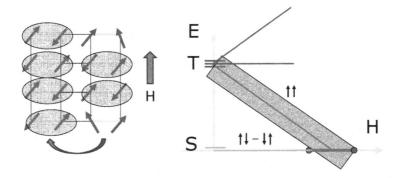

Fig. 5.   (Color online) Ladders under magnetic field. (left) The ground state of dimer systems (the strong rungs J_r denoted by the yellow lines) is made of singlets. This ground state is stable with respect to the weaker exchange J denoted by black lines. Under the magnetic field H some of the singlets are transformed into triplets, which can propagate from one rung to the next thanks to the exchange J. Using a mapping of the singlet to a vacuum of bosons, and the triplet to the presence of a hard core boson, this magnetic system realizes a tight binding model of interacting hard core bosons, whose density can be controlled by the magnetic field. (right) the energy spectrum of the system on the left. The dispersion of the triplets (only the lower dispersion is shown) leads to the presence of two quantum critical points when H is varied (denoted by red dots). At H_{c1} the first triplet ("boson") enters the system, while the band of triplets is full at H_{c2}.

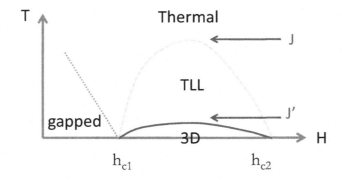

Fig. 6.   Phase diagram of Ladders under magnetic field. The phase below H_{c1} or above H_{c2} are gapped, (either the singlet phase or the fully polarized one). The gap (and hence the field h_{c1} is of the order of the rung coupling J_r. Between the two critical fields the band of triplets is partly filled. There is a coherence scale (corresponding roughly to the distance between the bottom or the top of the band, below which the systems is described by TLL physics. The typical scale is given by the triplet dispersion along the legs of the ladders, of the order of the exchange J. Due to the weak coupling J' between the ladders, there is an ordered antiferromagnetic phase in the XY plane, perpendicular to the magnetic field, that can take place.

5.2.   *Quantitative test of TLL —power laws*

Fortunately excellent experimental realization of such systems exist. One of them, that proved to be remarkably useful is provided by the BPCB compound, an organic ladder. The exchange constant can be obtained directly from neutron scattering measurements. They can also be directly extracted from a comparison of the measured magnetization and a computation using finite temperature Density

Matrix Renormalization group (DMRG) calculations. One can check, by comparing measured specific heat with the DMRG computed one that no important term was forgotten in the Hamiltonian. Having the exchange constants allows for a *quantitative* test of the TLL. The procedure is the following:

(1) Compute from the Hamiltonian the TLL parameters. The general procedure is explained in Ref. 1. Note that there are no adjustable parameters there. The values of u and K as a function of the magnetic field are totally determined. For an electronic system that would correspond going from an empty band to a full band.

(2) Compute several correlation functions of the system from the TLL theory and compare with the experimental ones. Note that now we have both a control parameter, which is the magnetic field, and the exponents are *not* adjustable parameters but known quantities.

This program can be implemented in practice for the BPCB. I will not give all the details here and refer the reader to Refs. 36, 42, 43 for more information. In short we used for the various correlation functions: (a) the NMR relaxation time $1/T_1$ which is related to the local spin–spin correlation function; (b) because of the interladder coupling, as shown in Fig. 6, there is a transition towards a three dimensional ordered phase. The $T_c(h)$ can be computed from the TLL theory; (c) the order parameter in the ordered phase. The list is of course not limitative, and other correlations can be computed as well, and compared to future experiments. An example is provided by the ESR resonance which was found to be in remarkable agreement.[44]

As an example the NMR $1/T_1$ is given for a TLL by:

$$
T_1^{-1} = \frac{1}{2}\gamma^2 \frac{\hbar}{k_B} A_\perp^2 \cdot \cos\left(\frac{\pi}{4K}\right) B\left(\frac{1}{4K}, 1 - \frac{1}{2K}\right) \cdot \frac{2A_0^x}{u}\left(\frac{2\pi T}{u}\right)^{\frac{1}{2K}-1}, \qquad (9)
$$

where all the parameters in the above formula, except the hyperfine coupling, are fully known as a function of the Hamiltonian and magnetic field (u, K, etc.). The *full* functional dependence of the relaxation time as a function of the temperature and magnetic field is thus known and can be quantitatively compared with the experiment. The result for the experimentally measured $1/T_1$ is shown in Fig. 7. The agreement is remarkable and provides a *quantitative* test of the TLL predictions. It is worthwhile to note that the description is not only valid in some distant asymptotic regime, but very efficient for practically all usable band filling. Such experiments demonstrates that the combination of analytic calculations based on the TLL theory and the numerical calculation of the TLL parameters provides an extremely powerful theoretical tool. Of course for the general experimental case, for which it is difficult to compute the TLL parameters *ab initio*, one can still measure them in one experiment and then use them for all the other experimental quantities.

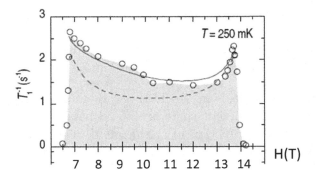

Fig. 7. (Color online) NMR $1/T_1$ measured for the ladder compound BPCB (dots). The red line is a theoretical calculation based on the TLL theory (see text). Note that the *only* adjustable parameter here is the hyperfine coupling constant (which is in addition roughly known). Since the $1/T_1$ is obtained for the whole range of magnetic field one could only shift up and down the theoretical curve by changing this parameter but not the overall shape, which is totally fixed by the calculation of the TLL parameters, amplitudes and general formulas. The remarkable agreement with the experimental curve thus constitute the first *quantitative* test of the TLL theory, where not only asymptotic behaviors with fitting exponents, but absolute the full dependence of observable quantities can be compared with experiments. After Ref. 42.

5.3. *Full dynamical correlations*

The TLL theory allows to compute the dynamical correlations. However in many cases, one can be interested in such correlations for energies comparable to the bandwidth. This for example the case with neutron scattering experiments for which the neutron energy can be much larger than the exchange energy J. It is normally extremely difficult to obtain such correlation functions directly. Exact diagonalization leads usually to small systems, for which a direct comparison with experiments is difficult because of the finite size effects. Quantum Monte Carlo can compute dynamical correlations, but obtains the results in imaginary time. It is then necessary to perform the analytic continuation to real time, using relatively uncontrolled methods such as maximum entropy. Fortunately in one dimension recent progress in the DMRG method have made it possible to compute directly the correlation functions in real time, with an extremely good accuracy. I will not go in detail on the corresponding techniques and relative merits of the various methods and refer the reader to the literature for more details on that point.[5]

The DMRG can be successfully applied in the case of the spin ladder systems, as shown in Fig. 8. The resolution is about $J/20$ in a energy and a comparable figure for the momentum resolution, better than the existing experiments. As a result one can clearly see the fractionalization of excitations and also predict additional features at high energy that can be directly tested experimentally and would be beyond the reach of an asymptotic theory. Of course the numerics fails at long time (low energy) but this is the regime where the analytic field theory (TLL) description is particularly efficient. The overlap between the two techniques is such that one does have a full description of the whole spectrum.

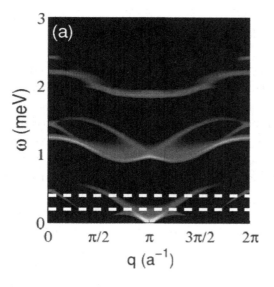

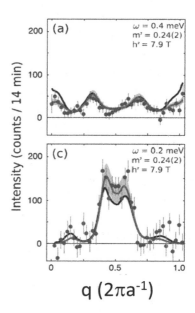

Fig. 8. (left) DMRG calculation of the dynamical spin–spin correlation for a spin ladder. The bottom part is the Des Cloiseaux–Pearson spectrum modified because of the incommensurability induced by the magnetic field. The higher parts of are reminiscicence of the other triplets, and are beyond the reach of the simple TLL description of the system. DMRG proves to be a very efficient way to obtain this part of the spectrum. (right) comparison with two scans of constant intensity ($E = 0.2$ and $E = 0.4$ meV) with neutron scattering experiments done one the compound BPCB. The agreement is excellent showing that we are now able to quantitatively compute such dynamical correlation functions. After Ref. 36.

The possibility to compute such dynamical correlations with an accuracy that is beyond the experimental one opens very interesting perspectives. In particular the neutron spectrum is very rich and one can thus use the comparison between the observed correlation functions and the essentially exact theoretical calculation to reconstruct the Hamiltonian of the system. This program has been successfully carried out for the DIMPY compound[45] for which it allowed to obtain very accurate values for the coupling constants. Such *ab initio* reconstruction of the Hamiltonian is clearly something that will become a useful technique in the future.

One can imagine using similar techniques to probe for fractionalization and spin-charge separation in other type of systems. For cold atomic gases proposals have been made but experiments still remain to be done.

6. Conclusions and Perspectives

One dimensional systems have proven to be extremely rich. It has seen in the recent years extremely important progress both on the theory side and on the experimental one. On the theory side, the development of novel methods both analytical and specially numerical have put us in good position to compute reliable many of the 1D properties that were elusive before. Likewise the explosion of experimental systems in which one dimensional physics is at play allowed for extremely stringent test of such TLL physics.

As a result several of the key features of TLL physics, namely the power-law behavior of the correlation functions and the fractionalization of the excitations have been observed in several experimental systems. Recently *quantitative* tests of the TLL physics could be performed in spin insulators. Such tests, which offered an extremely high level of control, both theoretically and experimentally fully confirmed the TLL predictions, not only for existence of power-laws and fractionalization but also as far as the universality of the dependence of the exponents in terms of the TLL parameters was concerned. Quite interestingly the TLL has proven not only to be accurate in asymptotic regimes but also showed a remarkable accuracy at intermediate energies. More generally the combination of TLL analytical calculations and of numerical DMRG calculation of dynamical correlation functions have given access to the full physics of one dimensional systems, for the whole energy range.

Such control on the theory side clearly opens the door to much more complex studies. In particular it allowed to do reconstruction of the Hamiltonian directly from neutron scattering spectra. Such combination of techniques must now be used to tackle problems which before were extremely hard, such as quansi-one dimensional systems and deconfinement, disordered or frustrated systems and of course out of equilibrium physics. There is no doubt that important progress should come, making it extremely exciting times to look at one dimensional physics.

Acknowledgments

My own understanding of the one dimensional systems and the several works described in these notes, have benefitted from so many people that it would be impossible to mention them all there. The recent works on spin insulators result from collaborations with P. Bouillot and C. Kollath. Several aspects of these works benefitted, on the theory side, from collaborations with S. Capponi, R. Chitra, R. Citro, S. Furuya, A. Laüchli, E. Orignac, B. Normand, M. Oshikawa, D. Poilblanc, A. Tsvelik and, on the experimental side, close interactions and constant discussions with, in particular, M. Klanjcek, M. Horvatic and C. Berthier for the NMR, B. Thielemann and C. Rüegg for Neutrons on BPCB and D. Schmidiger and A. Zheludev for neutrons on DIMPY. This work was supported by the Swiss NSF under MaNEP and Division II.

References

1. T. Giamarchi, in Quantum Physics in one Dimension, *International Series of Monographs on Physics*, Vol. **121** (Oxford University Press, Oxford, UK, 2004).
2. L. D. Landau, *J. Exp. Theor. Phys.* **8**, 70 (1958).

3. P. Nozieres, *Theory of Interacting Fermi Systems* (Benjamin, New York, 1961).
4. F. D. M. Haldane, *Phys. Rev. Lett.* **47**, 1840 (1981).
5. M. A. Cazalilla *et al.*, *Rev. Mod. Phys.* **83**, 1405 (2011).
6. A. Lebed (ed.) *The Physics of Organic Superconductors and Conductors* (Springer, Heidelberg, 2008).
7. D. Jérome, *Chem. Rev.* **104**, 5565 (2004).
8. C. Bourbonnais and D. Jerome, Interacting electrons in quasi-one-dimensional organic superconductors, in *Physics of Organic Superconductors and Conductors*, ed. A. G. Lebed (Springer, Heidelberg, 2008, p. 357).
9. T. Giamarchi, in *Physics of Organic Superconductors and Conductors*, ed. A. G. Lebed (Springer, Heidelberg, 2008), p. 357.
10. T. Giamarchi, *Phys. Rev. B* **44**, 2905 (1991).
11. T. Giamarchi, *Physica B* **230–232**, 975 (1997).
12. A. Schwartz *et al.*, *Phys. Rev. B* **58**, 1261 (1998).
13. W. Henderson *et al.*, *Eur. Phys. J. B* **11**, 365 (1999).
14. M. Dressel *et al.*, *Phys. Rev. B* **71**, 75104 (2005).
15. D. Pashkin, M. Dressel and C. A. Kuntscher, *Phys. Rev. B* **74**, 165118 (2006).
16. D. Pashkin *et al.*, *Phys. Rev. B* **81**, 125109 (2010).
17. Y.-S. Lee *et al.*, *Phys. Rev. Lett.* **94**, 137004 (2005).
18. M. Bockrath *et al.*, *Nature* **397**, 598 (1999).
19. M. Bockrath *et al.*, *Science* **291**, 283 (2001).
20. M. Grioni, S. Pons and E. Frantzeskakis, *J. Phys. C* **21**, 023201 (2009).
21. J. D. Denlinger *et al.*, *Phys. Rev. Lett.* **82**, 2540 (1999).
22. V. Vescoli *et al.*, *Eur. Phys. J. B* **13**, 503 (2000).
23. F. Wang *et al.*, *Phys. Rev. B* **74**, 113107 (2006).
24. J. Hager *et al.*, *Phys. Rev. Lett.* **95**, 186402 (2005).
25. F. Wang *et al.*, *Phys. Rev. Lett.* **103**, 136401 (2009).
26. T. Giamarchi, *Physics* **2**, 78 (2009).
27. A. M. Chang, *Rev. Mod. Phys.* **75**, 1449 (2003).
28. I. Bloch, J. Dalibard and W. Zwerger, *Rev. Mod. Phys.* **80**, 885 (2008).
29. S. Hofferberth *et al.*, *Nat. Phys.* **4**, 489 (2008).
30. L. Pitaevskii and S. Stringari, *Bose-Einstein Condensation* (Clarendon Press, Oxford, 2003).
31. E. Berg *et al.*, *Phys. Rev. B* **77**, 245119 (2009).
32. M. Endres *et al.*, *Science* **334**, 200 (2011).
33. D. A. Tennant *et al.*, *Phys. Rev. B* **52**, 13368 (1995).
34. M. B. Stone *et al.*, *Phys. Rev. Lett.* **91**, 037205 (2003).
35. B. Thielemann *et al.*, *Phys. Rev. Lett.* **102**, 107204 (2009).
36. P. Bouillot *et al.*, *Phys. Rev. B* **83**, 054407 (2011).
37. O. M. Auslaender *et al.*, *Science* **295**, 825 (2002).
38. Y. Tserkovnyak *et al.*, *Phys. Rev. Lett.* **89**, 136805 (2002).
39. A. Kleine *et al.*, *Phys. Rev. A* **77**, 013607 (2007).
40. A. Auerbach, *Interacting Electrons and Quantum Magnetism* (Springer, Berlin, 1998).
41. T. Giamarchi, C. Ruegg and O. Tchernyshyov, *Nat. Phys.* **4**, 198 (2008).
42. M. Klanjsek *et al.*, *Phys. Rev. Lett.* **101**, 137207 (2008).
43. B. Thielemann *et al.*, *Phys. Rev. B* **79**, 020408(R) (2009).
44. S. C. Furuya *et al.*, *Phys. Rev. Lett.* **108**, 037204 (2012).
45. D. Schmidiger *et al.*, *Phys. Rev. Lett.* **108**, 167201 (2012).

BOSONIZATION AND ITS APPLICATION TO TRANSPORT IN QUANTUM WIRES*

FEIFEI LI

*Department of Physics and Astronomy, Northwestern University,
2145 Sheridan Rd, Evanston, Illinois 60208, USA
feifei-li@northwestern.edu*

This paper reviews the method of bosonization for one-dimensional interacting fermions. Example of its application to mesoscopic systems is presented with an explicit calculation of the spin and charge currents in an interacting quantum wire subject to a uniform magnetic field and alternating voltage source. Special emphasis is given to the details of the calculation using bosonization and Keldysh techniques.

Keywords: Bosonization; mesoscopic systems; Luttinger liquids; quantum wires; spin current; rectification; Keldysh technique; nonequilibrium transport.

1. Introduction: The Breakdown of Fermi Liquid Theory in One Dimension

Landau's Fermi liquid theory is a successful theory that accounts for the low-energy behavior of a large class of conventional metals as well as other Fermi systems such as Helium-3 and nuclear matters.[1] One of the surprises that we learned from this theory is that, in three spatial dimensions, the low-energy physics of an interacting Fermi system can be essentially similar to a noninteracting Fermi gas, regardless of the strength and details of the interactions. In a Fermi liquid, the low-energy excitations are quasi-particles that can be classified in the same way as a Fermi gas. This picture relies on the long lifetime of the quasi-particles, which can be traced back to the fact that, in three dimensions, the simultaneous conservation of energy and momentum plus Pauli exclusion significantly suppresses the effectiveness of electron–electron interactions. This can be understood in the following phase space argument due to Landau. Let us begin with an excited electron in a reference noninteracting Fermi gas, and construct the fully interacting system by gradually switching-on the interactions. At the beginning, the excited electron has a momentum $\mathbf{k}_1$ above the Fermi surface ($k_1 > k_F$). During the switching-on process, it will collide with one of the electrons below the Fermi surface ($k_2 < k_F$). Momentum and energy of the initial state is changed as a result of the collision, hence the excitation has a finite lifetime. Let us be more precise. A general scattering process has to conserve

---

*This article first appeared in Modern Physics Letters B, Vol. 27, No. 11 (2013).

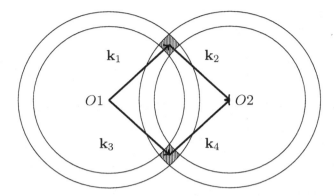

Fig. 1. Phase space argument for a general scattering process near the Fermi surface. The conservation of momentum and energy significantly limits the phase space of the scattering.

momentum and energy

$$\mathbf{k}_1 + \mathbf{k}_2 = \mathbf{k}_3 + \mathbf{k}_4, \tag{1}$$

$$k_1^2 + k_2^2 = k_3^2 + k_4^2, \tag{2}$$

where $\mathbf{k}_2$ and $\mathbf{k}_3$ are the final momentums of the two colliding electrons. The process is depicted in Fig. 1, where the conservation of momentum is manifest. In addition to the conservation laws, Paul exclusion requires that $k_3, k_4 > k_F$ because the states below the Fermi surface are already occupied. Now for a low-energy process that involves only the electrons near the Fermi surface $k_1, k_2, k_3, k_4 \sim k_F$, it is obvious from the geometry in Fig. 1 that the electrons cannot have arbitrary final momentums. Instead, there is only a small region (hatched in Fig. 1) for them to be scattered into. This region has a volume $\sim k_F(k_1 - k_F)^2 \sim q^2$ where $q = k_1 - k_F$, and it shrinks as one gets closer to the Fermi surface. Thus, from Fermi's Golden rule, the rate of scattering goes as q^2, hence the lifetime $\tau(q) \sim 1/q^2$. In addition, the excitation has an energy $E(q) \sim q$. Therefore, we expect that the wave function of the quasi-particle has a time-dependence $\exp[-iE(q)t - t/\tau(q)]$. In the limit that $q \to 0$, the lifetime $\tau(q) \sim 1/q^2$ becomes much larger than the oscillation period $1/E(q) \sim 1/q$. Thus, effectively, the excited electron $\mathbf{k}_1$ has a diverging lifetime, and the concept of quasi-particle is better defined as one gets closer and closer to the Fermi surface.

It is important to appreciate that the above phase space argument is a consequence of the tight constraints imposed by Fermi statistics and the conservation of momentum and energy in three spatial dimensions. In one dimension, however, situation is every different. The Fermi surface collapses to two Fermi points (see Fig. 2) near which the dispersion is linear $\Delta E \sim \Delta q$. Because there is no angle in one dimension, conservation of energy automatically implies conservation of momentum for such a linear dispersion. As a result, the phase space is much less constrained, and the scattering becomes so effective that a fermion-like single-particle excitation

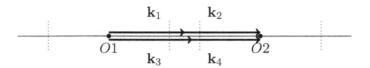

Fig. 2. Phase space for a scattering process in one dimension. In contrast to higher dimensions, the phase space is not highly constrained because simultaneous conservation of momentum and energy is guaranteed by the linear dispersion near the Fermi points.

has vanishing lifetime, in contrast to the situation in three dimensions. The differences between one dimension and higher dimensions is apparent in the real space. In three dimensions, it is possible for an electron to move without significantly affecting all the electrons in the system. But in one dimension, electrons can only move along a line. There is no way to move one without affecting all the others. Consequently, excitations in one dimension must be very different. They have to be collective ones, like density waves. Density wave is bosonic in nature. This observation suggests that one may understand a one-dimensional fermion system from a bosonic perspective. Indeed, the system can be solved by a method called bosonization, and we will review this technique in the next two sections.

2. Tomonaga–Luttinger Model

The surprise coming from the study of one-dimensional interacting fermion systems is that it is essentially solvable (for the low-energy physics), even in the presence of interactions. Historically, the investigation began with Tomonaga[2] and later Luttinger[3] who introduced the Tomonaga–Luttinger model. The correct solution of this model was first obtained, however, by Mattis and Lieb[4] in 1965. The method received further development during the 70s.[5] A systematic framework was eventually put together by Haldane,[6] who coined the term "Luttinger liquid" to describe the highly collective nature of one-dimensional fermions.

The Tomonaga–Luttinger model has been widely used to study one-dimensional electronic systems. The model shares the same low-energy physics of a realistic fermion system while having the advantage of being exactly solvable. To arrive at this model, we start with the ground state of a noninteracting fermion system described by

$$\psi_\sigma(x) = \frac{1}{\sqrt{L}} \sum_{k=-\infty}^{\infty} e^{ikx} C_\sigma(k), \qquad \text{with} \ \ k = \frac{2\pi}{L} n_k \ \text{ and } \ n_k \in \mathbb{Z}. \tag{3}$$

Here $\sigma = \uparrow, \downarrow$ is the spin index, L is the length of the system, and the momentum k takes discrete values in the finite system. $L \to \infty$ will be taken at the end of the calculation. Assuming the usual parabolic dispersion $E(k) = k^2/2m$ ($\hbar$ is set to 1),

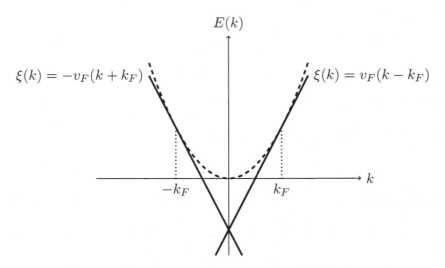

Fig. 3. The Tomonaga–Luttinger model replaces the original parabolic dispersion (dashed curve) by left and right moving branches (solid lines). The low-energy physics near the Fermi level is not affected by this extension.

the noninteracting system has the following Hamiltonian:

$$H = \sum_{\sigma=\uparrow,\downarrow} \sum_{k=-\infty}^{\infty} \xi(k) C_\sigma^\dagger(k) C_\sigma(k) \,. \tag{4}$$

Here the excitation spectrum $\xi(k) = E(k) - E(k_F)$ is measured from the Fermi level. It is linear near the left and right Fermi points

$$\xi(k) \approx \begin{cases} v_F q & k \approx k_F \\ -v_F q & k \approx -k_F \end{cases}, \tag{5}$$

where $v_F = k_F/m$ is the Fermi velocity, $q = k - k_F$ for $k \approx k_F$, and $q = k + k_F$ for $k \approx -k_F$. The Tomonaga–Luttinger model is obtained by extending this linear dispersion to infinity as shown in Fig. 3. This extension forces us to introduce two species of fermions for each spin σ: left movers $\psi_{L\sigma}(x)$ and right movers $\psi_{R\sigma}(x)$. In addition, we have to assume that, for each species, all states below the Fermi level are filled. Despite these differences, the low energy physics given by the Tomonaga–Luttinger model is the same as our original one, because the excitations far below the Fermi level are prohibited by Pauli exclusion, and the ones far above cannot be excited. Physically, the approximation corresponds to represent the original fermion field $\psi_\sigma(x)$ by left and right moving pieces

$$\psi_\sigma(x) \to \frac{1}{\sqrt{L}} \sum_{k \approx k_F} e^{ikx} C_\sigma(k) + \frac{1}{\sqrt{L}} \sum_{k \approx -k_F} e^{ikx} C_\sigma(k)$$

$$\to \psi_{R\sigma}(x) + \psi_{L\sigma}(x) \tag{6}$$

with

$$\psi_{r\sigma}(x) = \frac{1}{\sqrt{L}} \sum_{k=-\infty}^{\infty} e^{ikx} C_{r\sigma}(k)\,. \tag{7}$$

We have used $r = R$ for right movers and $r = L$ for left movers. Throughout this paper, we will also use the convention that $r = 1$ for $r = R$ and $r = -1$ for $r = L$ in a mathematical expression. The original Hamiltonian H is now replaced by the linearized Tomonaga–Luttinger Hamiltonian that describes four species of fermions,

$$H \rightarrow H_{\mathrm{TL}} = \sum_{r=R,L} \sum_{\sigma=\uparrow,\downarrow} H_{r\sigma} \tag{8}$$

with

$$H_{r\sigma} = \sum_{k=-\infty}^{\infty} r v_F k C_{r\sigma}^{\dagger}(k) C_{r\sigma}(k) \tag{9}$$

$$= \int_0^L dx \psi_{r\sigma}^{\dagger}(x)(-i r v_F \partial_x) \psi_{r\sigma}(x)\,. \tag{10}$$

Interaction can be easily introduced in this model. The general density-density interaction has the form

$$V = \sum_{r,r'} \sum_{\sigma,\sigma'} \int dx dx' V_{r\sigma,r'\sigma'}(x - x') \rho_{r\sigma}(x) \rho_{r'\sigma'}(x')\,, \tag{11}$$

where $\rho_{r\sigma}(x) = \psi_{r\sigma}^{\dagger}(x) \psi_{r\sigma}(x)$ is the density of the species $r\sigma$.

3. Bosonization

The Tomonaga–Luttinger model can be solved by the method of bosonization, even in the presence of strong interactions. Bosonization is a powerful tool for studying one-dimensional fermion systems. The basic idea is that, in one dimension, the Hamiltonian of a fermionic system can be mapped into a bosonic one. Quantities like the correlation functions that are difficult to calculate in the fermionic language turns out to be easy or even trivial to obtain in the bosonic language. There have been a number of reviews on this subject.[7–14] Here we provide an elementary introduction that allows to set stage for the calculation of transport in one-dimensional quantum wires. The interested reader is referred to the above-mentioned references for a thorough review.

Before proceeding to the details of bosonization, we need to address one subtle point in the Tomonaga–Luttinger model. The introduction of the left and right movers has forced us to allow the presence of infinitely many negative energy states.

This is necessary to make the model solvable. But it does come with a cost. The infinitely deep "vacuum" introduces various infinities. One needs to be very careful in defining the operators to avoid such infinities. Introducing the normal ordered product of operators $: ABC :$ does that. In a normal ordered product, the destruction operators are put to the right-hand side of the creation operators. For the case that A and B are linear combinations of creation and destruction operators (which is always the case here), normal ordering is equivalent to subtracting its vacuum average

$$: AB : = AB - \langle 0|AB|0 \rangle \,. \tag{12}$$

Normal ordered operators are bounded and well-behaved.

We are now in the position to motivate bosonization. Recall that only collective density-wave-like excitations can exist in one dimension, we start by looking at the normal ordered fermion density $: \rho_{r\sigma}(x) := : \psi_{r\sigma}^\dagger(x)\psi_{r\sigma}(x) :$ which can be written as a Fourier sum

$$
\begin{aligned}
: \rho_{r\sigma}(x) : &= \frac{1}{L} \sum_{k,k'=-\infty}^{\infty} e^{-i(k-k')x} : C_{r\sigma}^\dagger(k)C_{r\sigma}(k') : \\
&= \frac{1}{L} \sum_{p=-\infty}^{\infty} e^{irpx} \sum_{k=-\infty}^{\infty} : C_{r\sigma}^\dagger(k)C_{r\sigma}(k+rp) : \\
&= \frac{1}{L} \sum_{k=-\infty}^{\infty} : n_{r\sigma}(k) : + \frac{1}{L} \sum_{p>0} [e^{irpx}\rho_{r\sigma}(p) + e^{-irpx}\rho_{r\sigma}^\dagger(p)] , \tag{13}
\end{aligned}
$$

where we have defined, for $p > 0$, the Fourier component

$$\rho_{r\sigma}(p) = \sum_{k=-\infty}^{\infty} C_{r\sigma}^\dagger(k)C_{r\sigma}(k+rp) \quad (\text{for } p > 0) , \tag{14}$$

and the occupation number operator $n_{r\sigma}(k) = C_{r\sigma}^\dagger(k)C_{r\sigma}(k)$. Normal ordering has no effect on $\rho_{r\sigma}(p)$ since the vacuum average is zero for $p > 0$. Important insight can be obtained by looking at the commutator

$$
\begin{aligned}
\left[\rho_{r\sigma}(p), \rho_{r'\sigma'}^\dagger(p') \right] &= \sum_{k,k'=-\infty}^{\infty} \left[C_{r\sigma}^\dagger(k)C_{r\sigma}(k+rp), C_{r'\sigma'}^\dagger(k'+r'p')C_{r'\sigma'}(k') \right] \\
&= \delta_{r,r'}\delta_{\sigma,\sigma'} \sum_{k=-\infty}^{\infty} \left(C_{r\sigma}^\dagger(k)C_{r\sigma}(k+r(p-p')) \right. \\
&\quad \left. - C_{r\sigma}^\dagger(k+rp')C_{r\sigma}(k+rp) \right) . \tag{15}
\end{aligned}
$$

Usually, one would perform a change of variable $k \to k + rp'$ in the first term of Eq. (15) and finds that the commutator vanishes. In the Tomonaga–Luttinger model, this substitution is valid if $p \neq p'$ since the operator $C_{r\sigma}^\dagger(k)C_{r\sigma}(k+r(p-p'))$ is bounded. This is no longer the case when $p = p'$, as the first term becomes the occupation number operator $n_{r\sigma}(k)$ so that the commutator is the subtraction of two different infinities. To proceed, one needs to isolate these infinities using normal ordered operators and write

$$C_{r\sigma}^\dagger(k)C_{r\sigma}(k) = \; : C_{r\sigma}^\dagger(k)C_{r\sigma}(k) : \; + \langle 0 | C_{r\sigma}^\dagger(k)C_{r\sigma}(k) | 0 \rangle \tag{16}$$

$$= \; : C_{r\sigma}^\dagger(k)C_{r\sigma}(k) : \; + \Theta(k_F - rk) \,, \tag{17}$$

where $\Theta(x)$ is the Heaviside step function. The vacuum is filled from $-\infty$ to k_F for the right movers, and from $+\infty$ to $-k_F$ for the left movers. Change of variable can now be performed to cancel the normal ordered densities, and one discovers a nonvanishing commutator for $p = p'$

$$\left[\rho_{r\sigma}(p), \rho_{r'\sigma'}^\dagger(p') \right] = \delta_{r,r'}\delta_{\sigma,\sigma'}\delta_{p,p'} \sum_{k=-\infty}^{\infty} \left[\Theta(k_F - rk) - \Theta(k_F - r(k+rp)) \right]$$

$$= \delta_{r,r'}\delta_{\sigma,\sigma'}\delta_{p,p'} \frac{Lp}{2\pi} \,. \tag{18}$$

The fact that the fermion densities do not commute is the key to correctly solve the Tomonaga–Luttinger model. It suggests that we define, for $p > 0$, the boson creation and destruction operators:

$$b_{r\sigma}(p) = \sqrt{\frac{2\pi}{Lp}}\rho_{r\sigma}(p) = \sqrt{\frac{2\pi}{Lp}} \sum_{k=-\infty}^{\infty} C_{r\sigma}^\dagger(k)C_{r\sigma}(k+rp) \,, \tag{19}$$

$$b_{r\sigma}^\dagger(p) = \sqrt{\frac{2\pi}{Lp}}\rho_{r\sigma}^\dagger(p) = \sqrt{\frac{2\pi}{Lp}} \sum_{k=-\infty}^{\infty} C_{r\sigma}^\dagger(k+rp)C_{r\sigma}(k) \,, \tag{20}$$

which obviously satisfy the desired boson commutations

$$[b_{r\sigma}(p), b_{r'\sigma'}^\dagger(p')] = \delta_{r,r'}\delta_{\sigma,\sigma'}\delta_{p,p'} \,. \tag{21}$$

The bosonic nature of the one-dimensional fermion system can now be revealed. Simple calculation gives us

$$[b_{r\sigma}(p), H_{r\sigma}] = v_F p \, b_{r\sigma}(p) \,, \tag{22}$$

$$[b_{r\sigma}^\dagger(p), H_{r\sigma}] = -v_F p \, b_{r\sigma}^\dagger(p) \,. \tag{23}$$

But the only bosonic form of $H_{r\sigma}$ that can give the above commutators is the quadratic expression

$$H_{r\sigma} = \sum_{p>0} v_F p \, b_{r\sigma}^\dagger(p)b_{r\sigma}(p) \,. \tag{24}$$

This is a remarkable result. It is the reason behind the solvability of a one-dimensional interacting fermionic system. As we know, what makes an interacting problem so difficult is that the interaction terms are quartic in the fermion operators. However, expressed in the boson operators $b_{r\sigma}$, the interaction becomes quadratic. This of course does not help much in the usual situation because the kinetic part of the Hamiltonian will then become not solvable. What is special about one dimension is that the kinetic part is also quadratic in the boson operators, hence the full Hamiltonian can be trivially diagonalized.

We can even express the fermion field $\psi_{r\sigma}$ in terms of the boson operators. To see this, one calculate the commutator

$$[\psi_{r\sigma}(x), b_{r'\sigma'}(p)] = \sum_{k,k'=-\infty}^{\infty} \frac{1}{\sqrt{L}} e^{ikx} \sqrt{\frac{2\pi}{Lp}} [C_{r\sigma}(k), C_{r'\sigma'}^{\dagger}(k')C_{r'\sigma'}(k'+r'p)]$$

$$= \delta_{r,r'}\delta_{\sigma,\sigma'} \sqrt{\frac{2\pi}{Lp}} e^{-irpx}\psi_{r\sigma}(x). \tag{25}$$

Comparing to the operator identity

$$[e^A, B] = [A, B]e^A \quad \text{(if $[A, B]$ commutes with A and B)}, \tag{26}$$

Eq. (25) suggests that the fermion field can be written as the exponential of a bunch of boson operators. An expression that produces Eq. (25) is

$$\psi_{r\sigma}(x) = e^{J_{r\sigma}(x)} \tag{27}$$

with

$$J_{r\sigma}(x) = \sum_{p>0} \sqrt{\frac{2\pi}{Lp}} [e^{irpx}b_{r\sigma}(p) - e^{-irpx}b_{r\sigma}^{\dagger}(p)]. \tag{28}$$

This can be easily checked:

$$[e^{J_{r\sigma}(x)}, b_{r'\sigma'}(p)] = [J_{r\sigma}(x), b_{r'\sigma'}(p)]e^{J_{r\sigma}}(x)$$

$$= \left(\sum_{p'>0} \sqrt{\frac{2\pi}{Lp'}} e^{-irp'x}[-b_{r\sigma}^{\dagger}(p'), b_{r'\sigma'}(p)] \right) e^{J_{r\sigma}}(x)$$

$$= \delta_{r,r'}\delta_{\sigma,\sigma'} \sqrt{\frac{2\pi}{Lp}} e^{-irpx}e^{J_{r\sigma}(x)}. \tag{29}$$

However, Eq. (27) has two flaws. First, on the left-hand side, $\psi_{r\sigma}(x)$ changes the number of fermions of species $r\sigma$ by one. On the right-hand side, however, the expression contains $b_{r\sigma}$ and $b_{r\sigma}^{\dagger}$ which conserve the number of fermions. Therefore, we need an operator $\hat{\eta}_{r\sigma}$ that commutes with the boson operators and removes one fermion of species $r\sigma$. This operator is called Klein factor. In most situations, the

Klein factors drop out in perturbation calculations, so they are sometimes ignored in the literature. However, there are situations in which special care of these operators is needed.[15] In this paper, we will deal with the cases where the Klein factors do not play a role. Expression (27) has another problem. It can be shown that it does not reproduce the correct free fermion correlation function which should be

$$
\begin{aligned}
\langle \psi_{r\sigma}^\dagger(x)\psi_{r\sigma}(x')\rangle &= \frac{1}{L}\int_{-\infty}^{\infty}\frac{dk}{2\pi/L}e^{-ik(x-x')}\Theta(k_F - rk) \\
&= \int_{-\infty}^{k_F}\frac{dk}{2\pi}e^{-irk(x-x')+k\delta} \quad (\delta \to 0^+ \text{ added for convergence}) \\
&= \frac{1}{2\pi}\frac{e^{-irk_F(x-x')}}{\delta - ir(x-x')}.
\end{aligned}
\tag{30}
$$

Therefore, additional modification is needed. It was found that the following slightly modified expression does what we want:[5]

$$
\psi_{r\sigma}(x) = \hat{\eta}_{r\sigma}\frac{1}{\sqrt{2\pi\delta}}e^{irk_F x + J_{r\sigma}(\delta,x)},
\tag{31}
$$

where

$$
J_{r\sigma}(\delta, x) = \sum_{p>0}e^{-\delta p/2}\sqrt{\frac{2\pi}{pL}}[e^{irpx}b_{r\sigma}(p) - e^{-irpx}b_{r\sigma}^\dagger(p)].
\tag{32}
$$

Here, the same δ that appears in Eq. (30) also ensures the convergence of $J_{r\sigma}(\delta, x)$. The limit $\delta \to 0^+$ should be taken at the end of the calculation. We now verify that the new expression gives the correct free fermion correlation function. We have

$$
\begin{aligned}
\langle \psi_{r\sigma}^\dagger(x)\psi_{r\sigma}(x')\rangle &= \frac{e^{-irk_F(x-x')}}{2\pi\delta}\langle e^{J_{r\sigma}^\dagger(\delta,x)}e^{J_{r\sigma}(\delta,x')}\rangle \\
&= \frac{e^{-irk_F(x-x')}}{2\pi\delta} \\
&\quad \times \prod_{p>0}\langle e^{-\beta_p(x)b_{r\sigma}(p)+\beta_p^*(x)b_{r\sigma}^\dagger(p)}e^{\beta_p(x')b_{r\sigma}(p)-\beta_p^*(x')b_{r\sigma}^\dagger(p)}\rangle
\end{aligned}
\tag{33}
$$

with

$$
\beta_p(x) = e^{-\delta p/2}\sqrt{\frac{2\pi}{pL}}e^{irpx}.
\tag{34}
$$

We then use

$$
e^A e^B = e^{A+B}e^{[A,B]/2} \quad (\text{if } [A,B] \text{ commutes with } A \text{ and } B)
\tag{35}
$$

to move the destruction operators to the right and get

$$\langle \psi_{r\sigma}^\dagger(x)\psi_{r\sigma}(x)\rangle = \frac{e^{-irk_F(x-x')}}{2\pi\delta}\prod_{p>0}e^{\frac{1}{2}[-|\beta_p(x)|^2-|\beta_p(x')|^2+2\beta_p^*(x')\beta_p(x)]}$$

$$\times \underbrace{\langle e^{[\beta_p^*(x)-\beta_p^*(x')]b_{r\sigma}^\dagger(p)}e^{[\beta_p(x')-\beta_p(x)]b_{r\sigma}(p)}\rangle}_{=1}$$

$$= \frac{1}{2\pi\delta}e^{-irk_F(x-x')-G_{r\sigma}(x-x')}, \tag{36}$$

where

$$G_{r\sigma}(x-x') = \frac{1}{2}\sum_{p>0}\left[|\beta_p(x)|^2+|\beta_p(x')|^2-2\beta_p^*(x')\beta_p(x)\right]$$

$$= \sum_{p>0}\frac{2\pi}{pL}[e^{-p\delta}-e^{p[ir(x-x')-\delta]}]$$

$$= \sum_{n=1}^{\infty}\frac{1}{n}[(e^{-\frac{2\pi}{L}\delta})^n-(e^{\frac{2\pi}{L}[ir(x-x')-\delta]})^n]$$

$$= -\ln[1-e^{-\frac{2\pi}{L}\delta}]+\ln[1-e^{\frac{2\pi}{L}[ir(x-x')-\delta]}]$$

$$\stackrel{L\to\infty}{=} \ln\left[1-\frac{ir(x-x')}{\delta}\right]. \tag{37}$$

In Eq. (37), we have used the fact that the momentum is quantized $p = \frac{2\pi}{L}n$, and also used the Taylor expansion formula $\sum_{n=1}^{\infty}y^n/n = -\ln(1-y)$. As promised, Eq. (36) is the same as Eq. (30).

It is convenient to define a field

$$\phi_{r\sigma}(x) = -irJ_{r\sigma}(\delta,x)$$

$$= -ir\sum_{p>0}e^{-\delta p/2}\sqrt{\frac{2\pi}{pL}}[e^{irpx}b_{r\sigma}(p)-e^{-irpx}b_{r\sigma}^\dagger(p)] \tag{38}$$

in terms of which the bosonized fermion field is expressed as

$$\psi_{r\sigma}(x) = \frac{\hat{\eta}_{r\sigma}}{\sqrt{2\pi\delta}}e^{ir[k_Fx+\phi_{r\sigma}(x)]}. \tag{39}$$

Equation (39) is called *bosonization identity*. This identity establishes the mapping between the fermionic and bosonic representations in one dimension. It is possible to give a more rigorous proof of this identity, which requires a careful construction of the Klein factors as well as proving that the two representations give identical eigenvalues and degeneracies.[6,10,14] The physical meaning of the field $\phi_{r\sigma}(x)$ can be understood from its spatial gradient

$$\partial_x \phi_{r\sigma}(x) = \sum_{p>0} e^{-\delta p/2} \sqrt{\frac{2\pi}{pL}} p \left[e^{irpx} b_{r\sigma}(p) + e^{-irpx} b_{r\sigma}^\dagger(p) \right]$$

$$\stackrel{\delta \to 0}{=} \frac{2\pi}{L} \sum_{p>0} \left[e^{irpx} \rho_{r\sigma}(p) + e^{-irpx} \rho_{r\sigma}^\dagger(p) \right]. \tag{40}$$

Comparing it with Eq. (13), we find

$$: \rho_{r\sigma}(x) := \frac{1}{L} \sum_{k=-\infty}^{\infty} : n_{r\sigma}(k) : + \frac{1}{2\pi} \partial_x \phi_{r\sigma}(x)$$

$$\stackrel{L \to \infty}{=} \frac{1}{2\pi} \partial_x \phi_{r\sigma}(x). \tag{41}$$

Thus, $\partial_x \phi_{r\sigma}(x)/2\pi$ is just the fermion density at x. Further more, it is not difficult to show that the noninteracting Hamiltonian $H_{r\sigma}$ can also be expressed nicely in $\phi_{r\sigma}(x)$:

$$H_{r\sigma} = \frac{v_F}{4\pi} \int_0^L dx : (\partial_x \phi_{r\sigma}(x))^2 : . \tag{42}$$

Finally, applying the same trick that was used in Eq. (37) to sum over the momentums, we can calculate various quantities of interest, for example, the commutators for $\phi_{r\sigma}$

$$[\phi_{r\sigma}(x), \phi_{r'\sigma'}(x')] \stackrel{L \to \infty}{=} \delta_{r,r'} \delta_{\sigma,\sigma'} \ln \left[\frac{\delta + ir(x - x')}{\delta - ir(x - x')} \right]$$

$$\stackrel{\delta \to 0}{=} \delta_{r,r'} \delta_{\sigma,\sigma'} i\pi r \, \text{sign}(x - x'), \tag{43}$$

and the zero temperature correlation functions

$$\langle \phi_{r\sigma}(x) \phi_{r'\sigma'}(x') \rangle \stackrel{L \to \infty}{=} -\delta_{r,r'} \delta_{\sigma,\sigma'} \ln \left[\frac{2\pi}{L} \left[\delta - ir(x - x') \right] \right], \tag{44}$$

or, for a general space-time separation,

$$\langle \phi_{r\sigma}(x,t) \phi_{r'\sigma'}(x',t') \rangle \stackrel{L \to \infty}{=} -\delta_{r,r'} \delta_{\sigma,\sigma'} \ln \left[\frac{2\pi}{L} \left[\delta - ir(x - x') + iv_F(t - t') \right] \right], \tag{45}$$

which can be obtained directly from Eq. (44) with $x \to x - rv_F t$.

4. Example: One-Dimensional Spinless System

We now demonstrate the power of bosonization with a simple example of interacting spinless fermions. The kinetic energy is the sum of the left and right moving fermions, and by Eq. (42) (with normal ordering taken implicitly)

$$K = \sum_{r=R,L} \int_0^L dx\, \psi_r(x)^\dagger (-irv_F \partial_x) \psi_r(x)$$

$$= \frac{v_F}{4\pi} \int_0^L dx \left[(\partial_x \phi_R)^2 + (\partial_x \phi_L)^2 \right]. \tag{46}$$

Next, we introduce the linear combinations

$$\phi(x) = \frac{1}{2}[\phi_L(x) + \phi_R(x)], \tag{47}$$

$$\theta(x) = \frac{1}{2}[\phi_L(x) - \phi_R(x)]. \tag{48}$$

The reason we introduce these fields is that they satisfy the commutation relations

$$[\phi(x), \phi(x')] = [\theta(x), \theta(x')] = 0, \tag{49}$$

$$[\phi(x), \theta(x')] = -\frac{i\pi}{2} \operatorname{sign}(x - x'), \tag{50}$$

which is a consequence of Eq. (43). This means $\Pi(x) = \frac{1}{\pi}\partial_x\theta(x)$ is the conjugate momentum of $\phi(x)$

$$[\phi(x), \Pi(x')] = i\delta(x - x'). \tag{51}$$

The kinetic energy is already diagonalized in these new fields

$$K = \frac{v_F}{2\pi} \int_0^L dx [(\partial_x \phi)^2 + (\pi\Pi)^2]. \tag{52}$$

In order to calculate physical quantities, one needs correlation functions. We will focus on the zero temperature limit. The correlation function with respect to K $\langle \phi(x,t)\phi(x,t') \rangle$ can be directly calculated from Eq. (45). One finds

$$\langle \phi(x,t)\phi(x',t') \rangle = -\frac{1}{4}\ln\left[\left(\frac{2\pi}{L} \right)^2 ((\delta + iv_F(t - t'))^2 + (x - x')^2) \right], \tag{53}$$

and

$$\langle \theta(x,t)\theta(x',t') \rangle = -\frac{1}{4}\ln\left[\left(\frac{2\pi}{L} \right)^2 ((\delta + iv_F(t - t'))^2 + (x - x')^2) \right]. \tag{54}$$

We now introduce interactions. For illustration purpose, consider a simple short-ranged density–density interaction

$$V = \int_0^L dx \left[\frac{g_4}{2} \rho_R^2(x) + \frac{g_4}{2} \rho_L^2(x) + g_2 \rho_R(x)\rho_L(x) \right], \tag{55}$$

where g_4 and g_2 describe the interaction strength between the same and different species, respectively. The problem is no longer solvable in the fermionic language because V contains four fermion fields. However, Eq. (41) tells us that $\rho_r(x)$ is linear in $\phi_r(x)$ thus the full interacting Hamiltonian remains quadratic in $\phi(x)$ and $\theta(x)$

$$H_{\text{spinless}} = K + V$$

$$= \int_0^L dx \left\{ \left(\frac{v_F}{4\pi} + \frac{g_4}{8\pi^2} \right) [(\partial_x \phi_R)^2 + (\partial_x \phi_L)^2] + \frac{g_2}{4\pi^2} (\partial_x \phi_R)(\partial_x \phi_L) \right\}$$

$$= \frac{1}{2\pi} \int_0^L dx \left\{ \left(v_F + \frac{g_4}{2\pi} + \frac{g_2}{2\pi} \right) (\partial_x \phi)^2 + \left(v_F + \frac{g_4}{2\pi} - \frac{g_2}{2\pi} \right) (\partial_x \theta)^2 \right\}$$

$$= \frac{u}{2\pi} \int_0^L dx [g^{-1}(\partial_x \phi)^2 + g(\pi \Pi)^2], \tag{56}$$

where

$$u = \sqrt{\left(v_F + \frac{g_4}{2\pi} \right)^2 - \left(\frac{g_2}{2\pi} \right)^2}, \tag{57}$$

$$g = \sqrt{\frac{v_F + g_4/2\pi - g_2/2\pi}{v_F + g_4/2\pi + g_2/2\pi}}. \tag{58}$$

Here u has the dimension of velocity, and it is the velocity of the bosonic excitations in the interacting system. g characterizes the strength of the interaction. For repulsive interaction, $g < 1$, for attractive interaction, $g > 1$, and for a noninteracting system, $g = 1$.

There are many ways to obtain the correlation functions with respect to the ground state of Eq. (56) $\langle \phi(x,t)\phi(x',t') \rangle$ and $\langle \theta(x,t)\theta(x',t') \rangle$. Here, an easy way is to observe that the transformation $v_F \to u, \phi(x) \to g^{-1/2}\phi(x), \theta(x) \to g^{1/2}\theta(x)$ turns Eq. (52) into Eq. (56) and at the same time preserves the canonical commutation Eq. (51). This immediately translates Eqs. (53) and (54) into

$$\langle \phi(x,t)\phi(x',t') \rangle = -\frac{g}{4} \ln \left[\left(\frac{2\pi}{L} \right)^2 ((\delta + iv_F(t-t'))^2 + (x-x')^2) \right], \tag{59}$$

and

$$\langle \theta(x,t)\theta(x',t')\rangle = -\frac{1}{4g}\ln\left[\left(\frac{2\pi}{L}\right)^2((\delta + iv_F(t-t'))^2 + (x-x')^2)\right]. \quad (60)$$

Thus, the interacting system is solved.

5. Application: Charge and Spin Currents in Quantum Wires

We now move on to apply the bosonization technique to transport in one-dimensional electronic systems. The example given here is a calculation of the charge and spin current rectified by an asymmetric impurity potential in an interacting quantum wire subject to a uniform magnetic field and alternating voltage source. The system has been studied in Ref. 16 where the combination of bosonization and Keldysh technique allows us to conveniently calculate transport properties. The same calculation technique has also been used to study transport along quantum Hall edges.[17] The focus of the rest of this paper is to go over the essential steps involved in this type of calculations. Some of the results (mainly the renormalization group analysis that identifies interesting parameter regime and relevant operators) will be stated without proof. Interested reader may find the proofs in the original paper.[16]

5.1. *Model and physics of the problem*

The system we want to investigate is sketched in Fig. 4. It consists of a one-dimensional conductor (quantum wire) with a scatterer in the center of the system at $x = 0$. The scatterer creates an asymmetric potential $U(x) \neq U(-x)$. The size of the scatterer $a_U \sim 1/k_F$ is of the order of the electron wavelength. We also assume that the system is placed in a uniform magnetic field $\mathbf{H}$. The magnetic field defines the S_z direction of the electron spins. At its two ends, the quantum wire is connected to nonmagnetic electrodes, labeled by $i = 1, 2$. The left electrode, $i = 1$, is controlled by an AC voltage source V, while the right electrode, $i = 2$, is kept

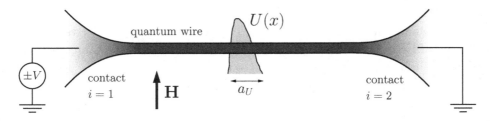

Fig. 4. Sketch of the one-dimensional conductor connected to two electrodes on both ends. Currents are driven through a voltage bias V that is applied to the left electrode while the right electrode is kept at ground. The system is magnetized by the field $\mathbf{H}$. Electrons are backscattered off the asymmetric potential $U(x)$. $U(x) \neq 0$ in the region of size $a_U \sim 1/k_F$.

on ground. Thus, an AC charge current $I_c(V)$ will flow through the quantum wire. In addition, an AC spin current $I_s(V)$ may be generated since the time-reversal symmetry is broken by the magnetic field. In a uniform wire, the magnetic field would not result in a net spin current since the conductances of the spin-up and -down channels would be the same,[18–20] e^2/h ($\hbar$ is restored from now on), and the spin currents of the spin-up and -down electrons would be opposite. In the presence of a potential barrier such a cancellation does not occur.[21] In a system with strong electron interaction, the spatial asymmetry of the wire leads to an asymmetric I–V curve,[22] $I(V) \neq I(-V)$. Thus, an AC voltage bias generates spin and charge DC currents, $I_s^r(V)$ and $I_c^r(V)$. We will focus on the low-frequency AC bias. We define the rectification current as the DC response to a low-frequency square voltage wave of amplitude V:

$$I_s^r(V) = [I_s(V) + I_s(-V)]/2\,, \tag{61}$$

$$I_c^r(V) = [I_c(V) + I_c(-V)]/2\,. \tag{62}$$

The above DC currents express via the currents of spin-up and -down electrons: $I_c^r = I_\uparrow^r + I_\downarrow^r$, $I_s^r = (\hbar/2e)[I_\uparrow^r - I_\downarrow^r]$. Equations (61) and (62) for the DC-currents do not contain the frequency ω of the AC-bias. They are valid as long as the frequency $\omega < eV/\hbar$ which for a typical system corresponds to $\omega < 1$ GHz.[16]

At $\omega < V$, the calculation of the rectification currents reduces to the calculation of the stationary contributions to the DC I–V curves $I_s(V)$ and $I_c(V)$ that are even in the voltage V. We assume that the Coulomb interaction between distant charges is screened by the gates. This will allow us to use the standard Tomonaga–Luttinger model with short range interactions.[23,24] Electric fields of external charges are also assumed to be screened. Thus, the applied voltage reveals itself only as the difference of the electrochemical potentials E_1 and E_2 of the particles injected from the left and right reservoirs. We assume that one lead is connected to the ground so that its electrochemical potential $E_2 = E_F$ is fixed. The electrochemical potential of the second lead $E_1 = E_F + eV$ is controlled by the voltage source (see Fig. 4). Since the Tomonaga–Luttinger model captures only low-energy physics, we assume that $eV < E_F$, where E_F is of the order of the bandwidth. Rectification occurs due to backscattering off the asymmetric potential $U(x)$. We will assume that the asymmetric potential is weak, $U(x) < E_F$. This will enable us to use perturbation theory.

For a narrow one-dimensional system, we can assume that the magnetic field $\mathbf{H}$ couples only to the electron spin ($k_{F\uparrow} \neq k_{F\downarrow}$) and we neglect the correction $-e\mathbf{A}/c$ to the momentum in the electron kinetic energy. Indeed, for a uniform field one can choose $\mathbf{A} \sim y$, where the y-axis is orthogonal to the wire, and y is small inside a narrow wire. In the absence of spin-orbit interaction, which is assumed here, the left and right movers still have the same Fermi velocity. So our slightly generalized bosonization identity is

$$\psi_{r\sigma}(x) = \frac{\hat{\eta}_{r\sigma}}{\sqrt{2\pi\delta}} e^{ir[k_{F\sigma}x + \phi_{r\sigma}(x)]}, \tag{63}$$

where we have allowed a spin dependent $k_{F\sigma}$. As shown in Ref. 21, the system in the absence of the asymmetric potential can be described by a quadratic bosonic Hamiltonian

$$H_0 = \sum_{r,r'=L,R} \sum_{\sigma,\sigma'=\uparrow,\downarrow} \int_0^L dx (\partial_x \phi_{r\sigma}) \mathcal{H}_{r\sigma,r'\sigma'} (\partial_x \phi_{r'\sigma'}) \tag{64}$$

with $\mathcal{H}_{r\sigma,r'\sigma'}$ a 4×4 matrix describing the interactions among the four species of fermions in the system. The quadratic Hamiltonian H_0 can be diagonalized by forming the new fields $\tilde{\phi}_c(x)$, $\tilde{\phi}_s(x)$ and their conjugate momentums $\tilde{\Pi}_c(x)$, $\tilde{\Pi}_s(x)$.[21] The new fields $\tilde{\phi}_{c,s}$ are related to the original fields $\phi_{r\sigma}$ through

$$\begin{pmatrix} \phi_\uparrow \\ \phi_\downarrow \end{pmatrix} = \begin{pmatrix} \sqrt{g_c}[1+\alpha] & \sqrt{g_s}[1+\beta] \\ \sqrt{g_c}[1-\alpha] & -\sqrt{g_s}[1-\beta] \end{pmatrix} \begin{pmatrix} \tilde{\phi}_c \\ \tilde{\phi}_s \end{pmatrix}, \tag{65}$$

where

$$\phi_\sigma = \phi_{L\sigma} + \phi_{R\sigma}, \tag{66}$$

and g_c, g_s, α, β are dimensionless parameters that depend on microscopic details and the magnetic field. For noninteracting electrons without a magnetic field, $g_c = g_s = 1/2$. $g_c < 1/2 \, (> 1/2)$ for repulsive (attractive) interactions. We have chosen the normalization of these fields such that the correlation function of $\tilde{\phi}_\nu$ fields with respect to H_0 is evaluated to

$$\langle \tilde{\phi}_c(t_1)\tilde{\phi}_c(t_2) \rangle = \langle \tilde{\phi}_s(t_1)\tilde{\phi}_s(t_2) \rangle = -2\ln[\delta + i(t_1 - t_2)/\tau_c] \tag{67}$$

with $\tau_c \sim \hbar/E_F$ the ultraviolet cutoff time. In the absence of the magnetic field, terms in Eq. (64) in the form of $\exp(\pm 2i\sqrt{g_s}\phi_s)$ may become relevant and open a spin gap for $g_s < 1/2$. Here they can be safely neglected since they are suppressed by the rapidly oscillating factors $\exp(\pm 2i[k_{F\uparrow} - k_{F\downarrow}]x)$. It is convenient to model the leads as the regions near the right and left ends of the wire without electron interaction.[18–20]

The effect of the central scatterer is described by

$$H' = \sum_\sigma \int_0^L dx U(x) \psi_\sigma^\dagger(x) \psi_\sigma(x). \tag{68}$$

Using Eq. (6), H' can be separated into two pieces: forward scattering

$$H_F = \sum_{r,\sigma} \int_0^L dx U(x) \psi_{r\sigma}^\dagger(x) \psi_{r\sigma}(x) = \sum_{r\sigma} \int_0^L dx U(x) \frac{1}{2\pi} \partial_x \phi_{r\sigma}(x), \qquad (69)$$

and backscattering

$$H_B = \sum_{r,\sigma} \int_0^L dx U(x) \psi_{r\sigma}^\dagger(x) \psi_{-r\sigma}(x) = \sum_{r,\sigma} \int_0^L dx U(x) e^{-ir[2k_{F\sigma}x + \phi_\sigma(x)]}. \qquad (70)$$

The forward scattering H_F has no effect on the current. Indeed, one can perform a redefinition $\phi_{r\sigma}(x) \to \phi_{r\sigma}(x) - \int^x dy f_{r\sigma}(y)$ in H_0, and choose $f_{r\sigma}(y)$ such that the resulting additional linear terms exactly cancel H_F. This redefinition simply shifts the phase of the terms in H_B and has no effect on the backscattering current. Therefore, we ignore H_F in the rest of the calculation. H_B describes the process that two electrons change their directions after being scattered by the impurity potential $U(x)$, hence it leads to backscattering current. Although Eq. (70) contains only two-particle processes of momentum transfer $\pm 2k_F$, electron–electron interaction renormalizes this term. In the renormalization group language, cumulant expansion generates multiple powers of H_B in an effective theory at low energies. Thus, the most general form of the backscattering Hamiltonian is[16,23]

$$H_B = \sum_{n_\uparrow, n_\downarrow} U(n_\uparrow, n_\downarrow) e^{i n_\uparrow \phi_\uparrow(0) + i n_\downarrow \phi_\downarrow(0)}. \qquad (71)$$

Here, the fields are evaluated at the impurity position $x = 0$ and $U(n_\uparrow, n_\downarrow) = U^*(-n_\uparrow, -n_\downarrow)$ since the Hamiltonian is Hermitian. The Klein factors are not written because they drop out in the perturbative expansion. $U(n_\uparrow, n_\downarrow)$ are the amplitudes of backscattering of $n_\uparrow$ spin-up and $n_\downarrow$ spin-down particles with $n_\sigma < 0$ for $L \to R$ and $n_\sigma > 0$ for $R \to L$ scattering. $U(n_\uparrow, n_\downarrow)$ can be estimated as[23,24] $U(n_\uparrow, n_\downarrow) \sim k_F \int dx U(x) e^{i n_\uparrow 2k_{F\uparrow}x + i n_\downarrow 2k_{F\downarrow}x} \sim U$, where U is the maximum of $U(x)$. In the case of a symmetric potential, $U(x) = U(-x)$, the coefficients $U(n_\uparrow, n_\downarrow)$ are real.

The spin and charge current can be expressed as

$$I_{s,c} = L_{s,c}^1 + R_{s,c}^1 = L_{s,c}^2 + R_{s,c}^2, \qquad (72)$$

where $L_{s,c}^i$ and $R_{s,c}^i$ denote the current of the left-and right-movers near electrode i. For a clean system ($U(x) = 0$), the currents obey[18-20] $R_{s,c}^1 = R_{s,c}^2$, $L_{s,c}^1 = L_{s,c}^2$ and $I_c = 2e^2 V/h$, $I_s = 0$. With backscattering off $U(x)$, particles are transferred between L and R in the wire, and hence $R_s^2 = R_s^1 + dS_R/dt$, $R_c^2 = R_c^1 + dQ_R/dt$, where Q_R and S_R denote the total charge and the z-projection of the spin of the right-moving electrons.[25] The currents $L_{s,c}^2$ and $R_{s,c}^1$ are determined by the leads (i.e. the regions without electron interaction in our model[18-20]) and remain the same as in the absence of the asymmetric potential. Thus, the spin and charge current can be represented as $I_c = 2e^2 V/h + I_c^{bs}$ and $I_s = I_s^{bs}$, where the backscattering current operators are[22–24,26]

$$\hat{I}_c^{bs} = d\hat{Q}_R/dt = i[H, \hat{Q}_R]/\hbar$$

$$= \frac{-ie}{\hbar} \sum_{n_\uparrow, n_\downarrow} (n_\uparrow + n_\downarrow) U(n_\uparrow, n_\downarrow) e^{in_\uparrow \phi_\uparrow(0) + in_\downarrow \phi_\downarrow(0)} , \qquad (73)$$

$$\hat{I}_s^{bs} = d\hat{S}_R/dt$$

$$= -\frac{i}{2} \sum_{n_\uparrow, n_\downarrow} (n_\uparrow - n_\downarrow) U(n_\uparrow, n_\downarrow) e^{in_\uparrow \phi_\uparrow(0) + in_\downarrow \phi_\downarrow(0)} . \qquad (74)$$

The calculation of the rectification currents reduces to the calculation of the currents Eqs. (73) and (74) at two opposite values of the DC voltage.

5.2. Calculation of the current

To find the backscattered current, we use the Keldysh technique.[27–30] We assume that at $t = -\infty$ there is no backscattering in the Hamiltonian ($U(x) = 0$), and then the backscattering is gradually turned on. Thus, at $t = -\infty$, the numbers N_L and N_R of the left- and right-moving electrons conserve separately: The system can be described by a partition function with two chemical potentials $E_1 = E_F + eV$ and $E_2 = E_F$ conjugated with the particle numbers N_R and N_L. This initial state determines the bare Keldysh Green functions.

We will consider only the zero temperature limit. The effect of voltage source is most conveniently handled by switching to the interaction representation $H_0 \to H_0 - E_1 N_R - E_2 N_L$. This transformation induces a time dependence in the electron creation and annihilation operators.[25,31] As a result, each exponent in Eqs. (71), (73) and (74) is multiplied by $\exp(ieVt[n_\uparrow + n_\downarrow]/\hbar)$. So we will use

$$H_B(t) = \sum_{n_\uparrow, n_\downarrow} U(n_\uparrow, n_\downarrow) e^{in_\uparrow \phi_\uparrow(t) + in_\downarrow \phi_\downarrow(t) + ieVt(n_\uparrow + n_\downarrow)/\hbar} , \qquad (75)$$

$$\hat{I}_c^{bs}(t) = \frac{-ie}{\hbar} \sum_{n_\uparrow, n_\downarrow} (n_\uparrow + n_\downarrow) U(n_\uparrow, n_\downarrow) e^{in_\uparrow \phi_\uparrow(t) + in_\downarrow \phi_\downarrow(t) + ieV(n_\uparrow + n_\downarrow)t/\hbar} , \qquad (76)$$

$$\hat{I}_s^{bs}(t) = -\frac{i}{2} \sum_{n_\uparrow, n_\downarrow} (n_\uparrow - n_\downarrow) U(n_\uparrow, n_\downarrow) e^{in_\uparrow \phi_\uparrow(t) + in_\downarrow \phi_\downarrow(t) + ieV(n_\uparrow + n_\downarrow)t/\hbar} , \qquad (77)$$

where we have used the abbreviations $\phi_\sigma(t) \equiv \phi_\sigma(x = 0, t)$. In the Keldysh formulation, the expectation value of the current I at t_0 is evaluated as

$$\langle I(t_0) \rangle = \left\langle \hat{T}_c \hat{I}(t_0) \exp\left[-\frac{i}{\hbar} \int_c dt H_B(t) \right] \right\rangle_0 , \qquad (78)$$

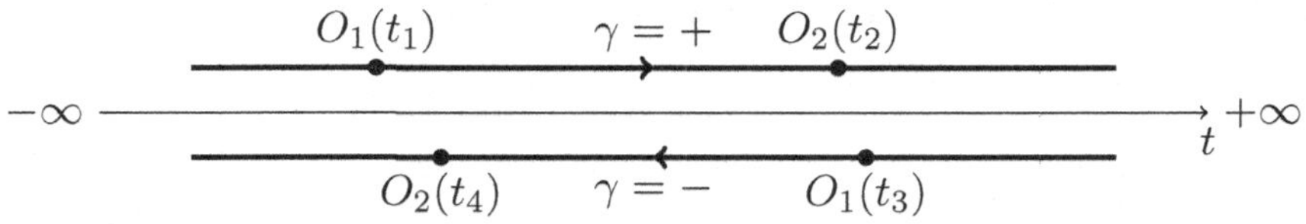

Fig. 5. The Keldysh contour. $O_1(t_1)$, $O_2(t_2)$ can lie on the forward branch ($\gamma = +$) or backward branch ($\gamma = -$).

where the integral is taken along the Keldysh contour denoted by c (see Fig. 5). In the Keldysh contour, the contour time t runs first forward from $-\infty$ to $+\infty$, then backward $+\infty \to -\infty$. $\hat{T}_c$ is the contour ordering symbol that orders the operators according to their contour time: operators with earlier contour times are put to the right. Unlike the usual time ordering, $\hat{T}_c$ may not respect the numerical value of t. For example, in Fig. 5, the operator $O(t_2)$ is earlier in contour time than $O(t_4)$ even though numerically $t_4 < t_2$. In fact, operators on the forward branch are always earlier than the ones on the backward branch. And for two operators that are both on the forward branch, contour ordering is the same as the usual time ordering, while for two operators that are both on the backward branch, contour ordering is anti-time ordering. The average $\langle \hat{T}_c \cdots \rangle_0$ is taken with respect to the ground state of H_0. Next, Eq. (78) is evaluated perturbatively by expanding the exponential

$$\exp\left[-\frac{i}{\hbar}\int_c dt H_B(t)\right] = \sum_{n=0}^{\infty} \frac{(-i)^n}{\hbar^n n!} \int_c dt_1 \cdots \int_c dt_n \left[\hat{T}_c H_B(t_1) \cdots H_B(t_n)\right]. \quad (79)$$

The leading contribution to the backscattering current emerges in the second order in $U(n,m)$. However, the rectification current results from the cancelation of the current at different signs of the voltage [see Eqs. (61) and (62)]. Hence the leading order contribution to the rectification current may emerge in the third order, depending on the interaction parameters g_c, g_s, α, and β. The condition for this to happen can be worked out using renormalization group analysis.[16] It turns out that, in such regiems, the spin rectification current may greatly exceed the charge rectification current.[16] We now demonstrate this interesting fact by an explicit calculation. The contribution 3rd order in $U(n,m)$ comes from the $n = 2$ terms in Eq. (79) since $I_{c,s}^{bs}$ also contains a $U(n,m)$

$$\langle I_{c,s}^{bs}(t_0) \rangle = \frac{(-i)^2}{\hbar^2 2!} \int_c dt_1 \int_c dt_2 \langle \hat{T}_c \hat{I}_{c,s}^{bs}(t_0) H_B(t_1) H_B(t_2) \rangle_0. \quad (80)$$

To evaluate the above integral, it is convenient to split the contour into forward ($\gamma = +$) and backward ($\gamma = -$) branches[32]

$$\int_c dt = \sum_{\gamma=\pm} \gamma \int_{-\infty}^{\infty} dt, \quad (81)$$

and attach a branch label to each operator in the expression:

$$O(t) \to O(\gamma, t). \tag{82}$$

The current operator $\hat{I}^{bs}_{c,s}(t_0)$ can be chosen to lie either on the forward or backward branch, and it is clear that the two choices give the same result. So in the new notation

$$\langle \hat{I}^{bs}_{c,s}(t_0) \rangle = \frac{-1}{2\hbar^2} \sum_{\gamma_1,\gamma_2} \gamma_1 \gamma_2 \int_{-\infty}^{\infty} dt_1 dt_2 \langle \hat{T}_c I^{bs}_{c,s}(+, t_0) H_B(\gamma_1, t_1) H_B(\gamma_2, t_2) \rangle_0$$

$$= \frac{-1}{4\hbar^2} \sum_{\gamma_0,\gamma_1,\gamma_2} \gamma_1 \gamma_2 \int_{-\infty}^{\infty} dt_1 dt_2 \langle \hat{T}_c \hat{I}^{bs}_{c,s}(\gamma_0, t_0) H_B(\gamma_1, t_1) H_B(\gamma_2, t_2) \rangle_0. \tag{83}$$

We will calculate charge current as an example. Spin current can be worked out in the same way. Substitute Eqs. (75) and (76) into Eq. (83), we have

$$\langle I^{bs}_c(t_0) \rangle = \frac{ie}{4\hbar^3} \sum_{n_\uparrow, n_\downarrow} \sum_{m_\uparrow, m_\downarrow} \sum_{l_\uparrow, l_\downarrow} (n_\uparrow + n_\downarrow) \sum_{\gamma_0, \gamma_1, \gamma_2} \gamma_1 \gamma_2 \int_{-\infty}^{\infty} dt_1 dt_2$$

$$\times U(n_\uparrow, n_\downarrow) U(m_\uparrow, m_\downarrow) U(l_\uparrow, l_\downarrow) e^{ieV[(n_\uparrow + n_\downarrow)t_0 + (m_\uparrow + m_\downarrow)t_1 + (l_\uparrow + l_\downarrow)t_2]/\hbar}$$

$$\times \langle \hat{T}_c e^{in_\uparrow \phi_\uparrow(0) + in_\downarrow \phi_\downarrow(0)} e^{im_\uparrow \phi_\uparrow(1) + im_\downarrow \phi_\downarrow(1)} e^{il_\uparrow \phi_\uparrow(2) + il_\downarrow \phi_\downarrow(2)} \rangle_0, \tag{84}$$

where for simplicity, we have used the abbreviations

$$\phi_\sigma(0) = \phi_\sigma(x = 0, \gamma_0, t_0), \tag{85}$$

$$\phi_\sigma(1) = \phi_\sigma(x = 0, \gamma_1, t_1), \tag{86}$$

$$\phi_\sigma(2) = \phi_\sigma(x = 0, \gamma_2, t_2). \tag{87}$$

The central quantity is now the multi-point correlation function

$$\langle \hat{T}_c e^{in_\uparrow \phi_\uparrow(0) + in_\downarrow \phi_\downarrow(0)} e^{im_\uparrow \phi_\uparrow(1) + im_\downarrow \phi_\downarrow(1)} e^{il_\uparrow \phi_\uparrow(2) + il_\downarrow \phi_\downarrow(2)} \rangle_0. \tag{88}$$

Note that the multi-point correlation function is nonzero only if

$$n_\uparrow + m_\uparrow + l_\uparrow = n_\downarrow + m_\downarrow + l_\downarrow = 0. \tag{89}$$

This can be understood from the following argument. The Hamiltonian H_0 is invariant under a constant shift $\phi_{r\sigma} \to \phi_{r\sigma} + F_\sigma/2$, thus we expect Eq. (88) is also invariant. However, it is easy to see that such a shift adds a nonzero phase factor $e^{i(n_\uparrow + m_\uparrow + l_\uparrow)F_\uparrow + (n_\downarrow + m_\downarrow + l_\downarrow)F_\downarrow}$ to Eq. (88), hence it must vanish unless the phase factor is 1. To evaluate Eq. (88), we use the decomposition[10]

$$\langle \hat{T}_c e^{O(1)} e^{O(2)} \cdots e^{O(n)} \rangle_0 = \prod_{i<j} e^{\langle \hat{T}_c O(i) O(j) \rangle_0}, \tag{90}$$

together with the contour ordered correlation function

$$\langle \hat{T}_c O(i) O(j) \rangle_0 = \begin{cases} \langle O(t_i) O(t_j) \rangle_0 & t_i \overset{\tau}{>} t_j \\ \langle O(t_j) O(t_i) \rangle_0 & t_i \overset{\tau}{<} t_j \end{cases}$$

$$= \begin{cases} \langle O(t_i - t_j) O(0) \rangle_0 & t_i \overset{\tau}{>} t_j \\ \langle O(t_j - t_i) O(0) \rangle_0 & t_i \overset{\tau}{<} t_j \end{cases}$$

$$= \langle O(\gamma_{ij}(t_i - t_j)) O(0)) \rangle_0 \,, \tag{91}$$

where $t_i \overset{\tau}{>} t_j$ means that $O(i) = O(\gamma_i, t_i)$ is further along the Keldysh contour c. The factor $\gamma_{ij} = \pm 1$ keeps track of the correct contour ordering and can be explicitly written as

$$\gamma_{ij} = \begin{cases} \gamma_i \, \text{sign}(t_i - t_j), & \gamma_i = \gamma_j \\ -\gamma_i, & \gamma_i = -\gamma_j \end{cases} . \tag{92}$$

Applying Eq. (91) to $\phi_\uparrow$ or $\phi_\downarrow$ in our problem, the following correlation functions are found:

$$\langle \hat{T}_c \phi_\uparrow(i) \phi_\uparrow(j) \rangle_0 = -2A \ln[\delta + i\gamma_{ij}(t_i - t_j)/\tau_c] \,, \tag{93}$$

$$\langle \hat{T}_c \phi_\downarrow(i) \phi_\downarrow(j) \rangle_0 = -2B \ln[\delta + i\gamma_{ij}(t_j - t_j)/\tau_c] \,, \tag{94}$$

$$\langle \hat{T}_c \phi_\uparrow(i) \phi_\downarrow(j) \rangle_0 = -2C \ln[\delta + i\gamma_{ij}(t_i - t_j)/\tau_c] \,, \tag{95}$$

where the constants A, B, C describes the interaction. They arises when we express ϕ_σ in terms of $\tilde{\phi}_{c,s}$ whose correlation functions are given by Eq. (67). Written explicitly,

$$A = g_c(1 + \alpha)^2 + g_s(1 + \beta)^2 \,, \tag{96}$$

$$B = g_c(1 - \alpha)^2 + g_s(1 - \beta)^2 \,, \tag{97}$$

$$C = g_c(1 - \alpha^2) - g_s(1 - \beta^2) \,. \tag{98}$$

As shown in Ref. 16, interesting results appear when the interaction strength A, B, and C are within the regime such that the most relevant operator is $U(1,0)$, the second most relevant operator is $U(0,-1)$, and the third most relevant operator is $U(-1,1)$. It was found that in this particular parameter regime, the leading contribution to the charge rectification current vanishes while the spin rectification current is finite for an asymmetric impurity potential. We now demonstrate this interesting fact. Collecting the terms in Eq. (84) that contain $U \equiv U(1,0)U(-1,1)U(0,-1)$ or its conjugate U^*, one has

$$I_c^{bs}(V) = 2\mathrm{Re}\left[\frac{ieU}{2\hbar^3}\sum_{\gamma_0,\gamma_1,\gamma_2}(\gamma_1-\gamma_0)\gamma_2\int_{-\infty}^{\infty}dt_1dt_2 e^{ieV(t_0-t_1)/\hbar}\right.$$

$$\left.\times\left(\delta+i\gamma_{01}\frac{(t_0-t_1)}{\tau_c}\right)^{-c}\left(\delta+i\gamma_{02}\frac{(t_0-t_2)}{\tau_c}\right)^{-a}\left(\delta+i\gamma_{12}\frac{(t_1-t_2)}{\tau_c}\right)^{-b}\right],$$

$$(99)$$

where $a = 2A - 2C$, $b = 2B - 2C$, $c = 2C$. We can then use Eq. (92) to separate the integral into different regions determined by the values of γ_0, γ_1, and γ_2. Each region can then be integrated into the product of Euler beta and gamma functions. After lengthy but straightforward calculations, we arrive at

$$I_c^{bs} = \frac{16e\tau_c^2}{\pi\hbar^3}\,\mathrm{sign}(eV)\left|\frac{eV\tau_c}{\hbar}\right|^{a+b+c-2}\mathrm{Re}[U(1,0)U(-1,1)U(0,-1)]$$

$$\times\,\Gamma(1-a)\Gamma(1-b)\Gamma(2-a-b-c)\Gamma(a+b-1)$$

$$\times\,\sin\frac{\pi a}{2}\sin\frac{\pi b}{2}\frac{\pi(a+b)}{2}\sin\pi(a+b+c).\qquad(100)$$

Spin backscattering current can be worked out in the same way. One finds

$$I_s^{bs} = \frac{16\tau_c^2}{\pi\hbar^2}\sin\frac{\pi a}{2}\sin\frac{\pi b}{2}\left|\frac{eV\tau_c}{\hbar}\right|^{a+b+c-2}$$

$$\times\,\Gamma(a+b-1)\Gamma(2-a-b-c)\Gamma(1-a)\Gamma(1-b)$$

$$\times\left\{\mathrm{Im}[U(1,0)U(-1,1)U(0,-1)]\cos\frac{\pi(a+b+c)}{2}\right.$$

$$\times\left[\sin\frac{\pi c}{2}+\cos\frac{\pi(a-b)}{2}\sin\frac{\pi(a+b+c)}{2}+\sin\frac{\pi(a+b)}{2}\cos\frac{\pi(a+b+c)}{2}\right]$$

$$\left.+\frac{1}{2}\mathrm{Re}[U(1,0)U(-1,1)U(0,-1)]\sin\frac{\pi(a-b)}{2}\sin\pi(a+b+c)\,\mathrm{sign}(eV)\right\}.$$

$$(101)$$

Indeed, the charge current Eq. (100) is an *odd* function of the voltage and hence does *not* contribute to the rectification effect. The spin current Eq. (101), however, contains both the *even* and *odd* parts, and the *even* part gives rise to a net spin rectification current when $\mathrm{Im}[U(1,0)U(-1,1)U(0,-1)] \neq 0$ which is satisfied for an asymmetric potential. Since these are the leading order results from perturbation theory, we expect that the charge rectification current in the system is not exactly zero, but is much smaller than the spin rectification current. It must be emphasized that the above result is valid within an interval of the applied voltage where perturbation theory is applicable, and the effect occurs only when the system is within a certain regimes of the parameter space: $(g_c, g_s, \alpha, \beta)$.[16] It would be interesting if this effect can be seen in an experimental system.

6. Summary

The effect of interaction in a one-dimensional fermion system is very different from the ones in higher dimensions. In one dimension, interacting fermions form Luttinger liquids whose behavior is best understood from the bosonic perspective. Combined with Keldysh technique, bosonization provides a powerful tool to analyse transport in one-dimensional quantum wires. The calculation technique presented in this brief review article can also be applied to other nonequilibrium transport problems in one dimension.

Acknowledgments

This work was supported by the NSF via Grant No. PHY-0854896. The author thanks D. Feldman, J. Koch, and A. Garg for many helpful discussions.

References

1. A. A. Abrikosov, L. G. Gor'kov and I. E. Dzyaloshinskii, *Methods of Quantum Field Theory in Statistical Physics* (Dover, New York, 1963).
2. S. Tomonaga, *Prog. Theor. Phys. (Kyoto)* **5** (1950) 544.
3. J. M. Luttinger, *J. Math. Phys.* **4** (1963) 1154.
4. D. C. Mattis and E. H. Lieb, *J. Math. Phys.* **6** (1965) 304.
5. A. Luther and I. Peschel, *Phys. Rev. B* **9** (1974) 2911.
6. F. D. M. Haldane, *J. Phys. C* **14** (1981) 2585.
7. V. J. Emery, in *Highly Conducting One-Dimensional Solids*, eds. J. T. Devreese, R. P. Evrard and V. E. Van Doren (Plenum, New York, 1979), pp. 247–303.
8. J. Sólyom, *Adv. Phys.* **28** (1979) 201.
9. H. J. Schulz, in *Proc. Les Houches Summer School LXI*, eds. E. Akkermans, G. Montambaux, J. Pichard and J. Zinn-Justin (Elsevier, Amsterdam, 1995), p. 533.
10. J. von Delft and H. Schoeller, *Ann. Phys.* **7** (1998) 225.
11. A. O. Gogolin, A. A. Nersesyan and A. M. Tsvelik, *Bosonization and Strongly Correlated Systems* (Cambridge University Press, Cambridge, 1999).
12. K. Schönhammer, *J. Phys.: Condens. Matter* **14** (2002) 12783.
13. D. Sénéchal, *Theoretical Methods for Strongly Correlated Electrons* (*CRM Series in Mathematical Physics*) (Springer, New York, 2003).
14. T. Giamarchi, *Quantum Physics in One Dimension* (Oxford University Press, Oxford, 2004).
15. K. T. Law, D. E. Feldman and Y. Gefen, *Phys. Rev. B* **74** (2006) 045319.
16. B. Braunecker, D. E. Feldman and F. Li, *Phys. Rev. B* **76** (2007) 085119.
17. D. E. Feldman and F. Li, *Phys. Rev. B* **78** (2008) 161304.
18. D. L. Maslov and M. Stone, *Phys. Rev. B* **52** (1995) R5539.
19. V. V. Ponomarenko, *Phys. Rev. B* **52** (1995) R8666.
20. I. Safi and H. J. Schulz, *Phys. Rev. B* **52** (1995) R17040.
21. T. Hikihara, A. Furusaki and K. A. Matveev, *Phys. Rev. B* **72** (2005) 035301.
22. D. E. Feldman, S. Scheidl and V. M. Vinokur, *Phys. Rev. Lett.* **94** (2005) 186809.

23. C. L. Kane and M. P. A. Fisher, *Phys. Rev. B* **46** (1992) 15233.
24. A. Furusaki and N. Nagaosa, *Phys. Rev. B* **47** (1993) 4631.
25. D. E. Feldman and Y. Gefen, *Phys. Rev. B* **67** (2003) 115337.
26. B. Braunecker, D. E. Feldman and J. B. Marston, *Phys. Rev. B* **72** (2005) 125311.
27. J. Rammer, *Quantum Field Theory of Nonequilibrium States* (Cambridge University Press, Cambridge, 2007).
28. A. Kamenev, *Field Theory of Nonequilibrium Systems* (Cambridge University Press, Cambridge, 2011).
29. L. V. Keldysh, *Sov. Phys. JETP* **20** (1965) 1018.
30. J. Rammer and H. Smith, *Rev. Mod. Phys.* **58** (1986) 323.
31. G. D. Mahan, *Many-Particle Physics* (Kluwer Academic/Plenum, New York, 2000), pp. 561–563.
32. C. L. Kane, *Phys. Rev. Lett.* **90** (2003) 226802.

Chapter IV

Generalizations to Higher Dimensions

FERMIONS IN TWO DIMENSIONS, BOSONIZATION, AND EXACTLY SOLVABLE MODELS*

JONAS DE WOUL* and EDWIN LANGMANN†

*Department of Theoretical Physics, Royal Institute of Technology KTH,
SE-106 91 Stockholm, Sweden*
**jodw02@kth.se*
†langmann@kth.se

We discuss interacting fermion models in two dimensions, and, in particular, such that can be solved exactly by bosonization. One solvable model of this kind was proposed by Mattis as an effective description of fermions on a square lattice. We review recent work on a specific relation between a variant of Mattis' model and such a lattice fermion system, as well as the exact solution of this model. The background for this work includes well-established results for one-dimensional systems and the high-T_c problem. We also mention exactly solvable extensions of Mattis' model.

Keywords: Bosonization; 2D correlated fermions; exactly solvable models.

1. Introduction

The theory of metals developed in the 1950–60s is the basis of an excellent qualitative description of many materials that exist in nature. This theory is based on the Fermi liquid picture of nearly-free quasiparticles and standard methods like diagrammatic perturbation theory; it is thoroughly discussed in most introductory textbooks on condensed matter physics (see e.g., Ref. 1).

Still, this otherwise so successful theory is known to fail for systems with strong correlation effects, as is typically observed in low-dimensional systems. For example, in the case of fermions confined to one dimension a different picture has been developed that radically departs from that of a Fermi liquid. This picture is largely based on prototype models that can be solved exactly by analytical techniques, and which nevertheless allow to take into account certain types of correlation effects. One important such technique is known as *bosonization*, which amounts to rewriting interacting fermions in terms of free bosons; we will return to this in later sections. For fermions confined to two dimensions, an equally satisfactory description has not yet been achieved. In this review, we describe a particular approach that uses a combination of mean field theory and bosonization to study such fermion systems.[2-7] In particular, we discuss a class of two-dimensional fermion models that can be solved exactly by bosonization, as well as their relation to models describing fermions on a square lattice. We also discuss the so-called high-T_c problem providing one physical motivation for this work.

---

*This article first appeared in International Journal of Modern Physics B, Vol. 26, No. 22 (2012).

Our plan is as follows. Section 1.1 introduces the models playing a central role in this paper, and Sec. 1.2 contains a biased overview of the literature on bosonization in higher dimensions.[a] Section 2 contains a discussion of lattice fermion models and their relation to the high-T_c problem (Sec. 2.1). As a pedagogical way of introducing the results and notation used in later parts, we also discuss the bosonization approach to one dimensional lattice fermion systems (Sec. 2.2 and 2.3). Section 3 reviews results for a simple prototype two dimensional lattice fermion system obtained in Refs. 2–5. Section 4 describes extensions of the latter results to more complicated models.[6,7] The final Sec. 5 contains a somewhat subjective discussion of our results from a broader perspective.

1.1. *Exactly solvable quantum field theory models*

The Luttinger model[8–11] is a prominent example of an exactly solvable quantum field theory (QFT[b]) model that has become a prototype for correlated fermion systems in one spatial dimension (1D).[12] This model describes two flavors of fermions, labeled by an index $r = \pm$ and by 1D momenta k, and is defined by the Hamiltonian

$$H_L = \sum_{r=\pm} \int dk\, v_F r k : \hat{\psi}_r^\dagger(k)\hat{\psi}_r(k): + \sum_{r,r'=\pm} \int \frac{dp}{2\pi} \hat{V}_{r,r'}(p)\hat{J}_r(-p)\hat{J}_{r'}(p) \qquad (1)$$

with $\hat{\psi}_r^{(\dagger)}(k)$ fermion field operators defined by the usual canonical anticommutator relations (CAR),

$$\hat{J}_r(p) = \int dk : \hat{\psi}_r^\dagger(k)\hat{\psi}_r(k+p): \qquad (2)$$

fermion density operators, $\hat{V}_{r,r'}(p)$ (suitable[c]) two-body potentials, $v_F > 0$ the Fermi velocity, and the colons indicating normal ordering. The technique to solve this model is based on mathematical results collectively known as bosonization; see (23)–(25) below. It is worth emphasizing that "exactly solvable" has a very strong meaning for this model: not only the eigenstates, but also all correlations functions can be computed analytically; see Ref. 13 and references therein.

As first pointed out by Mattis,[14] there exists a similar model describing fermions in 2D and which is also exactly solvable by bosonization. This model describes four flavors of fermions that are labeled by two indices $r, s = \pm$ and by 2D momenta $\mathbf{k} = (k_1, k_2)$; it is defined by the Hamiltonian

$$H_M = \sum_{r,s=\pm} \int d^2k\, v_F r k_s : \hat{\psi}_{r,s}^\dagger(\mathbf{k})\hat{\psi}_{r,s}(\mathbf{k}):$$

$$+ \sum_{r,s,r',s'=\pm} \int \frac{d^2p}{(2\pi)^2} \hat{V}_{r,s,r',s'}(\mathbf{p})\hat{J}_{r,s}(-\mathbf{p})\hat{J}_{r',s'}(\mathbf{p}), \qquad (3)$$

---

[a]The literature on higher-dimensional bosonization is huge, and our discussion is far from exhaustive.
[b]Note that by "QFT" we mean a quantum model with infinitely many degrees of freedom.
[c]See e.g., in Ref. 13.

with $k_s = (k_1 + sk_2)/\sqrt{2}$ and $\hat{J}_{r,s}(\mathbf{p}) = \int d^2k \; :\hat{\psi}^\dagger_{r,s}(\mathbf{k})\hat{\psi}_{r,s}(\mathbf{k}+\mathbf{p}):$ etc., similarly as above. Mattis, having the famous high-T_c problem of the cuprate superconductors[15] in mind, proposed this model as an effective description of fermions on a square lattice. We believe that the Mattis model deserves to be better known: First, exactly solvable models in 2D are rare, and any physically motivated such example deserves to be studied from different points of view. Second, as proposed and elaborated by us in recent work,[2-5] the exact solution of this model is a key part of a method to compute physical properties of 2D lattice fermion systems. Third, there exist exactly solvable extensions of the Mattis model that allow, for example, to study the effect of dynamical electromagnetic fields and spin on 2D interacting fermion systems by exact solutions.[6,7]

A main part of this review is on the relation of the Mattis model and a model of fermions on a square lattice. This relation is based on a method to derive effective QFT models for such lattice fermion systems. This method, which we call (partial) *continuum limit*, is well-established in 1D, and in the simplest case of spinless lattice fermions leads to the Luttinger model. The extension of this method to spinless fermions on a square lattice leads to a model that, in addition to the fermion fields $\hat{\psi}^{(\dagger)}_{r,s}(\mathbf{k})$, $r, s = \pm$, that describe the so-called *nodal* fermion degrees of freedom, also takes into account the so-called *antinodal* fermion degrees of freedom represented by operators $\hat{\psi}^{(\dagger)}_{r,0}(\mathbf{k})$, $r = \pm$. While the energy dispersion relations of the nodal fermions are linear in the momenta, $\epsilon_{r,s}(\mathbf{k}) = v_F r k_s$ for $r, s = \pm$, the antinodal fermions have hyperbolic dispersion relations, $\epsilon_{r,0}(\mathbf{k}) = r c_F k_+ k_- - \mu_0$ for $r = \pm$, with computable constants c_F and μ_0. The full Hamiltonian providing an effective description of 2D lattice fermions is the following extension of the Mattis model

$$H = \sum_{r,s=\pm} \int d^2k\, v_F r k_s :\hat{\psi}^\dagger_{r,s}(\mathbf{k})\hat{\psi}_{r,s}(\mathbf{k}):$$
$$+ \sum_{r=\pm} \int d^2k\, (c_F r k_+ k_- - \mu_0):\hat{\psi}^\dagger_{r,0}(\mathbf{k})\hat{\psi}_{r,0}(\mathbf{k}):$$
$$+ \sum_{r,r'=\pm}\sum_{s,s'=0,\pm} \int \frac{d^2p}{(2\pi)^2} \hat{V}_{r,s,r',s'}(\mathbf{p})\hat{J}_{r,s}(-\mathbf{p})\hat{J}_{r',s'}(\mathbf{p}) \quad (4)$$

with two-body interaction potentials $\hat{V}_{r,s,r',s'}$ determined by the underlying lattice fermion interactions.[3] The model defined by the Hamiltonian in (4) is a 2D analogue of the Luttinger model in the sense that it arises as a partial continuum limit of 2D lattice fermions, and that it is amenable to bosonization. We emphasize that this 2D Luttinger model is not exactly solvable; only the nodal fermion degrees of freedom can be mapped exactly to noninteracting bosons and thus be treated without approximations. Treating the antinodal fermions by mean field theory, one finds a significant parameter regime away from half filling for which the antinodal fermions are gapped.[4] We proposed that, in this partially gapped phase, the low-energy physics of the system can be described by the Mattis model.[4] This motivated us to study the Mattis model from a mathematical point of view, and we found that the Mattis model is indeed exactly solvable in the same strong sense as the Luttinger model.[5]

1.2. *Bosonization in higher dimensions (review)*

The literature on bosonization in dimensions higher than one is quite extensive. Below we mainly discuss work whose bosonization methods lie closest to ours, while references to other approaches are briefly mentioned at the end.

The first to apply bosonization to interacting, nonrelativistic fermions in higher dimension was Luther[16] (see also Ref. 13 for contemporary work). For a discretized spherical Fermi surface in three dimensions, one can bosonize in the radial direction at each point on the surface, while treating the discretized angles as flavor indices[16]; this is sometimes called radial- or tomographic bosonization. Using a generalized Kronig identity, one can then express the kinetic part of the Hamiltonian in terms of densities.[13,16] Similarly, the fermion fields can be written in terms of charge shift operators (also called Klein factors), which depend on the radial direction, and an exponential of boson operators.[13] However, in this early approach, only density operators with momenta in the radial direction behave as bosons, and thus interactions with transverse momentum exchange cannot be treated exactly.

Real interest in higher-dimensional bosonization came with the high-T_c problem and the prospect of finding non-Fermi liquid behavior in strongly correlated 2D models. In particular, Anderson suggested that the 2D Hubbard model on a square lattice has a Luttinger liquid phase away from half filling, and that this could be explored using bosonization methods.[17] Whether or not Luttinger-liquid behavior is possible in this model is still an open problem. Rigorous work on the renormalization group have shown that weakly coupled 2D fermions with Fermi surfaces of nonzero curvature are in general Fermi liquids[18] (see also Ref. 19 and references therein). Most of the attention has therefore been on Fermi surfaces with "flat parts", i.e., they contain portions that are straight (see for example Refs. 20 and 21).

Prior to Anderson's suggestion, Mattis proposed a 2D model of fermions with density–density interactions that could be solved exactly using bosonization.[14] The kinetic part contained four types of fermions with linearized tight-binding band relations on each side of a square Fermi surface [see (3)], which is reasonable for a Hubbard-type model with short-range hopping and near half filling. Unlike Ref. 16, the density operators were taken to be bosonic for all momentum exchange, although no details were provided. Rewriting the kinetic part in terms of densities, the Hamiltonian of the model could be diagonalized using a Bogoliubov transformation.

In more recent work, Luther[22] also studied fermions on the square Fermi surface with linear tight-binding band relations. Let $k_\parallel$ and $k_\perp$ denote the momenta parallel and perpendicular to a face of the square Fermi surface. Following Ref. 16, one would treat $k_\parallel$ as a flavor index, extend $k_\perp$ to plus and minus infinity, and introduce a Dirac sea in which all states $k_\perp < 0$ are filled. However, unlike the 1D case, one should note that there is a huge degeneracy in choosing the accompanying flavor index to the unbounded momenta. One can for example do a Fourier transformation (change of basis) in the $k_\parallel$-direction and then bosonize the fermions with a new flavor index $x_\parallel$. These can be interpreted as coordinates for a collection of parallel chains aligned in the $k_\perp$-direction. In this way, Luther was able to include boson-like interactions with momentum exchange also in the $k_\parallel$-direction.[22] The properties of this model were further investigated in Refs. 23 and 24.

Finally, we mention other work on rewriting interacting fermions as bosons, although these methods are somewhat different from that discussed here. Haldane[25] formulated a phenomenological approach in which density fluctuations corresponding to momentum exchange of states near an arbitrary D-dimensional Fermi surface are assumed bosonic. These ideas were further pursued in the works of Refs. 26–28. Bosonization methods using functional integration and Hubbard–Stratonovich transformations have been developed in Refs. 29 and 30. There is also an approach based on noncommutative geometry.[31]

2. Motivation and Background

2.1. *High-T_c problem and 2D lattice fermions*

The possible violation of Landau's Fermi liquid theory in models of strongly interacting fermions has been an actively researched problem for many years. Interest in this topic quickly grew with the discovery of high temperature superconductivity in the cuprates[15] and the realization that these materials display many properties not described by Fermi liquid theory; see e.g., Ref. 32 for review. Early on, it was suggested that 2D lattice fermion models of Hubbard-type capture the strongly correlated physics of cuprates.[33–36] These models are appealing due to their apparent simplicity. For example, the much-studied 2D Hubbard model can be defined by the Hamiltonian

$$H_{tU} = -t \sum_{\langle i,j \rangle} \sum_{\alpha=\uparrow,\downarrow} \left(c_{i,\alpha}^\dagger c_{j,\alpha} + c_{j,\alpha}^\dagger c_{i,\alpha} \right) + U \sum_i n_{i,\uparrow} n_{i,\downarrow}, \quad n_{i,\alpha} \equiv c_{i,\alpha}^\dagger c_{i,\alpha} \quad (5)$$

with fermion operators $c_{i,\alpha}^{(\dagger)}$ labeled by sites i of a 2D lattice and a spin index $\alpha = \uparrow, \downarrow$, and $\sum_{\langle i,j \rangle}$ a sum over all nearest-neighbor (nn) pairs on this lattice ($t > 0$ and $U > 0$ are the usual Hubbard parameters). The Hubbard model is conceptually simple since, if one restricts to a lattice with a finite number of sites, the Hamiltonian in (5) can be represented by a finite, albeit usually large, matrix.[d]

---

[d]For a lattice with $\mathcal{N}$ sites, the size of this matrix is $4^{\mathcal{N}} \times 4^{\mathcal{N}}$.

However, it has proven to be very difficult to do reliable computations for 2D Hubbard-like models for lattice sizes and intermediate coupling values of interest for the cuprates (i.e., for U/t in the range between 2 and 10, say). Thus, despite of much work over many years, no consensus has been reached on the physical properties of such models in large parts of the interesting parameter regime. To be more specific: one important parameter for Hubbard-like systems is *filling ν*, which is defined as the average fermion number per lattice site. For the Hubbard model this parameter is in the range $0 \leq \nu \leq 2$. There is consensus that, at half filling (i.e., for $\nu = 1$), the 2D Hubbard model describes a Mott insulator.

The challenge is to study the 2D Hubbard model close to, but away from, half filling, which is a regime that is much less understood. Our approach has been to look for other models that describe the same low energy physics but are more amenable to reliable computations. As we explain in Sec. 3.1, our search for such models was motivated by mean field results for the 2D Hubbard model.[37] Another important motivation and guide were 1D lattice fermion systems that, different from 2D, are well understood also by numerical and analytical methods: In Sec. 2.2 and 2.3 we review a particularly useful method in 1D, namely to perform a suitable continuum limit to obtain a QFT model that can be studied by bosonization. As already mentioned, this method is well-established in 1D, but we present it in a way that makes our generalization to 2D[3] natural. For simplicity, much of our discussion in this paper is on a spinless variant of the Hubbard model defined by the Hamiltonian

$$H_{tV} = -t \sum_{\langle i,j \rangle} \left(c_i^\dagger c_j + c_j^\dagger c_i \right) + \frac{V}{2} \sum_{\langle i,j \rangle} n_i n_j \,, \quad n_i \equiv c_i^\dagger c_i \tag{6}$$

with fermion operators $c_i^{(\dagger)}$ labeled only by lattice sites i; $t > 0$ and $V > 0$ are model parameters. For this so-called t-V model, the filling parameter is in the range $0 \leq \nu \leq 1$. Similarly as the Hubbard model, the 2D t-V model is believed to describe an insulator at half filling $\nu = 1/2$ and intermediate coupling values $V/t > 0$, and it has been a challenge to understand the model away from half filling. We will also shortly describe generalizations of our results for the 2D t-V model to the Hubbard model in Sec. 4.2.

2.2. *1D lattice fermions and the Luttinger model*

We now describe a method to derive the Luttinger model from the 1D variant of the so-called t-V model defined in (6).

In 1D, the lattice sites are $i = 1, 2, \ldots, \mathcal{N}$. Introducing a lattice constant $a > 0$ and the "length of space" $L \equiv \mathcal{N}a$, we define $\psi(x) \equiv a^{-1/2} c_j$ with the 1D spatial positions $x = ja$. Due to the finite L, possible fermion momenta k are in the set $\Lambda^* \equiv (2\pi/L)(\mathbb{Z} + 1/2)$ (we use antiperiodic boundary conditions). Using lattice

Fourier transform, we express the Hamiltonian in (6) in terms of fermion operators $\hat{\psi}^{(\dagger)}(k)$ on the Brillouin zone of the lattice[e]

$$\mathrm{BZ} = \left\{ k \in \Lambda^*; -\frac{\pi}{a} < k < \frac{\pi}{a} \right\} \tag{7}$$

as follows

$$\begin{aligned}
H_{tV} = &\int_{\mathrm{BZ}} \hat{d}k\, \epsilon(k) \hat{\psi}^\dagger(k) \hat{\psi}(k) \\
&+ \int_{\mathrm{BZ}} \hat{d}k_1 \cdots \int_{\mathrm{BZ}} \hat{d}k_4\, v(k_1, \ldots, k_4) \hat{\psi}^\dagger(k_1) \hat{\psi}(k_2) \hat{\psi}^\dagger(k_3) \hat{\psi}(k_4)
\end{aligned} \tag{8}$$

with the *band relation* $\epsilon(k) = -2t\cos(ka)$ and the *interaction vertex*

$$v(k_1, \ldots, k_4) = \hat{u}(k_1 - k_2) \sum_{n \in \mathbb{Z}} \hat{\delta}(k_1 - k_2 + k_3 - k_4 + (2\pi/a)n) \tag{9}$$

where $\hat{u}(p) = aV\cos(ap)/(2\pi)$; we write $\int_S \hat{d}k \equiv \sum_{k \in S}(2\pi/L)$ for scaled Riemann sums (S some subset of Λ^*) and $\hat{\delta}(k_1 - k_2) \equiv [L/(2\pi)]\delta_{k_1,k_2}$ for scaled Kronecker deltas. Our normalization of the CAR is such that $\{\hat{\psi}(k_1), \hat{\psi}^\dagger(k_2)\} = \hat{\delta}(k_1 - k_2)$. We note that these scalings are such that all formulas remain meaningful in the formal infinite volume limit $L \to \infty$, but it is convenient to keep L finite during computations. We use a particle physics jargon and refer to a and L as *UV- and IR-cutoff*, respectively.

Our aim is to modify the model so as to make it better amenable to analytical computations but not (much) change the low energy properties of the model. The strategy is to perform a continuum limit $a \to 0^+$, i.e., the lattice constant a is a small parameter, and subleading terms in a are assumed to be of lesser importance. However, this limit has to be taken with care: Physics folklore suggests that the low energy physics of a fermion system is dominated by momentum states close to the Fermi surface, i.e., one should only change the Hamiltonian so that only momentum states are affected that are far away from the Fermi surface. To be more specific: In the noninteracting case $V = 0$, the ground state of the Hamiltonian in (8) is

$$|\Omega\rangle = \prod_{k:\epsilon(k)-\mu \leq 0} \hat{\psi}^\dagger(k)|0\rangle \tag{10}$$

with the chemical potential μ determined by the fermion density and $|0\rangle$ the vacuum state such that $\hat{\psi}(k)|0\rangle = 0$ for all k. We now choose Q such that $\epsilon(rQ/a) = \mu$ for $r = \pm$, i.e., the filled momentum states are in the range $-Q/a < k < Q/a$, and the two points $\pm Q/a$ correspond to the Fermi surface. Note that filling in this situation is $\nu = Q/\pi$, i.e., half filling corresponds to $Q = \pi/2$. The ground state $|\Omega\rangle$ is then

---

[e]We use anti-periodic boundary conditions.

fully characterized by the following conditions,

$$\hat{\psi}_{\pm}(\pm k)|\Omega\rangle = \hat{\psi}_{\pm}^{\dagger}(\mp k)|\Omega\rangle = 0 \quad \forall\, k > 0\,, \tag{11}$$

where $\hat{\psi}_r(k) \equiv \hat{\psi}(K_r + k)$, $r = \pm$, are fermion operators labeled my momenta close to the Fermi surface points $K_r \equiv rQ/a$.

To prepare for this limit it is convenient to rewrite the t-V model as a model of two flavors of fermions. For that we divide the Brillouin zone in two subsets $\Lambda_r^* = \{k \in \mathrm{BZ} - K_r; r(k + K_r) > 0\}$ such that

$$\int_{\mathrm{BZ}} \hat{d}k\, f(k) = \sum_{r=\pm} \int_{\Lambda_r^*} \hat{d}k\, f(K_r + k) \tag{12}$$

for arbitrary functions f defined on the Brillouin zone. This shows that the fermion operators $\hat{\psi}_r(k)$ are to be labeled by momenta $k \in \Lambda_r^*$, $r = \pm$. Inserting this in (8) we obtain a Hamiltonian describing two flavors of fermion operators $\hat{\psi}_r(k)$, $k \in \Lambda_r^*$ and $r = \pm$, with dispersion relations $\epsilon_r(k) \equiv \epsilon(K_r + k)$ and interaction vertices

$$v_{r_1,\cdots,r_4}(k_1, \ldots, k_4) = \hat{u}(K_{r_1} - K_{r_2} + k_1 - k_2)$$
$$\times \sum_{n \in \mathbb{Z}} \hat{\delta}(K_{r_1} - K_{r_2} + K_{r_3} - K_{r_4} + k_1 - k_2$$
$$+ k_3 - k_4 + (2\pi/a)n)\,, \tag{13}$$

i.e., $H_{tV} = \sum_{r=\pm} \int_{\Lambda_r^*} \hat{d}k\, \epsilon_r(k)\hat{\psi}_r^{\dagger}(k)\hat{\psi}_r(k) + \cdots$ (the dots indicate obvious interaction terms).

We now modify the model as follows: First, we Taylor expand the dispersion relations $\epsilon_r(k) = \mu + r v_F k + O(a^2 k^2)$, $v_F = 2ta\sin(Q)$, and ignore the higher-order terms, i.e., we replace

$$\epsilon_r(k) \to \mu + r v_F k\,. \tag{14}$$

Second, we replace the interaction vertex by

$$v_{r_1,\cdots,r_4}(k_1, \ldots, k_4) \to \hat{u}(K_{r_1} - K_{r_2})\hat{\delta}(k_1 - k_2 + k_3 - k_4)$$
$$\sum_{n \in \mathbb{Z}} \hat{\delta}(K_{r_1} - K_{r_2} + K_{r_3} - K_{r_4} + (2\pi/a)n)\,, \tag{15}$$

i.e., we expand $\hat{u}(K_{r_1} - K_{r_2} + k_1 + k_2) = \hat{u}((r_1 - r_2)Q/a)[1 + O(a(k_1 - k_2))]$ and ignore the higher-order terms and, at the same time, modify allowed scattering terms in the interaction but only those where $a(k_1 - k_2 + k_3 - k_4)$ is "large". Since momenta k close to the Fermi surface are such that ka is "small", we expect that these changes are appropriate.

Inserting these changes in the Hamiltonian one obtains, after some computations,

$$H_{tV} \to \sum_{r=\pm} \int_{\Lambda_r^*} \hat{d}k \, r v_F k \hat{\psi}_r^\dagger(k) \hat{\psi}_r(k) + \int_{\tilde{\Lambda}^*} \frac{\hat{d}p}{2\pi} \, \hat{v}(p) \hat{\rho}_+(p) \hat{\rho}_-(-p) \tag{16}$$

with

$$\hat{\rho}_r(p) = \int_{\Lambda_r^*} \hat{d}k \, \hat{\psi}_r^\dagger(k) \hat{\psi}_r(k+p) \tag{17}$$

and[f]

$$\hat{v}(p) = g\theta(\pi - |p|a) \tag{18}$$

with $g = 2aV \sin^2(Q)/\pi$; here and in the following we use the notation $\tilde{\Lambda}^* \equiv (2\pi/a)\mathbb{Z}$. We note that, in these computations, one can ignore additive constants and chemical potential terms, i.e., terms proportional to $\sum_{r=\pm} \int_{\Lambda_r^*} \hat{d}k \hat{\psi}_r^\dagger(k) \hat{\psi}_r(k)$ (since the former terms amount to constant shifts of the energy, and the latter to a renormalization of the chemical potential, which both are irrelevant). The final modification is to replace on the r.h.s. in (16) and in (17)

$$\Lambda_r^* \to \Lambda^* \tag{19}$$

i.e., to remove the restriction on fermion momentum states by taking the limit $a \to 0^+$ in the sets Λ_r^*, $r = \pm$. We note that it is this approximation which turns the model into a QFT: Before we had a model with a finite number of fermion momentum states, but after (19) we have a quantum model with an infinite number of degrees of freedom. One finds that, in order to obtain a model that is mathematically well-defined, one needs to normal order *before* making the change in (19): defining

$$:A: \equiv A - \langle \Omega, A\Omega \rangle \tag{20}$$

for operators A and with the state $|\Omega\rangle$ fully characterized by (11), one can show that

$$\hat{J}_r(p) \equiv \int_{\Lambda^*} \hat{d}k \, :\hat{\psi}_r^\dagger(k) \hat{\psi}_r(k+p): \tag{21}$$

and

$$H_L \equiv \sum_{r=\pm} \int_{\Lambda^*} \hat{d}k \, r v_F k :\hat{\psi}_r^\dagger(k) \hat{\psi}_r(k): + \int_{\tilde{\Lambda}^*} \frac{\hat{d}p}{2\pi} \, \hat{V}(p) \hat{J}_+(p) \hat{J}_-(-p) \tag{22}$$

are mathematically well-defined. Note that normal ordering, before the replacement in (19), only amounts to dropping irrelevant additive constants and chemical potential terms.

---

[f]θ below is the Heaviside function.

The Hamiltonian in (22) makes mathematically precise a special case of the one in (1) (i.e. $\hat{V}_{r,r'} = \delta_{r,-r'}\hat{V}/2$). Note that the limit $a \to 0^+$ was only partial and, in particular, it is important to keep a finite in the interaction potential in (18): taking $a \to 0^+$ amounts to $\hat{V}(p) \to g$ (independent of p), which is exactly the limit where the Luttinger model becomes formally equal to the massless Thirring model.[9] However, as is well-known, this latter limit is delicate and requires a nontrivial multiplicative renormalization.[38]

It is important to note that, in the derivation above, one assumes $Q \neq \pi/2$. For $Q = \pi/2$ there are additional terms in the effective Hamiltonian that cannot be ignored and that, presumably, open a gap. This is consistent with $Q = \pi/2$ corresponding to half filling.

2.3. *Bosonization in 1D*

The exact solution of the Luttinger model relies on three mathematical facts: First, the normal ordered fermion densities do not commute (as one would naively expect) but obey the commutator relations

$$[\hat{J}_r(p), \hat{J}_{r'}(p')] = r\delta_{r,r'}p\hat{\delta}(p+p'),\tag{23}$$

where the right-hand side is an example of an *anomaly*. Second, the kinetic part of the Luttinger Hamiltonian can be expressed in term of the normal ordered fermion densities,

$$\int_{\Lambda^*} \hat{d}k\, rk:\hat{\psi}_r^\dagger(k)\hat{\psi}_r(k): = \pi \int_{\tilde{\Lambda}^*} \frac{\hat{d}p}{2\pi}:\hat{J}_r(p)\hat{J}_r(-p):,\tag{24}$$

which is known as *Kronig identity*. Third, the fermion field operators can be computed as

$$\hat{\psi}_\pm(k) = \lim_{\epsilon \to 0^+} \frac{1}{2\pi\sqrt{\epsilon}} \int_{-L/2}^{L/2} e^{\pm\pi i\hat{j}_\pm(0)x/L}(R_\pm)^{\mp 1} e^{\pm\pi i\hat{j}_\pm(0)x/L}$$

$$\times \exp\left(\pm \int_{\tilde{\Lambda}^*\backslash\{0\}} \hat{d}p\frac{1}{p}e^{ipx-\epsilon|p|}\hat{J}_\pm(p)\right) e^{-ikx}dx\tag{25}$$

with unitary operators $R_\pm$ obeying $R_+R_- = -R_-R_+$ and

$$R_r\hat{J}_{r'}(p)R_r^{-1} = \hat{J}_{r'}(p) - r\delta_{r,r'}\delta_{p,0}.\tag{26}$$

The commutator relations in (23) imply that the linear combinations

$$\hat{\Pi}(p) = \frac{1}{\sqrt{2}}\left(-\hat{J}_+(p) + \hat{J}_-(p)\right)$$

$$\hat{\Phi}(p) = \frac{1}{ip\sqrt{2}}\left(\hat{J}_+(p) + \hat{J}_-(p)\right) \qquad (p \neq 0)\tag{27}$$

are boson operators with the usual canonical commutator relations (CCR)

$$[\hat{\Pi}(p), \hat{\Phi}^\dagger(p')] = -i\hat{\delta}(p - p'), \quad [\hat{\Pi}(p), \hat{\Pi}^\dagger(p')] = [\hat{\Phi}(p), \hat{\Phi}^\dagger(p')] = 0 \quad (28)$$

with $\hat{\Pi}^\dagger(p) = \hat{\Pi}(-p)$ and similarly for $\hat{\Phi}$. Moreover, using (24), one can express the Luttinger Hamiltonian solely in terms of normalized density operators, and inserting (27) one finds

$$H_L = \frac{v_F}{2} \int_{\tilde{\Lambda}^* \setminus \{0\}} \hat{d}p \colon \left((1 - \gamma(p))\hat{\Pi}^\dagger(p)\hat{\Pi}(p) + (1 + \gamma(p))p^2\hat{\Phi}^\dagger(p)\hat{\Phi}(p) \right) \colon + \cdots,$$

$$(29)$$

where the dots indicate zero mode terms that are $O(1/L)$ and that we suppress, to not clutter our presentation. The formula in (29) shows that the Luttinger Hamiltonian is equivalent to a free boson model that can be diagonalized by a boson Bogoliubov transformation. Moreover, since (25) expresses the fermion field operators in terms of boson operators, it is possible to compute all fermion correlation functions exactly by analytical methods; see e.g., Ref. 13 for details.

3. Effective Model for a 2D Lattice Fermion System

In this section we review a series of papers where we propose[2,3] and study[4,5] an effective model for a 2D lattice fermion system of Hubbard type. In these papers we treat a generalization of the 2D t-V model where we also allow for a next-nearest neighbor hopping term proportional to t' but, for simplicity, we restrict ourselves here to the special case $t' = 0$.

3.1. Mean field results in 2D

Hartree–Fock theory is a variational method where the ground state of the interacting system is approximated by the one of noninteracting fermions in an external potential, and this potential is determined such that the variational energy is minimized.[39] In mean field theory one further restricts to variational states that are translationally invariant, which often is adequate for translationally invariant models. For the t-V model, there are three mean field parameters, and one, which we call Δ, is proportional to the magnitude of the density difference between neighboring sites, which is assumed to be constant.[4] A mean field solution with $\Delta > 0$ has the physical interpretation of a charge density wave (CDW) phase: the state is only invariant under translations by two sites, and it is usually insulating. A normal (N) phase corresponds to a state invariant under all lattice translations, i.e., $\Delta = 0$, and such a state is typically metallic. It is important to note that mean field theory allows to compare the variational energy of three types of states: First, the pure N state, second, the pure CDW state, and third, a phase separated state where parts of

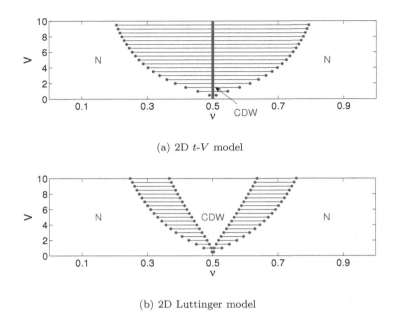

(a) 2D t-V model

(b) 2D Luttinger model

Fig. 1.   Comparison between the coupling (V) versus filling (ν) mean field phase diagrams of (a) the 2D t-V model and (b) the 2D Luttinger model for $t = 1$. Shown are the CDW and N phases as a function of coupling V and filling ν at zero temperature. The regime marked by horizontal lines is mixed, i.e., neither the CDW nor the N phase is thermodynamically stable. The CDW phase in (a) exists only at half filling $\nu = 0.5$. The CDW region in (b) corresponds to a partially gapped phase. The parameters used for (b) are $\kappa = 0.8$ and $Q/\pi = 0.45$. (Figure reprinted from Ref. 4 with kind permission from Springer Science+Business Media.)

the system are in the N and another part of the system in the CDW state.[4,40] If this third state is obtained it proves that the assumption underlying mean field theory is not correct: the true Hartree–Fock ground state is not translationally invariant, and this is an indication of unusual physical properties that cannot be accounted for by mean field theory.

Figure 1(a) shows the mean field phase diagram for the 2D t-V model as a function of filling ν and interaction strength V/t, setting $t = 1$. As shown, the pure CDW state occurs, but only at strictly half filling, and the N state only in a region quite far away from half filling: In the shaded regions between mean field theory fails. We believe that this failure of mean field theory in a large region away from half filling is due to treating all fermion degrees of freedom in the same way. In the 2D Luttinger model these degrees of freedom are disentangled and thus can be treated by different methods. As shown in Fig. 1(b), mean field theory is applicable in a much larger part of the doping regime if used only for the antinodal fermions. As we proposed,[4,5] the CDW region in Fig. 1(b) corresponds to a partially gapped phase where the antinodal fermions are gapped and thus do not contribute to the low energy properties of the system, and this region can be described by the Mattis model.

3.2. *Partial continuum limit of the 2D t-V model*

We consider the model defined by the Hamiltonian in (6), but now on a 2D diagonal square lattice with lattice constant a and size L, i.e., the lattice sites are $x = (x_1, x_2)$ with $x_{1,2}$ integer multiples of a such that the possible fermion momenta are $\mathbf{k} = (k_1, k_2)$ with $k_\pm = (k_1 \pm k_2)\sqrt{2}$ half-integer multiples of $2\pi/L$. We denote the set of all such fermion momenta as Λ^*.

The 2D Brillouin zone is a subset of a square (rather than of an interval in 1D),

$$\mathrm{BZ} = \left\{ \mathbf{k} \in \Lambda^*; -\frac{\pi}{a} < k_{1,2} < \frac{\pi}{a} \right\}, \tag{30}$$

and, similarly as in 1D, the Hamiltonian in Fourier space is fully characterized by the dispersion relation

$$\epsilon(\mathbf{k}) = -2t[\cos(ak_1) + \cos(ak_2)] \tag{31}$$

and the interaction vertex

$$v(\mathbf{k}_1, \ldots, \mathbf{k}_4) = \hat{u}(\mathbf{k}_1 - \mathbf{k}_2) \sum_{\mathbf{n} \in \mathbb{Z}^2} \hat{\delta}^2(\mathbf{k}_1 - \mathbf{k}_2 + \mathbf{k}_3 - \mathbf{k}_4 + (2\pi/a)\mathbf{n}) \tag{32}$$

with $\hat{u}(\mathbf{p}) = a^2 V[\cos(ap_1) + \cos(ap_2)]/(2\pi)^2$, i.e., the Hamiltonian can be written as in (8) etc. but with the obvious replacements $\hat{\psi}^{(\dagger)}(k) \to \hat{\psi}^{(\dagger)}(\mathbf{k})$,

$$\int \hat{d}k \to \int \hat{d}^2 k \equiv \sum_{\mathbf{k}} \left(\frac{2\pi}{L} \right)^2, \quad \hat{\delta}(k - k') \to \hat{\delta}^2(\mathbf{k} - \mathbf{k}') = \left(\frac{L}{2\pi} \right)^2 \delta_{\mathbf{k}, \mathbf{k}'}$$

etc.

The noninteracting Fermi surface at half filling coincides with the tilted square given by the four line segments $k_1 + sk_2 = r\pi/(2a)$, $r, s = \pm$, i.e., $\epsilon(\mathbf{k}) = 0$ on these lines. The midpoints of the sides of this square Fermi surface are $(rQ, rsQ)/a$ with $Q = \pi/2$. We are interested in the model close to half filling and assume that there is a Fermi surface containing four points

$$\mathbf{K}_{r,s} \equiv (rQ, rsQ)/a, \quad r, s = \pm \tag{33}$$

close to these midpoints, i.e., $Q \approx \pi/2$ is a parameter. Close to these four points, the dispersion relation has a linear behavior,

$$\epsilon(\mathbf{K}_{r,s} + \mathbf{k}) = -4t \cos(Q) + r v_F k_s + O(|a\mathbf{k}|^2), \quad r, s = \pm \tag{34}$$

with $v_F = 2\sqrt{2} \sin(Q)ta$. This suggests to approximate the dispersion relation in (31) by a linear one, keeping only the leading nontrivial term in (34). This leads to a Fermi surface that includes line segments through the points $\mathbf{K}_{r,s}$, as indicated by dashed lines in Fig. 2. However, there are momentum states close to the noninteracting Fermi surface at half filling where this approximation is qualitatively

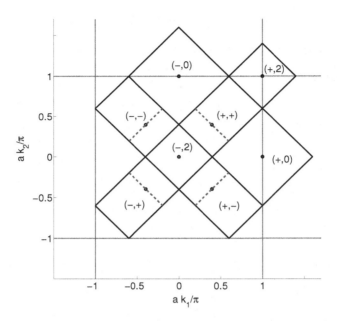

Fig. 2.   Division of the Brillouin zone BZ in eight regions $\mathbf{K}_{r,s} + \Lambda^*_{r,s}$ marked as (r,s) for $r = \pm$, $s = 0, \pm, 2$. The eight dots mark the points $\mathbf{K}_{r,s}$, and the parameters are $Q = 0.45\pi$ and $\kappa = 0.8$. Note that $k_{1,2}$ are only defined up to integer multiples of $2\pi/a$, and thus the union of these regions cover the BZ exactly once. (Figure reprinted from Ref. 3 with kind permission from Springer Science+Business Media.)

wrong: the points $\mathbf{K}_{+,0} \equiv (\pi, 0)/a$ and $\mathbf{K}_{-,0} \equiv (0, \pi)/a$ on this Fermi surface are saddle points of the dispersion relation in (31), and

$$\epsilon(\mathbf{K}_{r,0} + \mathbf{k}) = -r c_F (k_1^2 - k_2^2)/2 + O(|a\mathbf{k}|^3), \quad r = \pm \tag{35}$$

with $c_F = 2ta^2$. This hyperbolic behavior of the dispersion relation is important for the tendency of 2D lattice fermion system to develop a gap at half filling,[41] and it certainly cannot be ignored. There are two more special points, namely $\mathbf{K}_{+,2} = (\pi, \pi)/a$ and $\mathbf{K}_{-,2} = (0, 0)$, corresponding to a minimum and maximum of the dispersion relation, respectively. However, the regions close to the latter points are far away from the assumed Fermi surface, and we expect that they therefore can be ignored.

This suggests us to write the 2D t-V model as a model of eight flavors of fermion operators $\hat{\psi}_{r,s}(\mathbf{k}) \equiv \hat{\psi}(\mathbf{K}_{r,s} + \mathbf{k})$, $r = \pm$ and $s = 0, \pm, 2$, similarly as in 1D. For that we use a partition of the Brillouin zone in eight regions $\mathbf{K}_{r,s} + \Lambda^*_{r,s}$ such that

$$\int_{\mathrm{BZ}} \hat{d}^2 k\, f(\mathbf{k}) = \sum_{r=\pm} \sum_{s=0,\pm,2} \int_{\Lambda^*_{r,s}} \hat{d}^2 k\, f(\mathbf{K}_{r,s} + \mathbf{k}), \tag{36}$$

as indicated in Fig. 2. As already mentioned, we refer to the fermions with $s = \pm$ as *nodal* and the ones with $s = 0$ as *antinodal*. It is important to note that this partition depends on a parameter κ in the range $0 < \kappa < 1$: the antinodal regions are

$$\Lambda_{r,0}^* = \Lambda_0^* \equiv \left\{ \mathbf{k} \in \Lambda^*; -\kappa \frac{\sqrt{2}\pi}{a} < k_\pm < \kappa \frac{\sqrt{2}\pi}{a} \right\} \tag{37}$$

with $0 < \kappa < 1$, and this also determines the widths of the nodal regions: $\Lambda_{r,s=\pm}$ only contains momenta $\mathbf{k}$ such that $-\pi/\tilde{a} < k_{-s} < \pi/\tilde{a}$ with

$$\tilde{a} \equiv \frac{\sqrt{2}a}{1-\kappa}. \tag{38}$$

As discussed in more detail below, $\tilde{a}$ is important since it serves as UV cutoff for the nodal fermions.

We now modify the 2D t-V model, written as a eight flavor model, similarly as in 1D: First, the nodal and antinodal dispersion relations are modified by ignoring the $O(|a\mathbf{k}|^n)$-terms in (34) ($n = 2$) and (35) ($n = 3$), respectively. Second, the interaction vertices are modified by using the obvious generalization of (13) and (15) to 2D. Third, the restriction on the nodal momenta orthogonal to the assumed Fermi surface is removed, i.e.,

$$\Lambda_{r,s=\pm}^* \to \Lambda_s^* \equiv \left\{ \mathbf{k} \in \Lambda^*; -\frac{\pi}{\tilde{a}} < k_{-s} < \frac{\pi}{\tilde{a}} \right\} \tag{39}$$

with $\tilde{a}$ in (38), and, at the same time, all terms in the Hamiltonian involving fermion flavors with $s = 2$ are ignored. Note that this third change only modifies fermion degrees far away from the assumed Fermi surface. Similarly as in 1D, it is the change in (39) that leads to a QFT model, and this model is mathematically well-defined if this change is done after normal ordering.[3] By lengthy computations one finds a Hamiltonian which, remarkably, only contains interaction terms involving densities

$$:\hat{\rho}_{r,s}(\mathbf{p}): \equiv \int_{\Lambda_s^*} \tilde{d}^2 k_1 \int_{\Lambda_s^*} \tilde{d}^2 k_2 : \hat{\psi}_{r,s}^\dagger(\mathbf{k}_1)\hat{\psi}_{r,s}(\mathbf{k}_2): \hat{\delta}^2(\mathbf{k}_1 - \mathbf{k}_2 + \mathbf{p}) \tag{40}$$

for $r = \pm$, $s = 0, \pm$. A final change is needed to obtain the 2D Luttinger model: For reasons explained below, we replace in the Hamiltonian

$$:\hat{\rho}_{r,s=\pm}(\mathbf{p}): \to \hat{J}_{r,s}(\mathbf{p}) \equiv \int_{\Lambda_s^*} \tilde{d}^2 k_1 \int_{\Lambda_s^*} \tilde{d}^2 k_2 : \hat{\psi}_{r,s}^\dagger(\mathbf{k}_1)\hat{\psi}_{r,s}(\mathbf{k}_2):$$
$$\times \sum_{n\in\mathbb{Z}} \hat{\delta}^2(\mathbf{k}_1 - \mathbf{k}_2 + \mathbf{p} + 2\pi n \mathbf{e}_{-s}/\tilde{a}) \tag{41}$$

with $\mathbf{e}_\pm = (1, \pm 1)/\sqrt{2}$. Note that this last change amounts to adding umklapp terms to the nodal densities.

The changes described above lead to[3]

$$H_{tV} \to \mathcal{E}_0 + H_n + H_a + H_{na} \tag{42}$$

with the nodal part of the Hamiltonian

$$H_n = \sum_{r,s=\pm} \int_{\Lambda_s^*} \hat{d}^2 k v_F r k_s : \hat{\psi}_{r,s}^\dagger(\mathbf{k}) \hat{\psi}_{r,s}(\mathbf{k}) : + \int_{\tilde{\Lambda}^*} \frac{\hat{d}^2 p}{(2\pi)^2} \sum_{r,s,r',s'=\pm}$$

$$\times g(\delta_{s,s'}\delta_{r,-r'} + \delta_{s,-s'})\chi_s(\mathbf{p})\chi_{s'}(\mathbf{p})\hat{J}_{r,s}(-\mathbf{p})\hat{J}_{r',s'}(\mathbf{p}), \tag{43}$$

the antinodal part

$$H_a = \sum_{r=\pm} \int_{\Lambda_0^*} \hat{d}^2 k (-r c_F k_+ k_- - \mu_0) : \hat{\psi}_{r,0}^\dagger(\mathbf{k}) \hat{\psi}_{r,0}(\mathbf{k}) :$$

$$+ \int_{\tilde{\Lambda}^*} \frac{\hat{d}^2 p}{(2\pi)^2} \tilde{g} : \hat{\rho}_{+,0}(-\mathbf{p}) :: \hat{\rho}_{-,0}(\mathbf{p}) :, \tag{44}$$

and the nodal–antinodal interactions

$$H_{na} = \int_{\tilde{\Lambda}^*} \frac{\hat{d}^2 p}{(2\pi)^2} \sum_{r,r',s=\pm} \tilde{g}\chi_s(\mathbf{p}) : \hat{\rho}_{r,0}(-\mathbf{p}) : \hat{J}_{r',s}(\mathbf{p}), \tag{45}$$

with the coupling constants $g = 2V a^2 \sin^2(Q)$, $\tilde{g} = 2V a^2$, and the cutoff functions $\chi_s(\mathbf{p}) = \theta(\pi/\tilde{a} - |p_{-s}|)\theta(\kappa\pi/(\sqrt{2}a) - |p_s|)$; see Refs. 3 and 4 for further details, including explicit formulas for the constants $\mathcal{E}_0$ and μ_0, and how filling ν depends on Q and κ. The Hamiltonian $H = H_n + H_a + H_{na}$ makes mathematically precise a special case of the one in (4).

Similarly in 1D, one assumes $Q \neq \pi/2$ in the above derivation, and for $Q = \pi/2$ additional nodal interaction terms appear that cannot be bosonized in a simple manner and that, presumably, open a nodal gap. This is consistent with $Q = \pi/2$ corresponding to half filling.

3.3. *Bosonization in 2D*

The 2D analogues of the identities in (23) and (24) are[5]

$$[\hat{J}_{r,s}(\mathbf{p}), \hat{J}_{r',s'}(\mathbf{p}')] = r\delta_{r,r'}\delta_{s,s'} \frac{2\pi p_s}{\tilde{a}} \sum_{n\in\mathbb{Z}} \hat{\delta}^2(\mathbf{p} + \mathbf{p}' - 2\pi n \mathbf{e}_{-s}/\tilde{a}) \tag{46}$$

$$\int_{\Lambda_s^*} \hat{d}^2 k \, r k_s : \hat{\psi}_{r,s}^\dagger(\mathbf{k}) \hat{\psi}_{r,s}(\mathbf{k}) : = \tilde{a}\pi \int_{\tilde{\Lambda}} \frac{\hat{d}^2 p}{(2\pi)^2} : \hat{J}_{r,s}(-\mathbf{p})\hat{J}_{r,s}(\mathbf{p}) : . \tag{47}$$

with $r, s, r's,' = \pm$; see Ref. 3, Proposition 2.1 for a mathematically precise formulation. Moreover, there exists a generalization of (25) to 2D; see Ref. 3, Proposition 2.7.

Similarly as in 1D, the first of these identities implies that the operators

$$\hat{\Pi}_s(\mathbf{p}) = \sqrt{\frac{\tilde{a}}{4\pi}}\big(-\hat{J}_{+,s}(\mathbf{p}) + \hat{J}_{-,s}(\mathbf{p})\big)$$

$$(p_s \neq 0) \qquad (48)$$

$$\hat{\Phi}_s(\mathbf{p}) = \sqrt{\frac{\tilde{a}}{4\pi}}\frac{1}{ip_s}\big(\hat{J}_{+,s}(\mathbf{p}) + \hat{J}_{-,s}(\mathbf{p})\big)$$

are boson operators with the usual CCR, and using these operators, the nodal and mixed parts of the 2D Luttinger Hamiltonian can be written as[3]

$$H_n = \frac{v_F}{2}\sum_{s=\pm}\int_{\tilde{\Lambda}^*}\hat{d}^2p : \big((1-\gamma_s(\mathbf{p}))\hat{\Pi}_s^\dagger(\mathbf{p})\hat{\Pi}_s(\mathbf{p}) + (1+\gamma_s(\mathbf{p}))p_s^2\hat{\Phi}_s^\dagger(\mathbf{p})\hat{\Phi}_s(\mathbf{p})$$

$$+ \gamma(\mathbf{p})p_+p_-\hat{\Phi}_s^\dagger(\mathbf{p})\hat{\Phi}_s(\mathbf{p})\big) : + \cdots \qquad (49)$$

and

$$H_{na} = \int_{\tilde{\Lambda}^*}\frac{\hat{d}^2p}{(2\pi)^2}\sum_{r,s=\pm}\sqrt{\frac{4\pi}{\tilde{a}}}ip_s\tilde{g}\chi_s(\mathbf{p}):\hat{\rho}_{r,0}(-\mathbf{p}):\hat{\Phi}_s(\mathbf{p}), \qquad (50)$$

with $\gamma_s(\mathbf{p}) = \gamma\chi_s(\mathbf{p})$, $\gamma(\mathbf{p}) = \gamma\chi_+(\mathbf{p})\chi_-(\mathbf{p})$, and $\gamma = V(1-\kappa)\sin(Q)/(2\pi t)$; as before, the dots indicate zero mode terms that are $O(1/L)$.

3.4. *Relation to the Mattis model*

As described in the previous section, the 2D Luttinger Hamiltonian can be mapped exactly to a Hamiltonian describing noninteracting bosons coupled linearly to interacting antinodal fermions. It is possible to integrate out these nodal bosons exactly and thus obtain an effective model for the antinodal fermions only. In Ref. 4 we studied this latter model by conventional approximation methods including mean field theory, and we found a nontrivial region away from half filling where the antinodal fermions are gapped. We assume that the gapped fermions do not contribute to the low energy properties of the system, and we therefore proposed that the Mattis model describes the low energy physics of the 2D Luttinger model in this partially gapped phase.

3.5. *Solution of the Mattis model*

Since the nodal part of the 2D Luttinger Hamiltonian is quadratic in boson operators, it can be diagonalized by a boson Bogoliubov transformation. From this, we can compute the ground state and free energy of the nodal (Mattis) model.[5] Furthermore, there is a natural generalization of the boson-fermion correspondence formula in (25) to the two-dimensional case.[5] This formula allows to compute all fermion correlation functions exactly using standard results for free bosons. For example, we found that the fermion two-point function has algebraic decay for intermediate length scales, and with nontrivial exponents that depend on the coupling constants.

The result for the two-point function may be interpreted as a signature of Luttinger-liquid behavior in the Mattis model. We emphasize however that this does not automatically imply Luttinger-liquid behavior in the 2D t-V model of lattice fermions, even if the antinodal fermions are gapped. For this to hold true, one first needs to investigate the approximations introduced when deriving the 2D Luttinger model from the lattice system.

4. Extensions of the Mattis model

There are several ways to obtain new exactly solvable models of interacting fermions by extending the Mattis model. Below we shortly describe two such extensions.[6,7]

4.1. *A 2 + 1D quantum gauge theory with interacting fermions*

In Ref. 6, we consider the quantum gauge theory model obtained by minimally coupling the fermions in the Mattis model to a two-dimensional dynamical electromagnetic field. The gauged model shares many features with the so-called Schwinger model[42] of (1+1)D quantum electrodynamics with massless Dirac fermions; see also Refs. 43 and 44. The anomaly in (46) and the requirement of gauge invariance on the quantum level leads to a bare mass term proportional to the electric charge for the electromagnetic field, similarly as for the Schwinger model.

The Hamiltonian of the gauged model can be bosonized and subsequently diagonalized; one finds two gapped, or massive, boson modes and one gapless mode. We also computed the linear response of the magnetic field to an external current and found that there is a Meissner effect.

4.2. *Partial continuum limit of the 2D Hubbard model*

Our work on the 2D t-V model of spinless fermions can be extended to the Hubbard model [see (5)]. In Ref. 7, we applied the partial continuum limit to the 2D Hubbard model on a square lattice. This again leads to an effective QFT model of nodal fermions coupled to antinodal fermions. The nodal fermions can be bosonized using either abelian or nonabelian[45] methods, and in the latter case we obtain a natural 2D analogue of a Wess–Zumino–Witten model. Furthermore, by a specific truncation of the nodal part of the effective Hamiltonian, we obtain a spinfull variant of the Mattis model that is exactly solvable. The fundamental excitations of this model separate into independent spin and charge degrees of freedom.

5. Discussion

The high-T_c problem has been an outstanding challenge in theoretical physics for many years; see for example Ref. 46 for recent commentaries by leading experts on the status of this field. One important hypothesis underlying much of the work on the high-T_c problem is that *the 2D Hubbard model is an adequate prototype model for the cuprates.* From a theorist's point of view this hypothesis is highly appealing due to the (deceptively) simple and aesthetic form of the model. Furthermore, the model is considered "adequate" in the sense that, (i) it is possible to derive it using physical arguments from a more realistic description and, (ii) it contains enough physics to capture the qualitative properties of the cuprates, while better quantitative agreement can be obtained by straightforward extensions of the model. Still, while there seem to be little doubts about (i) and (ii), these criteria are not enough; an adequate model also requires that, (iii) it should be amenable to reliable computations and thus allow for experimental predictions. This criterion, which is obviously the most important one, remains to be fulfilled.

The results reviewed in this paper are part of a program aiming at finding a better balance between (i)–(iii) by modifying a 2D Hubbard-like lattice fermion model. Our emphasis has been on modifications that lead to exactly solvable models since, for such models, (iii) is always true. However, it is important to note that, even though we have proposed a derivation of such a model from a Hubbard-like model, we cannot claim to have shown that (i) is fulfilled: our derivation contains steps that need to be substantiated by mathematical arguments (these steps are spelled out in Refs. 3 and 5). We also stress that, up to now, (ii) remains open: we have not systematically investigated whether the Mattis model, or its various extensions, can describe the cuprates. Much remains to be done in our program.

Acknowledgments

This work was supported by the Göran Gustafsson Foundation and the Swedish Research Council (VR) under contract no. 621-2010-3708. We thank Farrokh Atai for carefully reading the manuscript.

References

1. G. D. Mahan, *Many Particle Physics* (Plenum Press, US, 1981).
2. E. Langmann, *Lett. Math. Phys.* **92**, 109 (2010).
3. E. Langmann, *J. Stat. Phys.* **141**, 17 (2010).
4. J. de Woul and E. Langmann, *J. Stat. Phys.* **139**, 1033 (2010).
5. J. de Woul and E. Langmann, Exact solution of a 2D interacting fermion model, arXiv:1011.1401 (to appear in *Comm. Math. Phys.*).
6. J. de Woul and E. Langmann, Gauge invariance, correlated fermion and Meissner effect in 2+1 dimensions, arXiv:1107.0891.

7. J. de Woul and E. Langmann, Partial continuum limit of the 2D Hubbard model, to appear.
8. S. Tomonaga, *Prog. Theor. Phys.* **5**, 544 (1950).
9. W. Thirring, *Ann. Phys.* **3**, 91 (1958).
10. J. M. Luttinger, *J. Math. Phys.* **4**, 1154 (1963).
11. D. C. Mattis and E. H. Lieb, *J. Math. Phys.* **6**, 304 (1965).
12. F. D. M. Haldane, *J. Phys. C* **14**, 2585 (1981).
13. R. Heidenreich, R. Seiler and D. A. Uhlenbrock, *J. Stat. Phys* **22**, 27 (1980).
14. D. C. Mattis, *Phys. Rev. B* **36**, 745 (1987).
15. J. G. Bednorz and K. A. Müller, *Z. Phys. B* **64**, 189 (1986).
16. A. Luther, *Phys. Rev. B* **19**, 320 (1979).
17. P. W. Anderson, *Phys. Rev. Lett.* **64**, 1839 (1990).
18. M. Disertori and V. Rivasseau, *Phys. Rev. Lett.* **85**, 361 (2000).
19. M. Salmhofer, *Renormalization: An Introduction* (Springer, New York, 1999).
20. A. T. Zheleznyak, V. M. Yakovenko and I. E. Dzyaloshinskii, *Phys. Rev. B* **55**, 3200 (1997).
21. V. Mastropietro, *Phys. Rev. B* **77**, 195106 (2008).
22. A. Luther, *Phys. Rev. B* **50**, 11446 (1994).
23. J. O. Fjærestad, A. Sudbo and A. Luther, *Phys. Rev. B* **60**, 13361 (1999).
24. O. F. Syljuåsen and A. Luther, *Phys. Rev. B* **72**, 165105 (2005).
25. F. D. M. Haldane, Luttinger's Theorem and bosonization of the Fermi Surface, in *Perspectives in Many-Particle Physics — Proceedings of the International School of Physics Enrico Fermi, Course CXXI*, eds. R. Broglia and J. R. Schrieffer (North Holland, Netherlands, 1994), p. 530.
26. A. H. Castro Neto and E. Fradkin, *Phys. Rev. Lett.* **72**, 1393 (1994).
27. A. Houghton and J. B. Marston, *Phys. Rev. B* **48**, 7790 (1993).
28. R. Hlubina, *Phys. Rev. B* **50**, 8252 (1994).
29. P. Kopietz and K. Schönhammer, *Z. Phys. B* **100**, 259 (1996).
30. J. Fröhlich, R. Götschmann and P. A. Marchetti, *J. Phys. A* **28**, 1169 (1995).
31. A. P. Polychronakos, *Phys. Rev. Lett.* **96**, 186401 (2006).
32. D. A. Bonn, *Nat. Phys.* **2**, 159 (2006).
33. P. W. Anderson, *Science* **235**, 1196 (1987).
34. V. J. Emery, *Phys. Rev. Lett.* **58**, 2794 (1987).
35. C. M. Varma, S. Schmitt-Rink and E. Abrahams, *Solid State Commun.* **62**, 681 (1987).
36. F. C. Zhang and T. M. Rice, *Phys. Rev. B* **37**, 3759 (1988).
37. E. Langmann and M. Wallin, *J. Stat. Phys.* **127**, 825 (2007).
38. K. G. Wilson, *Phys. Rev. D* **2**, 1473 (1970).
39. V. Bach, E. H. Lieb and J. P. Solovej, *J. Stat. Phys.* **76**, 3 (1994).
40. E. Langmann and M. Wallin, *Phys. Rev. B* **55**, 9439 (1997).
41. H. J. Schulz, *Phys. Rev. B* **39**, 2940 (1989).
42. J. Schwinger, *Phys. Rev.* **128**, 2425 (1962).
43. N. S. Manton, *Ann. Phys. (NY)* **159**, 220 (1985).
44. H. Grosse, E. Langmann and E. Raschhofer, *Ann. Phys. (NY)* **253**, 310 (1997).
45. E. Witten, *Commun. Math. Phys.* **92**, 455 (1984).
46. J. Zaanen *et al.*, *Nat. Phys.* **2**, 138 (2006).

LUTTINGER LIQUID, SINGULAR INTERACTION AND QUANTUM CRITICALITY IN CUPRATE MATERIALS*

C. DI CASTRO and S. CAPRARA

Dipartimento di Fisica, Università di Roma "La Sapienza"
Piazzale Aldo Moro, 2, Rome I-00185, Italy

With particular reference to the role of the renormalization group (RG) approach and Ward identities (WI's), we start by recalling some old features of the one-dimensional Luttinger liquid as the prototype of non-Fermi-liquid behavior. Its dimensional crossover to the Landau normal Fermi liquid implies that a non-Fermi liquid, as, e.g., the normal phase of the cuprate high temperature superconductors, can be maintained in $d > 1$ only in the presence of a sufficiently singular effective interaction among the charge carriers. This is the case when, nearby an instability, the interaction is mediated by critical fluctuations. We are then led to introduce the specific case of superconductivity in cuprates as an example of avoided quantum criticality. We will disentangle the fluctuations which act as mediators of singular electron–electron interaction, enlightening the possible order competing with superconductivity and a mechanism for the non-Fermi-liquid behavior of the metallic phase.

This paper is not meant to be a comprehensive review. Many important contributions will not be considered. We will also avoid using extensive technicalities and making full calculations for which we refer to the original papers and to the many good available reviews. We will here only follow one line of reasoning which guided our research activity in this field.

Keywords: Renormalization group; Luttinger liquid; quantum criticality; cuprate superconductors.

1. Introduction

1.1. *Criticality and renormalization group*

Up to the sixties of the last century the entire world of condensed matter or, more specifically, every N-body system in a stable phase was considered to be reducible to a collection of quasiparticles.

At the phenomenological level, a system of strongly interacting particles was considered for sufficiently low temperature as a gas of quasiparticles (the quasiparticles of the normal Fermi liquid theory for metals or liquid ^{3}He, the phonons and rotons of superfluid ^{4}He, gapped excitations of superconductors, spin waves in magnets ...), leaving — at most — a weak residual Hartree-type interaction.

Theoretically this approach privileged the development of techniques apt to obtain the simplified dynamics of quasiparticles for each system, starting from the microscopic description. The statistical part was trivial, a gas of excitations with at most a normal distribution of fluctuations vanishing in the thermodynamic limit, according the $1/N^{1/2}$ law.

*This article first appeared in International Journal of Modern Physics B, Vol. 26, No. 22 (2012).

Out of this scheme, a puzzling behavior arises in the proximity of criticality (the first example of what is now called "complexity") where the collective phenomena do not appear as a simple superposition of single microscopic events and the laws of great numbers are modified (violation of the $1/N^{1/2}$ law). It is not true in this case that each sufficiently large portion of the system has an average behavior independent from the rest.

At zero temperature criticality is related to the competition between two ground states, rather than between two phases. Quantum fluctuations of the zero-point motion lead to quantum critical behavior near a Quantum Critical Point (QCP), which is obtained by tuning a parameter to balance the relative energies of the two states. Anomalous finite temperature behavior arises even though the transition is driven by non-thermal parameters, as applied and/or chemical pressure, doping, magnetic field, in fragile structures like liquid-solid helium, antiferromagnets and so on.

A common aspect of critical phenomena is the singular power-law behavior of the physical responses, like the spin susceptibility in ferromagnets near the Curie point, or the compressibility near the gas-liquid critical point. The power laws are characterized by sets of so-called critical exponents, which are equal for apparently different transitions, like the para-ferromagnet or gas-liquid, provided the corresponding model systems, when expressed in terms of the order parameter φ, share the same symmetry. The order parameter characterizes the difference of the two phases (the spontaneous magnetization in the ferromagnet or the density diffference near the gas-liquid critical point) and becomes the classical field of the field-theoretic models representing the system near criticality. A prototype of these models is the Landau–Wilson φ^4-model, which is a generalization of the mean-field Landau theory of second-order phase transitions. This universal behavior observed near the critical point of phase transitions results from the fact that the correlation of the fluctuations of the order parameter extends to infinity and the system becomes scale invariant: Near a critical point, a change of the length scale leaves the physics unchanged, provided each parameter of the model is rescaled. The physical parameters μ_l are then characterized by their scaling dimension x_l in terms of an inverse length.

The RG approach[1] implements these new universal simplifying concepts and asymptotically near criticality transforms into scaling transformation thus reproducing the phenomenology of criticality. The statistical aspects become prevalent with respect to the previous procedures in many body theory of approximately solving the dynamics, thus reducing each system to a gas of quasiparticles.

Two major RG transformations were introduced, each one implementing a distinct aspect of universality.

While approaching the critical point, the details specifying a particular system do not matter.[2] The great (infinite) number of degrees of freedom involved as the correlation length ξ of the order parameter goes to infinity, is well-accounted for by the small subset of parameters, which are considered to be relevant. Once the proper choice of the relevant parameters is made, e.g., the deviation from the critical temperature and the order parameter φ (or its conjugate field h), we can change the other parameters of the Hamiltonian, specifying the details of each system, and maintain the physics unchanged, provided we renormalize the relevant parameters correspondingly.

In 1969 the field theoretical RG approach was introduced in critical phenomena simultaneously in Russia[3–6] and in Rome.[7–9] The field theoretical RG implements this first form of universality and relates one model system to another by varying the coupling and suitably renormalizing the other variables and the correlation functions (vertices and propagators) to take care of the singularities in perturbation theory.

Since the correlation length ξ goes to infinity, the degrees of freedom at short distance do not influence the critical behavior. The original system and the system expressed in terms of block-variables, related to larger and larger cells, should have the same critical behavior. In 1971, Wilson[10–12] implements this second idea of universality by introducing his new RG transformation with the elimination of degrees of freedom at large wavevectors, i.e., at short distances.

Both RG transformations aim to describe the system near the critical point in terms of few renormalized parameters, which correspond to a finite set of relevant directions with respect to a given fixed point. Relevant and renormalized parameters coincide asymptotically in the infrared region.

When the model parameters do not change anymore by iterating the transformation (fixed point), scaling and expressions of critical indices follow. Universality classes (yielding sets of critical indices common to different systems with the same symmetry) appear as domains of attraction of different fixed points.

The eigenvectors of the linearized RG transformation around each fixed point with negative scaling dimensions ($x_l < 0$) define the tangent plane to the critical surface and correspond to the irrelevant variables. The eigenvectors with positive scaling dimensions ($x_l > 0$) define the directions of escape from critical surface and are the relevant variables. The so-called marginal eigenvectors with scaling dimensions $x_l = 0$ may occur. In this case, the critical indices or the scaling dimensions depend on the marginal parameters of the model, as in the Luttinger liquid (see Sec. 2.4).

1.2. *Use of renormalization group for systems in stable phases and singular perturbation theory*

The most successful use of RG is indeed in critical phenomena with the asymptotic summation of infrared *singular* perturbative terms to give, for $d < 4$, the singular power law behavior of physical response functions.

A different, more recent, application of RG deals with cases when perturbation theory is singular even in stable phases with *finite* physical response functions, e.g., for the two cases of interacting fermions at $d = 1$[13–15] and interacting bosons with Bose–Einstein condensation for $d \leq 3$.[16,17] The case of interacting fermions at $d = 1$ was the first case to call into question the picture of a collection of quasiparticles given by the Landau theory of normal Fermi liquid.

Both cases deal with stable liquid phases far from criticality and finite response functions, which therefore require exact cancellation of singularities to all orders in perturbation theory, instead of their summation into a power law singular behavior as in critical phenomena.

In critical phenomena, a general problem in the renormalization is to initialize the action to follow, in the parameter space of the Hamiltonian, a renormalized trajectory with a small number of flowing parameters compared to the enormous number of degrees of freedom strongly correlated within a coherence distance ($\xi \to \infty$). This subtle procedure of filtering out those variables appropriate to the description of critical systems and of their self organization cannot be carried out in general without a knowledge of the fundamental symmetry inherent to each specific problem, and in particular of the order parameter, to make the proper choice of the basic variables entering the field theoretic model, e.g., the Landau–Wilson model, on which the RG transformation acts.

Moreover, symmetry properties can be translated into Ward Identities (WI's), which establish relations among the various terms of the skeleton structure of the problem and their connection to physical quantities, thus in general simplifying the RG treatment (see, for example, the case of disordered electron systems.[18–20]

In the case of singular perturbation theory in stable phases, additional symmetries and related WI's have to be present to implement the cancellation of singularities in physical responses, giving finite results despite a singular perturbation theory. In the cases of the Bose liquid (not treated here)[16,17] and of interacting fermions in $d = 1$ (Luttinger model), the additional WI controlling this cancellation allows for a closure of the hierarchical equations of the response functions,[13,14,21] thus providing the exact low-energy description of these systems. Exact solution of one-dimensional systems with forward scattering, the Luttinger liquid, was also previously obtained via bosonization in Refs. 22–26.

The crossover of the Luttinger liquid to the Fermi liquid in $d > 1$[15,27] and the non-Fermi-liquid behavior of an interacting electron system with singular forward interaction in $d > 1$[28–30] are also controlled by specific WI's and RG transformation. Extension of the bosonization technique to $d > 1$[31,32] was also used in this last case.[33–37]

1.3. *Normal Fermi liquid and its breakdown in cuprates*

In the absence of symmetry breaking, the description of the low energy behavior of metals in terms of a small set of parameters can be achieved via the RG scheme by an iterated elimination of high-energy degrees of freedom far from the Fermi surface (see, e.g., Ref. 38).

The concept of a normal Fermi liquid, which successfully applies to liquid ^{3}He and to ordinary metals in $d = 3$, relies on the existence of a fixed point Hamiltonian with asymptotically free low-lying excitations, i.e., the quasiparticles, with discontinuous occupation number in momentum space n_k at $T = 0$ at the Fermi surface ($k = k_F$),

$$|n_{k_{F+}} - n_{k_{F-}}| = Z < 1\,,$$

where $k_{F\pm}$ indicates the limit $k \to k_F^{\pm}$ from above or below. This discontinuity still marks the Fermi surface in the interacting system and its *finite* reduction with respect to the Fermi gas ($Z = 1$) is given by the finite "wavefunction renormalization" Z, renormalizing in this case the Fermi field operators. The presence of the discontinuity at the Fermi surface, together with the Pauli principle, compel the inverse quasiparticle lifetime to be $\tau^{-1} \approx \max(T^2, \varepsilon^2)$, where ε is the deviation of the energy of the quasiparticle from the Fermi energy. The resistivity at finite temperature due to the electron–electron interaction is then proportional to T^2. The energy uncertainty $h\tau^{-1}$ due to the finite lifetime of a quasiparticle near the Fermi surface is small compared to its energy ε and the quasiparticle concept is well-defined.

All the momentum transferring scattering processes become asymptotically ineffective and the expressions of the specific heat, the spin susceptibility and the compressibility or the Drude peak of a Fermi gas are still valid but for *finite* multiplicative renormalizations due to the residual Hartree type interaction among the quasiparticles.

The Fermi liquid breaks down if the quasiparticle spectral weight at the Fermi surface is suppressed by the presence of an interaction-induced anomalous scaling dimension $\eta > 0$ in $Z \approx |k - k_F|^{\eta}$, where $k - k_F$ measures the deviation of the quasiparticle momentum from the Fermi momentum. Z vanishes at the Fermi surface, the low-lying single particle excitations are suppressed and the quasiparticle concept loses its validity. The low-energy behavior of the system is dominated by the charge and spin collective modes. These modes propagate with different velocities, leading to the so-called charge and spin separation, as opposed to the behavior of a normal Fermi liquid, where quasiparticle excitations simultaneously carry charge and spin. This non-Fermi-liquid behavior is achieved in one-dimensional interacting electron system. The metallic phase is the so-called Luttinger liquid with a non-universal index η, finite charge and spin density responses in the long wavelength limit and the corresponding propagating low-lying collective modes.

The main motivation for discussing non-Fermi-liquid behavior in general comes from numerous experimental evidences in several systems. In particular the metallic phase of superconducting cuprates is not a Fermi liquid (see, e.g., Ref. 39). These materials are insulating with antiferromagnetic (AFM) long-range order when stoichiometric, and become strange metals and then superconducting upon chemical doping (see Sec. 4, Fig. 7). The cuprates are strongly anisotropic materials, and their structure consists of copper–oxygen planes intercalated with rare-earth slabs. The doped charge carriers are introduced in the copper–oxygen planes giving rise to quasi two-dimensional systems. When the chemical doping x is such that the critical temperature T_c for the onset of superconductivity reaches its maximum value (optimal doping), the in-plane resistivity is linear in temperature (rather than quadratic, as in Fermi liquids) over a wide temperature range. Below the optimal doping, in the so-called underdoped region, the anomalous metallic behavior is even stronger and a pseudogap opens below a doping dependent crossover temperature $T^*(x)$, affecting thermodynamic, spectroscopic, and transport properties. Furthermore, in contrast to the Fermi-liquid behavior, the inverse quasiparticle scattering time around optimal doping is linear in temperature or frequency, the energy uncertainty becomes of the same order of the energy, and the quasiparticle concept loses its validity. The appearance of a novel metallic phase, at least as relevant as the appearance of high temperature superconductivity itself, has opened a wide theoretical debate.

Among the various suggestions for a breakdown of the Fermi-liquid behavior,[39,40] the possibility that a Luttinger-liquid-like phase could be extended to two dimensions has been considered.[41,42] We will indicate the constraints that have to be satisfied by the effective electron–electron interaction to extend the non-Fermi-liquid behavior to dimension higher than one. This led to consider as a further possibility for the occurrence of non-Fermi-liquid behavior that the system is close to an instability.[39,43–48] The nearby critical fluctuations couple to the charge carriers giving rise to an effective interaction among the fermionic quasiparticles singular enough to destroy the Fermi-liquid behavior.

Critical fluctuations can be due to a charge instability [incommensurate charge density wave[43,44] (CDW)] of the Fermi liquid coming from the markedly metallic overdoped region, to a magnetic instability[45–48] at low doping, or to a combination of the two, with a stripe like modulation of the charge and spin density.[49,50] We will deal with the quantum criticality related to the CDW as the onset of stripe phase in high-T_c superconducting cuprates. Unveiling the critical modes that act as mediators of the effective interaction may also be a key to understand the elusive electronic order, which competes with superconductivity in the underdoped region.[51]

2. The Luttinger Liquid and its Crossover to the Fermi Liquid

2.1. *The Tomonaga-Luttinger (TL) model*

In one-dimensional systems the Fermi surface consists of two points αk_F, where $\alpha = \pm$ refer to right- and left-moving fermions, respectively. The peculiar Fermi surface leads in general to consider few scattering processes that are incorporated in the continuous low-energy model known as the "g-ology model".[52] This model includes the small momentum transfer scattering processes (g_2 and g_4), the back scattering (g_1) and Umklapp scattering (g_3), as depicted in Fig. 1. Many features of the g-ology model were already clear within perturbative renormalization calculations.[52]

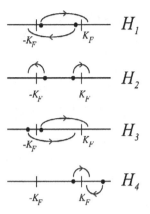

Fig. 1.   The relevant scattering processes considered within the g-ology model. The dots indicate the initial states of two scattering quasiparticles and the arrows point to their final states.

For the sake of simplicity, we will only refer to the Tomonaga–Luttinger (TL)[53,54] model with linear spectrum ($\varepsilon_k = v_F \mathbf{k}$), a bandwidth cut-off Λ and the couplings g_4 and g_2 for the two forward scattering processes depicted in Fig. 1. The Hamiltonian $H = H_0 + H_2 + H_4$ (H_0 is the free particle Hamiltonian) is specified by

$$H_0 = \sum_{|\mathbf{k}|<\Lambda} \alpha v_F \mathbf{k} a^+_{\alpha,\sigma}(\mathbf{k}) a_{\alpha,\sigma}(\mathbf{k}) \,,$$

$$H_2 = \frac{1}{V} \sum_{\mathbf{q}} \sum_{\sigma\sigma'} g_2^{\sigma\sigma'} \rho_{+,\sigma}(\mathbf{q}) \rho_{-,\sigma'}(-\mathbf{q}) \,, \tag{1}$$

$$H_4 = \frac{1}{2V} \sum_{\mathbf{q}} \sum_{\sigma\sigma'} g_4^{\sigma\sigma'} \sum_{\alpha} \rho_{\alpha,\sigma}(\mathbf{q}) \rho_{\alpha,\sigma'}(-\mathbf{q}) \,.$$

Henceforth, v_F indicates the Fermi velocity. We adopt the boldface notation for the momentum $\mathbf{k}$, measured relative to $\pm k_F$, even in $d = 1$, to distinguish from bivectors $k = (\mathbf{k}, \varepsilon)$, which include the energy variable ε. The couplings may be spin dependent,

$$g_i^{\sigma\sigma'} = g_{i\parallel} \delta_{\sigma,\sigma'} + g_{i\perp} \delta_{\sigma,-\sigma'} \,,$$

and

$$\rho_{\alpha,\sigma}(\mathbf{q}) = \sum_{\mathbf{k}} a^+_{\alpha,\sigma}(\mathbf{k} - \mathbf{q}) a_{\alpha,\sigma}(\mathbf{k}) \tag{2}$$

are the density operators for right ($\alpha = 1$) and left ($\alpha = -1$) movers in terms of creation and destruction operators of spin $1/2$ fermions with spin projection σ, $a^+_{\alpha,\sigma}(\mathbf{k})$ and $a_{\alpha,\sigma}(\mathbf{k})$. The discrete structure of the Fermi surface of one-dimensional fermions in the presence of forward scattering implies the separate charge and spin conservation in the low energy processes for particle near the left and right Fermi points respectively. The related symmetry[13–15,21] allows for the exact solution of the model, yielding the Luttinger liquid behavior. We will exploit the WI's following from the above symmetry to constrain the structure of the RG equations to have a line of fixed points with an anomalous dimension η in Z depending on the couplings g_2 and g_4. The same symmetry ensures the finiteness of the compressibility and the spin susceptibility, thus making possible a metallic phase with no quasiparticles as low-lying excitations.

We give now few technical definitions. The physical content of a system resides in the Green's functions defined as the ground-state expectation value (angular brackets) of the time-ordered operator products:

$$
\begin{aligned}
G^{(2n,l)}&(\mathbf{k}_1, t_1, \ldots; \mathbf{k}_1', t_1', \ldots; \mathbf{k}_1'', t_1'', \ldots) \\
&= (-i)^{n+l} \langle \tau_t a_{\alpha_1 \sigma_1}(\mathbf{k}_1, t_1) \cdots a_{\alpha_n \sigma_n}(\mathbf{k}_n, t_{n1}) \times a^+_{\alpha_1' \sigma_1'}(\mathbf{k}_1', t_1') \cdots a^+_{\alpha_n' \sigma_n'}(\mathbf{k}_n', t_n') \\
&\quad \times \rho_{\alpha_1'' \sigma_1''}(\mathbf{k}_1'' t_1'') \cdots \rho_{\alpha_l'' \sigma_l''}(\mathbf{k}_l'' t_l'') \rangle \,,
\end{aligned}
\tag{3}
$$

where a^+, a and ρ are the operators in the Heisenberg representation and τ_t is the time-ordering operator, and analogously for the spin density operators. The single-particle Green's function (the propagator) is given by:

$$
G_{\alpha\sigma} \equiv G^{(2,0)}(\mathbf{k}, t; \mathbf{k}, 0) = -i \langle \tau_t a_{\alpha\sigma}(\mathbf{k}, t) a^+_{\alpha\sigma}(\mathbf{k}, 0) \rangle \,,
$$

where we used the fact that the expectation value vanishes for different α, σ, and $\mathbf{k}$. The Green's functions in the frequency representation will be denoted by $G^{(2n,l)}(\{k\})$ and are obtained as a Fourier transform of Eq. (3) with respect to the time arguments. The single-particle Green's function for the free case is $G_0 = [\varepsilon - \alpha v_F k + \alpha \mathrm{sgn}(\mathbf{k})i0^+]^{-1}$, with $Z = 1$. In the diagrammatic representation of perturbation theory, the single-particle Green's function is represented by a solid line and the interaction by a dashed line. In a given diagram, external lines correspond to ingoing and outgoing particles. The diagrams for $G^{(2n,l)}(\{k\})$ are classified in diagrams that can or cannot be divided into disjoint parts with cutting an internal line. The latter are called one-particle irreducible. By truncating the external lines of the one-particle irreducible diagrams for $G^{(2n,l)}(\{k\})$ one obtains the corresponding vertices $\Gamma^{(2n,l)}(\{k\})$. The four-point vertex $\Gamma^{(4)} \equiv \Gamma^{(4,0)}(\{k\})$ represents the fully dressed interaction. In order to extract the physical properties from the Green's functions, one should in principle solve an infinite set of hierarchical equations, since the equations of motion couple a given Green's function with higher order Green's functions and vertices. The additional WI's present in the TL model allow for a closure of this set and for its full solution.

Let us recall that for ordinary critical phenomena, the paradigmatic model is the Landau–Wilson φ^4-model:

$$
H = \int (c|\nabla\varphi|^2 + r\varphi^2 + u\varphi^4) d^d x = \sum_k [(ck^2 + r)\varphi^2 + u\varphi^4] \,.
$$

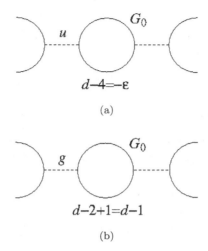

Fig. 2. (a) Power counting to determine the dimension of the effective coupling in critical phenomena. (b) Power counting to determine the dimension of the effective coupling in the TL model.

The field φ represents the order parameter of the broken-symmetry phase, $r \approx T - T_c$ is the deviation from the critical temperature, and u is the coupling of the fields with respect to the free (φ^2) Gaussian case. Given H to be dimensionless, when measured in units of the finite characteristic energy $k_B T_c$, it turns out that r has bare canonical dimensions in inverse length equal to 2, the field φ has bare dimensions $(d-2)/2$ and the coupling u has bare dimension $4-d$. The perturbation theory with respect to the bare dimensionless coupling $u/|r|^{(4-d)/2}$ becomes meaningless at criticality $(r \to 0)$ for $d < 4$, and a renormalized theory is required. The infrared singularities for $d < 4$ are manifest already in the first perturbative correction to the coupling u, i.e., to the four point vertex function depicted in Fig. 2(a). The loop correction consists of a momentum integration over two free Gaussian propagators (the two lines) $G_0 = (ck^2 + r)^{-1}$ and a simple power counting gives four as upper critical dimensionality, where logarithmic singularities appear. Below $d = 4$, stronger infrared singularities persist in perturbation theory and one needs the use of RG to take care of these singularities.

The singular behavior of perturbation theory in terms of the couplings g_i for the TL model appears again by power counting at $T = 0$. Indeed, a dimensional analysis of the model yields the canonical dimensions in inverse length: $[H] = 1$, $[a(k)] = -1/2$ and $[g_i] = 0$, i.e., g_i's are marginal. In the one loop correction to the four-point vertex function [see Fig. 2(b)], one has d for the momentum and 1 for the frequency integration (at $T = 0$, the time or its conjugate variable the frequency appear as additional variables), -2 for the two free propagators (the single particle Green function has in this case a linear dependence on momenta, rather then quadratic, and $d + 1 - 2 = d - 1$). This shifts the logarithmic infrared singularities from $d = 4$ (in the case of ordinary critical phenomena) to $d = 1$ in the TL model.

We will show that in this case one parameter Z, renormalizing the inverse single particle Green's function (the propagator), is enough to fully renormalize the TL model. In this case, Z vanishes at the Fermi points, driving the non-Fermi-liquid behavior.

2.2. *The total charge and spin conservation and the related Ward identities*

The total charge ($a = c$) and spin ($a = s$) density operators $\rho^a(\mathbf{q})$ and the right and left charge and spin density operators $\rho^a_\alpha(\mathbf{q})$ are given by:

$$\rho^a(\mathbf{q}) = \sum_\alpha \rho^a_\alpha(\mathbf{q}),$$

$$\rho^c_\alpha = \rho_{\alpha\uparrow}(\mathbf{q}) + \rho_{\alpha\downarrow}(\mathbf{q}), \qquad (4)$$

$$\rho^s_\alpha = \rho_{\alpha\uparrow}(\mathbf{q}) - \rho_{\alpha\downarrow}(\mathbf{q}).$$

The right and left density operators ρ^a_α satisfy the commutation relation[13]:

$$[\rho^a_\alpha(\mathbf{q}), \rho^a_{\alpha'}(\mathbf{q}')] = \frac{V}{\pi} \delta_{\alpha,\alpha'} \delta_{\mathbf{q},-\mathbf{q}'} \alpha \mathbf{q}.$$

From the above commutation relation, the continuity equation related to the total charge (spin) conservation is obtained:

$$i\partial_t \rho^a(\mathbf{q}, t) = [\rho^a(\mathbf{q}, t), H] = \mathbf{q} v^a j^a(\mathbf{q}, t),$$

$$v^a = v_F + \frac{1}{\pi}(g^a_4 - g^a_2), \qquad (5)$$

where $j^a = \sum_\alpha \alpha \rho^a_\alpha$ is the current operator, which gives the physical current when multiplied by the interaction-dependent velocity v^a, and $g^{c,s}_i = 1/2(g_{i\|} \pm g_{i\perp})$.

These conservations can be transformed into WI's by applying $i\partial_t$ to any correlation function and using the continuity equation. Without making explicit derivations, we summarize the main results[13–15]: The total conservations connects vertices and correlations, e.g., the one particle Green's function G is related with the charge (spin) density ($\mu = 0$) and current ($\mu = 1$) vertices $\Lambda^a_{\alpha,\sigma;\mu}(k, q)$:

$$\omega \Lambda^a_{\alpha,\sigma;0}(p, q) - v_F \mathbf{q} \Lambda^a_{\alpha,\sigma;1}(p, q) = \varepsilon^a_\sigma [G^{-1}_{\alpha,\sigma}(p - q/2) - G^{-1}_{\alpha,\sigma}(p + q/2)], \qquad (6)$$

with $\varepsilon^a_\sigma = -1$ if $a = s$ and $\sigma = \downarrow$ and $\varepsilon^a_\sigma = +1$ otherwise. The truncated vertices $\Lambda^a_{\alpha,\sigma;\mu}(k, q)$ are obtained from the one-particle irreducible part of $G^{(2,1)}$ with $\rho_{\alpha\sigma}$ substituted by ρ^a ($\mu = 0$) and j^a ($\mu = 1$) in Eq. (3), respectively. These vertices are irreducible also with respect to cutting an interaction line. Notice that v_F, instead of v^a, appears in the WI, Eq. (6), for the irreducible vertices.[13,14] This WI is valid for all systems with total conservation. By itself is of course not enough to solve any model.

2.3. *Left and right conservation laws and the additional Ward identities*

In addition to the usual total charge and spin conservation, the discrete structure of the Fermi systems in $d = 1$ allows for more stringent conservation laws when large momentum scattering processes are absent, as in the TL model: charge (spin) density near the left and right Fermi points is conserved separately.

Let us introduce the so-called axial charge (spin) density as the difference operator $\tilde{\rho}^a = \rho^a_+ - \rho^a_- = j^a$. These quantities also are now conserved and obey the continuity equation:

$$i\partial_t \tilde{\rho}^a(\mathbf{q}, t) = [\tilde{\rho}^a(\mathbf{q}, t), H] = \mathbf{q}\tilde{v}^a \tilde{j}^a(\mathbf{q}, t), \qquad (7)$$

with the axial current $\tilde{j}^a = \rho^a_+ + \rho^a_- = \rho^a$ and the interaction-dependent velocity $\tilde{v}^a = v_F + (g^a_4 + g^a_2)/\pi$.

One can define the correlation functions and the vertices for the axial density and current and, in complete analogy with the total charge ($a = c$) (spin, $a = s$) conservation, additional WI's are derived with v^a substituted by $\tilde{v}^a$. The main consequences are:

(1) Combining the WI's from global and axial conservation laws, the low frequency and momentum limits of the density–density and current–current correlation functions are now determined in terms of v^a and $\tilde{v}^a$. The finite compressibility κ, spin susceptibility χ and the Drude weight $\mathrm{Re}\sigma(\omega)$ (σ is the conductivity) are:

$$\kappa = 2/(\pi\tilde{v}^c), \quad \chi = 2/(\pi\tilde{v}^s), \quad \mathrm{Re}\sigma(\omega) = 2v^c\delta(\omega),$$

(2) The continuity Eqs. (5) and (7) for ρ^a and $\tilde{\rho}^a$ are easily combined into a harmonic-oscillator equation for ρ^a:

$$i\partial_t \rho^a(\mathbf{q}, t) + v^a \tilde{v}^a |\mathbf{q}|^2 \rho^a(\mathbf{q}, t) = 0,$$

hence it follows that the undamped collective charge and spin modes move with different velocities $u^a = (v^a \tilde{v}^a)^{1/2}$ and the specific heat is linear in temperature:

$$C_V = \frac{\pi}{6} \frac{u^c + u^s}{u^c u^s} T.$$

The contribution to C_V comes here from collective modes and not from quasiparticles, as in the Fermi liquid.

(3) It is clear from the separate right and left conservation that the current vertex and the density vertex differ only from the sign of the particle velocity:

$$\Lambda^a_{\alpha,\sigma;1}(p, q) = \alpha\Lambda^a_{\alpha,\sigma;0}(p, q). \qquad (8)$$

From Eq. (6) we can now eliminate the current vertex and obtain:

$$(\omega - \alpha v_F \mathbf{q}) \Lambda^a_{\alpha,\sigma;0}(p,q) = \varepsilon^a_\sigma [G^{-1}_{\alpha,\sigma}(p-q/2) - G^{-1}_{\alpha,\sigma}(p+q/2)] \,. \tag{9}$$

This WI can be formally solved to express the density vertex, shortly named Λ_0 from now on, in terms of G only. This is the key point to produce a closed solution for the single particle Green's function G, see Sec. 2.5. We henceforth drop the index α to simplify the notation, making the dependence on α of the various quantities explicit when necessary.

2.4. The Luttinger liquid line of fixed points

The only dramatic change introduced by the Luttinger liquid solution is in the single particle Green's function G, which maintains all the singular aspects of the problem. We have to show that no other renormalization is required besides the renormalization Z for G. The bare couplings g_i are dimensionless. From the perturbative corrections to the four-point vertex $\Gamma^{(4)}$, one sees that, at each order, a vertex and an integration over two propagators G are added. This is also the skeleton structure of the renormalized theory. If in the renormalized quantities the renormalization Z^{-2} for the two propagators and the vertex renormalization Z_4 compensate each other, the RG transformation would not act on the couplings and the system would have therefore a line of fixed points in terms of their bare values, providing an anomalous behavior in G only.

The WI, Eq. (9) allows us to eliminate the density vertex Λ^a_0 in favor of G^{-1}. We should therefore introduce Λ^a_0 in the argument instead of $\Gamma^{(4)}$. In analogy with the skeleton diagrammatic structure of quantum electrodynamics, where the role of the interaction is played by the photon propagator, any Green's or vertex function can be expressed in terms of the propagator G, the irreducible charge (spin) vertex Λ^a_0 and the effective interaction D^a, obtained as a resummation of the couplings g^a (analogous to the photon propagator in quantum electrodynamics). In the skeleton structure of the four point vertex $\Gamma^{(4)}$, Λ^a and D^a appear in the combination $(\Lambda^a)^2 D^a$ (see Fig. 3). For small q, D^a can be obtained as RPA summation of *bare* bubbles Π^a_0 for each coupling g^a, according to Fig. 4.

It is enough to show that the renormalized polarization bubble Π^a coincides with Π^a_0 for the non-interacting system. The vertex and self-energy corrections of the renormalized bubbles indeed cancel each other due to Eq. (9), as depicted in Fig. 5.

In conclusion, it is important to notice that the effective interaction D^a, resulting from the resummation in Fig. 4, has zero dimensionality, i.e., due to the WI, Eq. (9), does not require to be renormalized and is marginal. On the other hand, the renormalization of the vertex Λ^a, again according to Eq. (9), is equal to the renormalization Z of G^{-1}. Therefore, the four point vertex $\Gamma^{(4)}$ (see Fig. 3) renormalizes as $(\Lambda^a)^2$ i.e., $Z_4 = Z^2$, which therefore cancels the renormalization Z^{-2} of the two

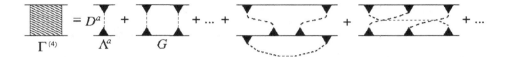

Fig. 3. Perturbative corrections to the four point vertex $\Gamma^{(4)}$ (filled square): the vertex Λ^a (black triangle) and the effective interaction D^a (dashed line) always appear in the combination $(\Lambda^a)^2 D^a$.

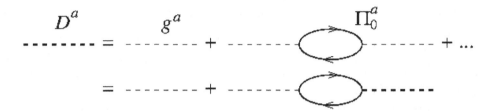

Fig. 4. The effective interaction D^a (thick dashed line) results from the RPA resummation of the bare interaction g^a (thin dashed line) dressed by the bare fermion polarization bubble Π_0^a (thin solid line).

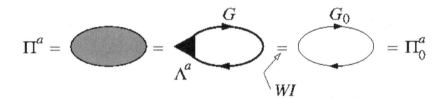

Fig. 5. The cancellation of self-energy and vertex corrections to the polarization bubble is enforced by the WI, Eq. (9). Here G is the full propagator and G_0 is the free-particle propagator.

propagators. This is the requirement to obtain a line of fixed points in terms of the bare couplings g_i. It can be further shown that no other renormalizations are required in the TL model.[13–15] This result assigns a different universality class to the Luttinger liquid with respect to the Fermi liquid.

2.5. *Solution for the fermion Green's function*

Once the existence of the line of fixed points has been ascertained, a RG perturbative calculation of Z, and thus of the asymptotic behavior of G, could be obtained. For our purposes, it is however more convenient to proceed to calculating G directly from its Dyson equation with the help of the WI, Eq. (9).[15,21]

The effective interaction D, entering into the Dyson equation, resummed according to Fig. 4 and with respect to the index a, contains all the information of the charge and spin collective modes and all the couplings via their velocities[15]

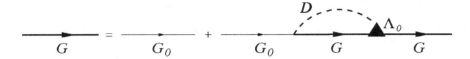

Fig. 6. Dyson equation for the fermion Green's function.

$$u^a = (v^a \tilde{v}^a)^{1/2},$$

$$D(q) = \pi(\omega - \alpha v_F \mathbf{q}) \sum_{a=c,s} \left[\frac{(2 - \eta^a)(u^a - v_F)}{\omega - u^a \mathbf{q}} + \frac{\eta^a(u^a + v_F)}{\omega + u^a \mathbf{q}} \right], \quad (10)$$

where $\eta^a = [(v^a/\tilde{v}^a)^{1/2} + (\tilde{v}^a/v^a)^{1/2} - 2]/4$ and $\eta = \eta^c + \eta^s$, as it is shown in the next section, turns out to be the anomalous index of G, which depends on the bare couplings through the velocities.

Once D is known, since the density vertex can be eliminated by means of the WI, Eq. (9), the Dyson equation for the fermion propagator (Fig. 6) becomes a closed integral equation

$$(\omega - \alpha v_F \mathbf{p})G(p) = 1 - \int \frac{d\mathbf{q}d\omega'}{(2\pi)^2} \frac{D(q)G(p-q)}{\omega - \omega' - \alpha v_F \mathbf{q}}. \quad (11)$$

The solution of Eq. (11) can be obtained after a Fourier transform to real space and time, and has the form:

$$G(\mathbf{r}, t) = e^{L(\mathbf{r},t) - L(0,0)} G_0(\mathbf{r}, t), \quad (12)$$

where $L(\mathbf{r}, t)$ is the Fourier transform of $iD(q)[\omega - \alpha v_F \mathbf{q} + i0^+ \text{sgn}(\omega)]^{-2}$, and the free-particle Green's function is $G_0(\mathbf{r}, t) = (1/2\pi)[|\mathbf{r}| - v_F t + i0^+ \text{sgn}(t)]^{-1}$. Using Eq. (10) for D, it is found that $L(\mathbf{r}, t)$ behaves logarithmically in $|\mathbf{r}|$ and t. Since $L(\mathbf{r}, t)$ appears in the exponent of Eq. (12), G behaves as a power law, with an anomalous exponent that depends on g_2 and g_4 through v^a and $\tilde{v}^a$. The single-particle excitation density and the discontinuity in the momentum distribution of the quasiparticles vanish at the Fermi surface with the anomalous exponent η.

The single-particle propagation is realized in a complex way, and results from the superposition of collective charge and spin modes, which propagate with different velocities (charge and spin separation).

3. Crossover of the Luttinger Liquid to the Fermi and Anomalous Liquids

3.1. *Crossover with short-range interactions*

The effect of increasing the space dimensionality on the Luttinger liquid can be evaluated by extending the WI approach to continuous dimension $1 < d < 2$.[15,27] As generalization of the TL model in $d > 1$, we assume a model Hamiltonian for the low-lying excitations close to the Fermi surface with dominating forward small q scattering processes ($|\mathbf{q}| < \Lambda \ll k_F$).

The extension of the theory to non-integer dimension $1 < d < 2$ is accomplished, as customary, by analytical continuation of the Feynman diagrams to the complex d plane. It is sufficient to continue momentum integrals of functions f that depend on the momentum $\mathbf{k}$ only via $|\mathbf{k}|$ and the angle θ between $\mathbf{k}$ and another fixed momentum $\mathbf{p}$. Then

$$\int d^d\mathbf{k} f(\mathbf{k}) = S_{d-1} \int d|\mathbf{k}||\mathbf{k}|^{d-1} \int_0^\pi d\theta (\sin\theta)^{d-2} f(|\mathbf{k}|, |\mathbf{k}||\mathbf{p}|\cos\theta), \qquad (13)$$

where S_d is the surface of the unit sphere in d dimensions. When $d \to 1$, $S_{d-1} \approx d-1$, and $S_{d-1}(\sin\theta)^{d-2} \to \delta(\theta) + \delta(\pi - \theta)$. The steps leading in $d = 1$ to the additional WI's, Eqs. (8) or (9), are no longer strictly valid for $d > 1$. Nonetheless, for $d < 2$, both equations are still asymptotically valid, when the exchanged momenta $\mathbf{q}$ are small, since the typical integrals in Eq. (13) are peaked at $\theta = 0$, π. In particular, since the relevant $\mathbf{k}$ vectors are asymptotically bound to be either parallel or antiparallel, near the Fermi surface we can still express the current vertex in terms of the density vertex, like in Eq. (8),

$$\mathbf{\Lambda}(p, q) = v_F \hat{\mathbf{p}} \Lambda_0(p, q), \qquad (14)$$

where $\mathbf{\Lambda}$ and Λ_0 are the current and density vertex, respectively, $\hat{\mathbf{p}} = \mathbf{p}/p_F$, and

$$\Lambda_0(p, q) = \frac{G^{-1}(p + q/2) - G^{-1}(p - q/2)}{\omega - v_F \hat{\mathbf{p}} \cdot \mathbf{q}}. \qquad (15)$$

Inserting Eq. (15) into the Dyson equation, one obtains again the Green's function in a form similar to Eq. (12), where however $L(\mathbf{r}, t)$ is now the Fourier transform of $iD(q)[\omega - v_F q_r + i0^+ \text{sgn}(\omega)]^{-2}$ and $q_r = \mathbf{q} \cdot \hat{\mathbf{p}}$ is the *radial* component of $\mathbf{q}$. The expression for $L(\mathbf{r}, t)$ involves now an average of the effective interaction $D(q)$ over the $d - 1$ components of the transverse momentum,

$$\bar{D}_\Lambda(q_r, \omega) = \frac{S_{d-1}}{(2\pi)^{d-1}} \int_0^{\sqrt{\Lambda^2 - q_r^2}} dq_t q_t^{d-2} D\left(\frac{q_r}{\omega}, \frac{q_t}{\omega}\right). \qquad (16)$$

When $d \to 1$ one recovers the exact expression for D in the TL. When $1 < d < 2$, the effective interaction, which was marginal for $d = 1$, scales to zero in the infrared, since Eq. (16) implies the scaling relation $\bar{D}(q_r, \omega) \approx \omega^{d-1} \bar{D}(q_r/\omega)$. This result clearly illustrates the marginality of scattering processes with small transferred momenta in $d = 1$, and their irrelevance in higher dimensions, in the case of regular (i.e., short ranged) scattering, as in the normal Fermi liquid.

3.2. *Singular interactions*

Singular interactions can compensate the vanishing of the effective interaction $\bar{D}$ as ω goes to zero, and extend in this way the non-Fermi-liquid behavior to dimension greater than one.

According to our previous discussion, in $1 < d < 2$ forward processes still dominate the scattering at low energy, and the additional WI is asymptotically valid. In systems with short-range e–e interaction, however, when the e–e interaction is dressed by the collective modes, the integral over the transverse momentum leads to a suppression of the mixing between the single particle and the collective modes. The system crosses over to a FL behavior.

A non-FL behavior can be achieved in the presence of a sufficiently singular long-range (LR) e–e interaction,[28–30,33–37] which we write in the form

$$V_{\mathrm{LR}} = \frac{(g_{\mathrm{LR}})^2}{2} \frac{1}{q^\alpha} \,.$$

This interaction strongly depends on the exchanged wavevector q (as a power law with exponent $-\alpha$); its strength is given in terms of the coupling g_{LR}. As we have seen in Sec. 2.4, the bare effective interaction V_{LR} (playing here the role of the g's in the g-ology model), must be dressed by the RPA series. Again, as in the $d = 1$ case, the irrelevance of self-energy and vertex corrections to the RPA resummation in the presence of singular interaction can be confirmed by means of a RG analysis.[30] The RPA resummation with the bare bubbles acquires a complicated momentum and frequency dependence. For energies $\omega < v_F q$, the effective interaction becomes short-ranged, due to the particle-hole screening. However, in the opposite limit, $\omega > v_F q$, the effective interaction becomes singular,

$$D_{\mathrm{LR}}(\mathbf{q}, \omega) = \frac{1}{q^\alpha} \frac{\omega^2}{\omega^2 - c^2 q^{2-\alpha}} \,, \tag{17}$$

with $c \propto g_{\mathrm{LR}}$. The poles of D_{LR} describe undamped collective plasmon excitations, which stay gapless for arbitrary d, provided $\alpha < 2$, as it will be assumed henceforth. The dimension of momenta remains $[q] = 1$, whereas the dimension of frequency is now dictated by the pole of the Eq. (17), and is $[\omega] \equiv z = 1 - \alpha/2$. We notice that the bare fermion propagator has dimension $-z$ instead of -1 in the dominating regime, where D_{LR} has dimension $-\alpha$. The power counting of the four-point vertex correction in Fig. 1(b) gives now $d + z - \alpha - 2z \equiv d - d_C$, and the marginality is shifted to $d_C \equiv 1 + \alpha/2 > 1$. Indeed, if we substitute D with D_{LR} in the integral Eq. (11) for the one particle Green's function, continued to d dimensions, $L(\mathbf{r}, t)$ appearing in its solution (12) behaves logarithmically (as in the Luttinger liquid for $d = 1$) when $d = d_C$. As in the Luttinger liquid, thanks to the WI, Eq. (9), Z^2 and the four-point vertex renormalization Z_4 compensate each other and the effective fermion-fermion coupling $u \propto (g_{\mathrm{LR}})^2$ scales with its bare dimension $d_C - d$. An even stronger violation of the Fermi-liquid picture is found when $d < d_C$, where $L(\mathbf{r}, t)$ has a power law singularity and the wavefunction renormalization Z vanishes exponentially while u scales to strong coupling. For $d > d_C$, u vanishes in the infrared and the Fermi liquid is recovered.

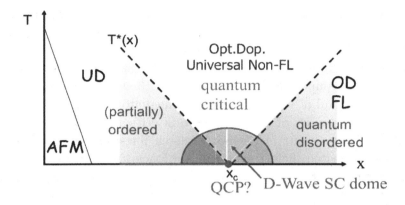

Fig. 7. Schematic phase diagram of high-T_c superconducting cuprates in the temperature T versus doping x plane. The stoichiometric ($x = 0$) and slightly doped cuprates are AFM insulators. An avoided QCP at $x = x_c$, hidden underneath the superconducting dome, naturally divides the phase diagram into three regions: the underdoped (UD) (partially) ordered region, for $T < T^*(x)$; the quantum critical region, around optimal doping (Opt. Dop.), where the only energy scale is the temperature; the overdoped (OD) quantum disordered region, where the system behaves as a Fermi liquid in the metallic phase.

4. Avoided Quantum Criticality in Cuprates

As we have seen in the previous section, a normal Fermi liquid is obtained as soon as $d > 1$ for generic interacting Fermi particles, unless a singular potential compensates the reduction of mixing of the single particle with the collective modes. This is the case when the interaction is mediated by critical collective modes near instability. This is one of the reasons that led to look in the cuprates for a QCP separating two states underlying superconductivity. This QCP would indeed occur as a transition point at $T = 0$ between two competing ground states, if superconductivity would not take over and avoid this quantum criticality (see Fig. 7).

In quantum criticality[55,56] not only the symmetry of the local order parameter enters, but also its dynamics and therefore the collective modes are inextricably mixed in the generalized Landau–Wilson functional. In general, near a quantum phase transition, abundant dynamical long-range and long-time fluctuations exists which may naturally dress the fermionic quasiparticles, mediating an effective retarded e–e interactions which strongly depends on momentum and frequency, as well as on the control parameters (temperature and doping) ruling the proximity to criticality.

The most studied example of avoided QCP and non-Fermi-liquid behavior in the normal phase is the case of heavy fermion metals like $CePd_2Si_2$ (see, e.g., Ref. 57). In this case, the competing states are a Kondo-compensated Fermi-liquid state and an AFM state. Spin-density-wave fluctuations, the relevant collective modes, dress the fermionic quasiparticles leading to non-Fermi-liquid behavior and also act as the glue for the onset of superconductivity. Electrons reorganize themselves to give the superconducting state and avoid the QCP.

A similar situation has been suggested to occur in cuprates by several experimental results, both in thermodynamic and transport measurements (see, e.g., Ref. 58–60). In particular, the crossover line $T^*(x)$, mentioned in Sec. 1.3, which separates the anomalous metallic phase from the pseudogap phase, extrapolates to $T^* = 0$ K at around optimal doping ($x \approx 0.15$–0.20). However, the nature of the competing order and of the related critical mediators is still debated.[43–48,61–69]

Unraveling the dynamics of specific quantum criticality, and the eventual competitions leading to it, is one of the hot research lines and the starting point for future, more sophisticated, field theoretical approaches. In particular, in cuprates, disentangling the relevant modes of the underlying state on which high-temperature superconductivity establishes, is a relevant current issue, as well as how these modes, coupling to fermionic quasiparticles, may account for the violation of the Fermi liquid and act as the glue for pairing.

Along the years, in our group, we elaborated on the idea that strongly correlated electron systems are on the verge of an instability towards phase separation in high- and low-density regions. Strong correlation, forbidding double occupancy of two fermions on the same site, reduces the homogenizing kinetic energy contribution. This, in turn, enhances the charge susceptibility and favors phase separation. In the absence of long-range Coulomb interactions, a wealth of mechanisms, which are ineffective in ordinary metals (e.g., short-range exchange, or coupling to phonons), can make the system unstable. Indeed, phase separation is observed in cuprates, whenever mobile ions are present to compensate for the electron charge unbalance. In phase separation, near criticality, the effective interaction is singular and stronger than Coulomb, and spoils the Fermi-liquid behavior.[30] Its occurrence should establish the onset of the heterogeneous phase below $T^*(x)$. However, in most cases, long-range Coulomb interactions frustrate macroscopic phase separation. As a compromise, mesoscopic charge segregation or an incommensurate CDW may occur in the system. This has indeed been found in the Hubbard–Holstein model with long-range Coulomb interaction.[43,44] The modulation of the CDW was characterized by a finite wavevector $\mathbf{Q}_c \approx (\pm\pi/2, 0);\ (0, \pm\pi/2)$, and the avoided QCP was located near optimal doping, $x_{QCP} \approx 0.19$. According to our proposal, the avoided incommensurate charge-density-wave instability line $T_{CO}(x)$ should be identified with the line $T^*(x)$.[70]

Upon underdoping, the CDW should gradually evolve into the highly anharmonic charge profile of the stripe phase observed in some cases by neutron scattering.[49,50] Moreover, we recall that the metallic phase of cuprates is obtained from the AFM undoped parent compound by chemical doping. Persistent incommensurate spin fluctuations with characteristic wavevector close to the ordering wavevector of the AFM phase $\mathbf{Q}_s \approx (\pi, \pi)$ are expected to be enhanced within the charge-poor domains promoted by CDW formation. The region of existence of nearly critical spin fluctuations is in this way extended even in the optimally and slightly over-doped regions of the phase diagram, far from the AFM QCP, located at low doping

($x_{\mathrm{AFM}} \approx 0.05$). A continuous evolution upon doping of the relative importance of charge and spin fluctuations is then present, the former dominating around optimal doping, the latter dominating in the underdoped region.

We will not recall here the complete scenario of high-T_c superconductors stemming from the proposal of quantum criticality related to the onset of a heterogeneous phase. According to the line followed in this work, we only address the question if we can detect the relevant modes, which would be implied in building up the effective singular interaction. This was recently achieved[51] by suitably interpreting the experiments of Raman spectroscopy in the $La_{2-x}Sr_xCuO_4$ family of cuprates. Strongly interconnected nearly critical charge and spin fluctuations are present, as implied by the above picture of incommensurate charge-density-wave quantum criticality.

To be unbiased, we assume that both charge and spin collective mode propagators take the form of a dynamical Ornstein–Zernicke propagator within the Gaussian approximation,

$$D_a(\mathbf{q}, \omega_n) = -\frac{1}{m_a + \nu_a(\mathbf{q} - \mathbf{Q}_a)^2 + |\omega_n| + \dfrac{\omega_n^2}{\bar{\Omega}_a}}, \qquad (18)$$

where henceforth the label $a = c, s$. The mass m_a is proportional to the inverse square correlation length for charge (spin) fluctuations, ν_a is a scale controlling the dispersion of the collective mode, the scale $\bar{\Omega}_a$ separates the low-frequency Landau-damped diffusive regime from the high-frequency propagating regime and ω_n is the bosonic Matsubara frequency. Notice that the effective interaction D in the extension of Luttinger liquid to singular interaction (see Sec. 3.2) was gapped when $\alpha \geq 2$. Here, the role of the gap is played by the mass m_a, which vanishes at criticality and the effective interaction is sufficiently singular to destroy the Fermi liquid, at least in some regions of the Fermi surface.[71–73] Indeed, when the fermion quasiparticles are coupled to critical collective modes with finite characteristic wavevectors $\mathbf{Q}_a$, the renormalization Z of the single particle Green's function vanishes at the so-called hot spots, i.e., the points of the Fermi surface connected by $\mathbf{Q}_a$ (see Fig. 8). The resulting violation of the Fermi liquid cannot however be inferred directly from the scheme of Sec. 3.2, valid for forward singular interaction, since the modulating $\mathbf{Q}_a$ vectors introduce dominant scattering with large momentum transfer. The question whether a violation of the Fermi-liquid behavior at some points of the Fermi surface only would be enough to account for the non-Fermi-liquid properties of cuprates has been posed.[74] Indeed, one could argue that transport properties are dominated by quasiparticles with momenta away from the hot spots, described by a Fermi-liquid propagator with finite Z. However, in the present theory with both charge and spin quasi-critical collective modes, the Fermi surface is disseminated with hot spots (24 hot spots due to spin and 16 hot spots due to charge collective modes are found for typical values of $\mathbf{Q}_s$ and $\mathbf{Q}_c$, and for the Fermi surface suited to $La_{2-x}Sr_xCuO_4$ systems[51]). Then, the argument about the dominance of Fermi-liquid quasiparticles

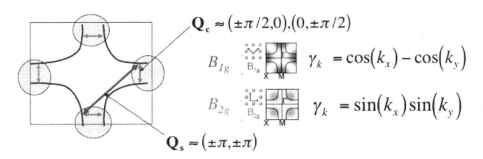

Fig. 8. (Color online) Sketch of the Fermi surface typical of the $La_{2-x}Sr_xCuO_4$ cuprate family, with examples of hot spots associated to the characteristic wavevectors of charge ($\mathbf{Q}_c$) and spin ($\mathbf{Q}_s$) fluctuations. The structure of the Raman vertices γ_k in the B_{1g} and B_{2g} channels is also shown in the small insets: the red (green) areas are the regions where the vertex in the B_{1g} (B_{2g}) channel is large in absolute value and the white areas are the region where the vertex changes sign. The wavevector $\mathbf{Q}_c$ connects region of the Fermi surface where γ_k has the same sign in the B_{1g} channel and changes sign in the B_{2g} channel. The opposite happens for the wavevector $\mathbf{Q}_s$.

is by no means conclusive. As we discuss below, the present theory, although with a completely different microscopic realization, yields a violation of the Fermi liquid that closely resembles the marginal-Fermi-liquid scenario.[75,76] This scenario was phenomenologically introduced to describe a non-Fermi-liquid quasiparticle inverse lifetime linear in energy and temperature.

In Ref. 51 the crucial observation was made that the effect of collective modes with finite characteristic wavevector on the fermionic quasiparticles could be investigated and resolved by means of Raman spectroscopy. On the one hand, Raman spectroscopy probes different regions of the Fermi surface by selecting the polarization of the ingoing and outgoing photons. On the other hand, due to the specific value of the characteristic wavevector $\mathbf{Q}_a$ connecting different regions of the Fermi surface, specific asymptotic cancellations occur in the Raman response, depending on the relative sign of the Raman vertices involved (see Fig. 8). From this dependence it will follow that, at low frequency, the charge collective mode affects the Raman response in the so-called B_{2g} channel [with Raman vertex $\gamma_k = \sin(k_x)\sin(k_y)$] whereas the spin collective mode contributes to the B_{1g} channel [with Raman vertex $\gamma_k = \cos(k_x) - \cos(k_y)$].

Indeed, within a standard memory function approach,[77] the Raman response function can be written as

$$\chi(\omega) = \frac{\chi_0 \omega}{\omega + M(\omega)},$$

where χ_0 is a real constant and $M(\omega)$ is the complex memory function resulting from scattering processes mediated by the propagator D_a, Eq. (18). Within a perturbative approach, the first contributions to $M(\omega)$ come from self-energy and vertex corrections associated with the exchange of one collective mode, dressing the free-electron Raman bubble (see Fig. 9).

Fig. 9. Self-energy (S) and vertex (V) corrections to the Raman response, due to the exchange of a collective mode propagator, Eq. (18), (dashed line). The solid line represents the bare fermion propagator and the black circle represents the Raman vertex γ_k.

At this lowest order in perturbation theory, the charge and spin collective modes do not mix and contribute independently, i.e., $M(\omega) = M_c(\omega) + M_s(\omega)$. Assuming that close to criticality the perturbative corrections are dominated by the pole of the collective mode propagator, Eq. (18), one finds

$$\text{Im}\, M_a(\omega) = \frac{1}{\omega} \int_0^\infty dz [\alpha^2 F(z)] \sum_{p=\pm 1} \left[\omega \coth\left(\frac{z}{2T}\right) - p(z + p\omega) \coth\left(\frac{z + p\omega}{2T}\right) \right],$$

which has the standard form of an inverse lifetime due to the scattering with a bosonic collective mode, and $\alpha^2 F_a$ is the spectral function of the charge ($a = c$) and spin ($a = s$) collective modes. In the present case, direct calculation of the diagrams in Fig. 9 gives:

$$\alpha^2 F_a(\omega) = g_a \left[\arctan\left(\frac{\bar{E}_a}{\omega} - \frac{\omega}{\bar{\Omega}_a}\right) - \arctan\left(\frac{m_a}{\omega} - \frac{\omega}{\bar{\Omega}_a}\right) \right], \tag{19}$$

where $\bar{E}_a$ is the energy cutoff of the collective mode dispersion and the dimensionless coupling

$$g_a \propto \gamma_{\text{HS}}(\gamma_{\text{HS}} - \gamma_{\text{HS}'})$$

depends on the Raman vertices calculated at two hot spots connected by the characteristic wavevector $\mathbf{Q}_a$. This expression enforces the cancellation quoted above: Due to the specific values of $\mathbf{Q}_s$ and $\mathbf{Q}_c$, $\gamma_{\text{HS}} - \gamma_{\text{HS}'}$ vanishes in the B_{1g} channel for the charge collective mode and in the B_{2g} channel for the spin collective mode (see Fig. 8), at least at low frequency, where the dominant pole approximation is asymptotically valid. Therefore, the analysis of the Raman response in the two channels allows disentangling the contribution of the two collective modes, which would instead contribute jointly, e.g., in optical and ARPES spectra.

Equation (19), resulting from coupling the fermionic quasiparticles with the charge and spin modes, provides a generalization of the so-called gapped marginal-Fermi-liquid spectral function,[78] and formally reduces to it in the limit $\bar{\Omega}_a \to 0$ (with $\bar{E}_a\bar{\Omega}_a$ and $m_a\bar{\Omega}_a$ finite). In this limit, the spectral function assumes a box-like shape, and this implies that the fermionic quasiparticle inverse lifetime increases linearly with frequency as long as the frequency is inside the box and saturates to a constant outside the box. The uncertainty of the quasiparticle is equal to its energy value, showing a "marginal" violation of Fermi-liquid behavior.

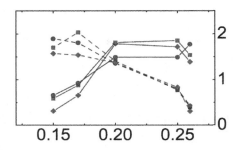

Fig. 10. (Color online) Spin (dashed lines) and charge (full lines) spectral weight W_s and W_c (in units of 10^3 cm^{-1}) as a function of doping x, for $T = 50$ K (blue line, circles), $T = 100$ K (green line, squares), and $T = 200$ K (red line, diamonds).

By a systematic fit of the Raman data for various temperatures (50 K $< T <$ 200 K) and doping levels (0.15 $< x <$ 0.26), the parameters of the two collective modes and their evolution were extracted. The overall spectral weight W_a of each collective mode is obtained integrating Eq. (19) over all frequencies. W_c increases with increasing doping and tends to saturate in the most overdoped samples, whereas W_s is larger in the less doped sample and decreases with increasing doping (see Fig. 10).

The spectral intensity of the two modes is equal at doping $x \approx 0.19$. Thus, the fermionic quasiparticles are more strongly scattered by spin collective modes for $x < 0.19$ and by charge collective modes for $x > 0.19$. At the doping $x \approx 0.19$, separating the two regions, we also obtained a mass of the charge collective mode linearly vanishing in temperature, making the temperature the only relevant energy scale, as it would implied by the presence a QCP at this doping.

Thus, our analysis supports the proposal that the phase competing with superconductivity in the underdoped region is characterized by (nearly) ordered stripe-like textures, whose onset on the overdoped side is marked by an incommensurate CDW instability. The related charge and spin fluctuations mediate a singular retarded effective e–e interaction, which accounts for anomalous physical responses. As in the case of heavy fermions, at low temperature the charge carriers reorganize themselves into a superconducting phase and avoid the occurrence of the QCP for charge ordering.

References

1. C. Domb and M. S. Green (eds.), *Phase Transitions and Critical Phenomena*, Vol. VI (Academic Press, New York, 1976).
2. L. P. Kadanoff *et al.*, *Rev. Mod. Phys.* **39**, 395 (1967).
3. A. A. Migdal, *Sov. Phys. JETP* **28**, 1036 (1969).
4. A. M. Polyakov, *Sov. Phys. JETP* **28**, 533 (1969).
5. A. I. Larkin and D. E. Khmel'nitskii, *Sov. Phys. JETP* **29**, 1123 (1969).
6. A. M. Polyakov, *Sov. Phys. JETP* **3**, 151 (1970).

7. C. Di Castro and G. Jona-Lasinio, *Phys. Lett. A* **29**, 322 (1969).

8. F. De Pasquale, C. Di Castro and G. Jona-Lasinio, in *Proceedings of the International School of Physics 'Enrico Fermi'*, Course LI, ed. M. S. Green (Academic Press, New York, 1971),

9. C. Di Castro and G. Jona-Lasinio, in *Phase Transitions and Critical Phenomena*, Vol. VI, eds. C. Domb and M. S. Green (Academic Press, New York, 1976).

10. K. G. Wilson, *Phys. Rev. B* **4**, 3174 (1971).

11. K. G. Wilson, *Phys. Rev. B* **4**, 3184 (1971).

12. K. G. Wilson and J. Kogut, *Phys. Rep. C* **12**, 75 (1974).

13. C. Di Castro and W. Metzner, *Phys. Rev. Lett.* **67**, 3852 (1991).

14. W. Metzner and C. Di Castro, *Phys. Rev. B* **47**, 16107 (1993).

15. W. Metzner, C. Castellani and C. Di Castro, *Adv. Phys.* **47**, 317 (1998).

16. C. Castellani *et al.*, *Phys. Rev. Lett.* **78**, 1612 (1997).

17. F. Pistolesi *et al.*, *Phys. Rev.* **69**, 024513 (2004).

18. C. Castellani *et al.*, *Phys. Rev. B* **30**, 527 (1984).

19. C. Castellani and C. Di Castro, *Phys. Rev. B* **34**, 5935 (1986).

20. A. M. Finkel'stein, in *Physics Reviews*, Vol. 14, Part 2, ed. I. M. Khalatnikov (Harwood Academic Publishers, Glasgow, 1990).

21. I. E. Dzyaloshinskii and A. I. Larkin, *Sov. Phys. JETP* **38**, 202 (1974).

22. D. C. Mattis and E. H. Lieb, *J. Math. Phys.* **6**, 304 (1965).

23. A. Luther and E. Peschel, *Phys. Rev. B* **9**, 2911 (1974).

24. F. D. M. Haldane, *Phys. Rev. Lett.* **45**, 1358 (1980).

25. F. D. M. Haldane, *Phys. Rev. Lett.* **47**, 1840 (1981).

26. J. Voit, *Rep. Prog. Phys.* **58**, 977 (1995).

27. C. Castellani, C. Di Castro and W. Metzner, *Phys. Rev. Lett.* **72**, 316 (1994).

28. P. Bares and X. G. Wen, *Phys. Rev. B* **48**, 8636 (1993).

29. C. Castellani and C. Di Castro, *Physica C* **235–240**, 99 (1994).

30. C. Castellani *et al.*, *Nucl. Phys. B* **594** 747 (2001).

31. A. Luther, *Phys. Rev. B* **19**, 320 (1979).

32. F. D. M. Haldane, *Helv. Phys. Acta* **65**, 152 (1992).

33. A. Houghton and J. B. Marston, *Phys. Rev. B* **48**, 7790 (1993).

34. A. Houghton, H.-J. Kwon and J. B. Marston, *Phys. Rev. B* **50**, 1351 (1994).

35. A. Houghton *et al.*, *J. Phys. C* **6**, 4909 (1994).

36. H.-J. Kwon, A. Houghton and J. B. Marston, *Phys. Rev. B* **52**, 8002 (1995).

37. P. Kopietz and K. Schönhammer, *Z. Phys. B* **100**, 561 (1996).

38. R. Shankar, *Rev. Mod. Phys.* **66**, 129 (1994).

39. C. M. Varma, Z. Nussinov and W. van Saarloos, *Phys. Rep.* **361**, 267 (2002).

40. C. M. Varma *et al.*, *Phys. Rev. Lett.* **63**, 1996 (1989).

41. P. W. Anderson, *Phys. Rev. Lett.* **64**, 1839 (1990).

42. P. W. Anderson, *Phys. Rev. Lett.* **65**, 2306 (1990).

43. C. Castellani, C. Di Castro and M. Grilli, *Phys. Rev. Lett.* **75**, 4650 (1995).

44. C. Castellani, C. Di Castro and M. Grilli, *Z. Phys. B* **103**, 137 (1997).

45. P. Monthoux, A. V. Balatsky and D. Pines, *Phys. Rev. B* **46**, 14803 (1992).

46. A. Sokol and D. Pines, *Phys. Rev. Lett.* **71**, 2813 (1993).

47. P. Monthoux and D. Pines, *Phys. Rev. B* **50**, 16015 (1994).

48. A. V. Chubukov, P. Monthoux and D. K. Morr, *Phys. Rev. B* **56**, 7789 (1997).

49. M. Tranquada *et al.*, *Nature* **375**, 561 (1995).

50. J. M. Tranquada *et al.*, *Phys. Rev. B* **54**, 7489 (1996).

51. S. Caprara *et al.*, *Phys. Rev. B* **84**, 054508 (2011).

52. J. Sólyom, *Adv. Phys.* **28**, 201 (1979).

53. S. Tomonaga, *Progr. Thor. Phys.* **5**, 544 (1950).

54. J. M. Luttinger, *J. Math. Phys.* **4**, 1154 (1963).

55. J. A. Hertz, *Phys. Rev. B* **14**, 1165 (1976).

56. A. J. Millis, *Phys. Rev. B* **48**, 7183 (1993).

57. P. Gegenwart, Q. Si and F. Steglich, *Nat. Phys.* **4**, 186 (2008).
58. J. L. Tallon and J. W. Loram, *Physica C* **349**, 53 (2001).
59. G. S. Boebinger *et al.*, *Phys. Rev. Lett.* **77**, 5417 (1996).
60. S. H. Naqib *et al.*, *Phys. Rev. B* **71**, 054502 (2005).
61. A. J. Millis, H. Monien and D. Pines, *Phys. Rev. B* **42**, 167 (1990).
62. C. M. Varma, *Phys. Rev. Lett.* **83**, 3538 (1999).
63. G. Baskaran and P. W. Anderson, *Phys. Rev. B* **37**, 580 (1988).
64. L. B. Ioffe and A. I. Larkin, *Phys. Rev. B* **39**, 8988 (1989).
65. P. A. Lee and N. Nagaosa, *Phys. Rev. B* **46**, 5621 (1992).
66. L. Benfatto, S. Caprara and C. Di Castro, *Eur. Phys. J. B* **17**, 95 (2000).
67. S. Chakravarty *et al.*, *Phys. Rev. B* **63**, 094503 (2001).
68. W. Metzner, D. Rohe and S. Andergassen, *Phys. Rev. Lett.* **91**, 066402 (2003).
69. B. L. Altshuler, L. B. Ioffe and A. J. Millis, *Phys. Rev. B* **50**, 14048 (1994).
70. S. Andergassen *et al.*, *Phys. Rev. Lett.* **87**, 056401 (2001).
71. A. V. Chubukov, D. K. Morr and A. Shakhnovich, *Philos. Mag. B* **74**, 563 (1996).
72. A. V. Chubukov and D. K. Morr, *Phys. Rep.* **288**, 347 (1998).
73. S. Caprara *et al.*, *Phys. Rev. B* **59**, 14980 (1999).
74. R. Hlubina and T. M. Rice, *Phys. Rev. B* **51**, 9253 (1995).
75. C. M. Varma, *Int. J. Mod. Phys. B* **3**, 2083 (1989).
76. C. M. Varma *et al.*, *Phys. Rev. Lett.* **63**, 1996 (1989).
77. W. Götze and P. Wölfle, *Phys. Rev. B* **6**, 1226 (1972).
78. M. R. Norman and A. V. Chubukov, *Phys. Rev. B* **73**, 140501(R) (2006).

LUTTINGER MODEL IN DIMENSIONS $d > 1$

DANIEL C. MATTIS

Department of Physics and Astronomy, University of Utah,
Salt Lake City, UT 84112 USA
mattis@physics.utah.edu

In the present work the Luttinger model is formally extended to $d > 1$ dimensions. For purposes of comparison with the original model only *short-range* forces linking the fermions have been considered (as in the Hubbard model). The Fermi sea is decomposed into sectors labeled by wave-vectors $\mathbf{q}$, each containing a set of orthogonal, independent normal modes. Operators that create or destroy normal modes within any given sector are shown to commute with those originating in other sectors. Because the Hamiltonian within each sector is a quadratic form in these normal modes it can be diagonalized in a multitude of ways. One convenient method consists of mapping it onto a model first introduced elsewhere in this volume, the $d = 1$ *expanded Luttinger model*. It too can be diagonalized, albeit not so trivially as the original, but by constructing and then diagonalizing a *boson string*. A product of eigenstates — one from each sector — is used to construct an exact eigenstate of the full many-body problem if the algebra connecting all the normal modes is Abelian. However, some form of non-Abelian algebra governs the normal modes in the presence of any sort of long-range order (LRO), whether ferromagnetic, antiferromagnetic or superconducting. (Note that the Mermin-Wagner theorem does preclude spontaneous LRO in dimensions $d \leq 2$). Although it is also important to understand the nature of *dynamical exchange corrections* to arbitrary-range interactions — including the physically important Coulomb interaction — a detailed derivation, together with such other conundrum as the calculation of one-fermion distribution functions from "first principles," all are topics relegated to future publications.

Contents

Figures

1. Introduction

In the second half of the twentieth century theoretical physics became largely concerned with the properties of $N \to \infty$ interacting particles in an infinite volume $V \to \infty$, known as "the many-body problem." Field theory, Feynman diagrams and diagrammatic many-body perturbation theory were all invented to solve it, but each brought along its own peculiar set of difficulties. Luttinger's model[1] as originally constructed was arguably the first to yield an exact solution to the many-fermion problem, albeit in the highly specialized universe of spinless two-component fermions in $d = 1$ spatial dimension ($d = 2$ space–time). In this model, particles move without dispersion (at constant velocities, either $\pm v_F$) and interact *via* delta-function two-body forces. Its solution, as originally elaborated by E. Lieb and the present author,[1] was both *simple* and *explicit*, was based on a transformation that replaced the usual *quadratic* + *quartic* form in the fermion operators by a simple quadratic form in bosons.[1] But despite many efforts by numerous authors this model proved surprisingly difficult to generalize. Once nontrivial dispersion is introduced or non-point-like interactions are considered, the procedure ceases to be simple *even in* 1D. Nevertheless, explicit solutions to generalized models exist. One such generalization is detailed elsewhere in this volume.[2,a] It requires replacing the boson operators $a(q)$ and $a^\dagger(q)$ by infinitely long boson *strings*, $a_j(q)$ and $a_j^\dagger(q)$, $j = 0, 1, 2 \ldots$ Because the Hamiltonian of the strings in this *expanded Luttinger model* remains quadratic in the enlarged set of bosonic operators, it can be diagonalized.

But however much this generalization may have enlarged the universe of solvable problems in mathematical physics, still it only concerns motion in $d = 1$ spatial

---

[a]The reference is reprinted on p. 105 of the present volume — it is a paper generalizing Ref. 1 to arbitrary dispersion and solving it in $d = 1$ spatial dimension by analogy with a boson string. Other earlier investigations generalizing Ref. 1 include Refs. 3 and 4.

dimension and merely gains us an incremental advance in understanding the physical world.

But what if the original Luttinger model *could* be expanded to higher dimensions? If in d dimensions the Fermi surface $e(\mathbf{k}_F)$ were somehow *special* — supposing, for example, that it contained a finite number of "flat" and parallel portions as, for example, at $e(k) = 0$ in the tight-binding band: $e(k) = -t/2 \, (\cos k_x + \cos k_y)$ — the mapping of fermionic models onto a small set of separable 1D Luttinger models could then be envisaged.[5] This "Luttingerization" has even been rendered mathematically rigorous.[6,7,b]

However, a far more ambitious idea than just the approximation of circles by polygons has been floating around for over three decades, according to which fermionic degrees of freedom might be well represented by a bosonic quadratic form capable of being exactly diagonalized in all dimensions $d > 1$, *regardless* of dispersion $e(k)$.

Such idealized generalizations of the "random phase *approximation*" are typically found under the rubric of *bosonization*. Books or book-length reviews have been consecrated to this end.[8,9] But it was only recently, in connection with the aforementioned generalizations of the Luttinger model[2–4,a] in 1D, that bosonization in dimensions $d > 1$ based on Luttinger's model became both clear and compelling.[10] The present text shows how to apply this discovery to the Hubbard model and points to work in progress and to future applications.

The principal concepts are easily stated: that an ideal gas of N fermions in any spatial dimension sustains a number $O(N)$ of dynamically independent bosonic *normal modes*, just as does any extended molecule or solid, whether it is initially comprised of classical or even *quantum* particles. In the presence of momentum-conserving two-body interactions that perturb the ground state, the bosonic degrees of freedom continue to be grouped into distinct sectors and, in the absence of long-range order (LRO), to satisfy an Abelian algebra.

For simplicity the presentation in this chapter is mostly restricted to short-range forces (what is generally known as the "Hubbard model,") and generally assumes a circularly or spherically isotropic band structure (e.g., $e(\mathbf{k}) = e(|\mathbf{k}|)$). At the end, in Secs. 21–24, we broach advanced topics — such as the complications that are the result of arbitrary dispersions or arbitrary-range potentials and their effects on "exchange forces" and spin waves. We leave to future analysis the non-Abelian algebras that control the internal degrees of freedom in the presence of various types of LRO.

2. First-Principles Fermionic Formulation

Consider the prototype conservative system of N ($N \to \infty$) SU(2) fermions in a volume $V = L^d$ ($L^d \to \infty$) in spatial dimension $d > 1$ at finite particle density N/V.

---

[b]Reference 7 is reprinted on p. 200 in the present volume.

Particles interact *via* two-body forces — the Coulomb repulsion being one example — at temperatures low compared to the degeneracy temperature. With periodic boundary conditions, there is momentum conservation at each collision. The total Hamiltonian describing this conservative system is the sum of kinetic energy H_1 and interactions H_2. Using anticommuting fermion operators $c_\sigma^\dagger(\mathbf{k})$ and $c_\sigma(\mathbf{k})$ that create or destroy fermions labeled by $(\mathbf{k}, \sigma)$, subject to anti-commutation relations, e.g., $c_\sigma^\dagger(\mathbf{k})c_{\sigma'}(\mathbf{k}') + c_{\sigma'}(\mathbf{k}')c_\sigma^\dagger(\mathbf{k}) \equiv \{c_\sigma^\dagger(\mathbf{k}), c_{\sigma'}(\mathbf{k}')\} = \delta_{\sigma,\sigma'}\delta_{\mathbf{k},\mathbf{k}'}$, the kinetic energy H_1 is:

$$H_1 = \sum_k \sum_\sigma (e(\mathbf{k}) - e_F)c_\sigma^\dagger(\mathbf{k})c_\sigma(\mathbf{k}) \equiv \sum_k \sum_\sigma \varepsilon(\mathbf{k})\tilde{n}_\sigma(\mathbf{k}), \tag{1}$$

where $\varepsilon(\mathbf{k}) = e(\mathbf{k}) - e_F$ is the energy measured from the Fermi level (e_F); σ (actually, σ_z) refers to spin "up/down". The "effective mass" approximation[c] $e(\mathbf{k}) = \hbar^2 k^2/2m^*$ and units $\frac{\hbar^2}{2m^*} = 1$ and $k_F = 1$ are assumed for definiteness. The Fermi level e_F is chosen such that $1/2N$ states have $\varepsilon(\mathbf{k}) < 0$ and are therefore doubly occupied at $T = 0$ in the absence of interactions, whereas all states with $\varepsilon(\mathbf{k}) > 0$ are unoccupied. In the ground state of N free fermions the occupation-number operator $\tilde{n}_\sigma(\mathbf{k}) = c_\sigma^\dagger(\mathbf{k})c_\sigma(\mathbf{k})$ has eigenvalue 1 if $\varepsilon(\mathbf{k}) < 0$ and 0 for $\varepsilon(\mathbf{k}) > 0$ for either value of σ, $\pm 1/2$. We know that in the presence of interactions the averaged discontinuity in $\tilde{n}_\sigma(\mathbf{k})$ (from just below the Fermi level to just above it) is a number Z_F in the range $0 \le Z_F \le 1$ that decreases with increasing two-body interactions or increasing T. In principle it could vanish even at $T = 0$ if interactions are repulsive and sufficiently strong, or if the forces are weak, if they have a structure capable of sustaining Cooper pairs. In some respects, such as their tendency to cause Z_F to shrink, thermal fluctuations and two-body interactions have qualitatively similar consequences: both deplete states below e_F of their particles while causing formerly empty states above e_F to become occupied.[d]

Two-body interactions are governed by an expression explicitly quartic in the fermions,

$$H_2 = \frac{1}{2!\text{Vol}} \sum_{\mathbf{k},\mathbf{k}',\mathbf{q}} V(\mathbf{q}) \sum_{\sigma,\sigma'} c_\sigma^\dagger\left(\mathbf{k} + \frac{\mathbf{q}}{2}\right) c_{\sigma'}^\dagger\left(\mathbf{k}' - \frac{\mathbf{q}}{2}\right) c_{\sigma'}\left(\mathbf{k}' + \frac{\mathbf{q}}{2}\right) c_\sigma\left(\mathbf{k} - \frac{\mathbf{q}}{2}\right)$$

$$= \frac{1}{2!\text{Vol}} \sum_{\mathbf{q}} V(\mathbf{q}) : \left\{\sum_{\sigma'} \sum_\sigma \rho_{\sigma'}(\mathbf{q})\rho_\sigma(-\mathbf{q})\right\} : \tag{2}$$

where $\rho_\sigma(\mathbf{q}) = \sum_\mathbf{k} c_\sigma^\dagger\left(\mathbf{k} + \frac{\mathbf{q}}{2}\right)c_\sigma\left(\mathbf{k} - \frac{\mathbf{q}}{2}\right)$ and colons $:\dots:$ indicates that the product of fermion operators in the curly brackets is to be "normally ordered" (annihilation operators to the right, creation operators to the left).

---

[c]Many of the present results remain qualitatively valid for differing laws of dispersion, including $e(\mathbf{k}) = \nu_F|\mathbf{k}|$ at low energies in graphene, given that only the magnitude of the coefficients and not much else, can be affected by the dispersion.

[d]Coherently in the case of dynamic redistributions, incoherently in the case of thermal fluctuations.

We shall specialize to just the following: $V(\mathbf{q}) = U = $ constant for $q < q_0$, zero otherwise. This point-like *Hubbard interaction* is governed by two independent parameters: the coupling constant U and the cutoff q_0, a wave-vector inversely proportional to the spacing d_0 of atoms in the physical lattice containing the fermions.

Although the interaction H_2 in Eq. (2) is relevant to *all* nontrivial properties of the Fermi gas it will not be invoked prior to Sec. 5, because our first and most important task is to derive a methodology by which to reëxpress the *kinetic energy* of Eq. (1) in 2D and 3D (including its ground-state and *low-lying excitations*) in a set of boson normal modes. In this we use a methodology previously crafted to solve our *expanded* Luttinger model in 1D.[2]

3. The Fermi Sea and Its Bosonic Excitations

Traditionally, collective vibrations of individual particles about presumed positions of equilibrium are known as "normal modes." After being orthogonalized and quantized, the associated creation or annihilation operators satisfy Bose–Einstein commutation relations *ab initio*. Notable examples include photons in electrodynamics and phonons in solid state physics. The magnons in Heisenberg's model of magnetism are less obviously so because of issues with their commutation relations (the spin wave operators are non-Abelian).

But here the premises are different. For, starting from what is already highly quantum-mechanical *fermion* physics, we propose to construct linear combinations of low-lying excitations that *behave* as bosons — but then only in the "thermodynamic limit" (this last meaning that N, the effective number of degrees of freedom and L, the characteristic length of the system, must both tend to ∞ in such a way that N/L^d remains finite in d dimensions). The system's dynamics are governed by an Hamiltonian that is quadratic in these bosons. Once diagonalized, its eigenstates yield a complete picture of the energies and correlation functions in the eigenstates of the many-body system.

The *plasma mode* of a charged electron gas or the *zero sound* of an uncharged Fermi liquid are bosons, as is well known. It is less widely known that such fluids made up of identical fermions can support an infinite number of *additional*, distinct, collective excitations, including velocity-dependent density- or spin-fluctuations as well as *two distinct virtual branches* of propagating normal modes associated with superconductivity. Some have very short lifetimes. Omitting spin indices, let us start by considering a bare-bones prototype *annihilation* operator $a_0(\mathbf{q})$ (a quantity of as yet unknown symmetry) constructed as a linear combination of elementary excitations of a free (noninteracting) Fermi gas, as follows:

$$a_0(\mathbf{q}) \equiv \frac{C(\mathbf{q})}{\sqrt{q \times \text{Vol}}} \sum_{\mathbf{k}} \theta(\mathbf{k} \cdot \mathbf{q}) c^\dagger(\mathbf{k} - \mathbf{q}/2) c(\mathbf{k} + \mathbf{q}/2) \,. \tag{3}$$

Here and throughout, $\theta(\mathbf{k} \cdot \mathbf{q})$ is the Heaviside function ($\theta = 1$ if its argument is positive, zero otherwise). Vol $= L^d$ in d dimensions. Although no single constituent of Eq. (3) such as $c^\dagger(\mathbf{k}-\mathbf{q}/2)c(\mathbf{k}+\mathbf{q}/2)$ satisfies Bose–Einstein commutation relations with the others, *each* does lower the momentum of any state to which it is applied by a constant $\mathbf{q}$ and lowers its energy by $\Delta \varepsilon = ((\mathbf{k} + \mathbf{q}/2)^2 - (\mathbf{k} - \mathbf{q}/2)^2) = (2\mathbf{k} \cdot \mathbf{q})$, which is positive due to the factor $\theta(\mathbf{k} \cdot \mathbf{q})$. The units are $\frac{\hbar^2}{2m^*} = 1 = k_F$. Thus $a_0(\mathbf{q})$ is a "lowering" operator of sorts, albeit having a *spectrum* of energies ranging from 0 (if $\mathbf{k} \perp \mathbf{q}$) to maximum $2k \cdot q$ (if $\mathbf{k}//\mathbf{q}$). Let us, moreover, postulate that the commutator $[a_0(\mathbf{q}), a_0^\dagger(\mathbf{q})] \equiv a_0(\mathbf{q})q_0^\dagger(\mathbf{q}) - a_0^\dagger(\mathbf{q})a_0(\mathbf{q})$ is equal to 1, what we shall call the "normalization" condition. It is what *determines* the over-all factor $C(\mathbf{q})$ in (3).

Thus,

$$1 = [a_0(\mathbf{q}), a_0^\dagger(\mathbf{q})] = \frac{C^2(\mathbf{q})}{|\mathbf{q}|\text{Vol}} \sum_k \theta(\mathbf{k} \cdot \mathbf{q})(\tilde{n}(\mathbf{k} - \mathbf{q}/2) - \tilde{n}(\mathbf{k} + \mathbf{q}/2))$$

$$= \frac{C^2(\mathbf{q})}{|\mathbf{q}|(2\pi)^d} \int d^d k\, \theta(\mathbf{k} \cdot \mathbf{q})(\tilde{n}(\mathbf{k} - \mathbf{q}/2) - \tilde{n}(\mathbf{k} + \mathbf{q}/2))$$

$$= \frac{C^2(\mathbf{q})}{|\mathbf{q}|(2\pi)^d} \int d^d k\, \theta(\mathbf{k} \cdot \mathbf{q})(f(\mathbf{k} - \mathbf{q}/2) - f(\mathbf{k} + \mathbf{q}/2)). \tag{4}$$

In the last line of (4) we replaced each operator in the sum by its thermodynamic and/or quantum average, that is, by an ordinary function f. Although this may appear to be a poor starting approximation, such a substitution is in fact *mathematically exact* and is, moreover, essential, as it allows the coefficient $C(\mathbf{q})$ to be just an ordinary, calculable, function of its argument and not some undecipherable operator! Invoking the "central limit theorem" of statistics, we find there is no error in replacement of a variable or operator in an integral by its average.[e] The "distribution function" f that needs to be used here is the average of $\tilde{n}(\mathbf{k}) = c^\dagger(\mathbf{k})c(\mathbf{k})$, a function of k, of the temperature T and of the strength and nature of the two-body interactions. In other words, the average $f(\mathbf{k}) = \langle \tilde{n}(\mathbf{k}) \rangle$ must be taken, not just with respect to thermal fluctuations, but also subject to *all* the internal dynamical forces (interactions). This is discussed further in Sec. 23.

Although there might be several ways to go about it, *the calculation of a fermion one-body correlation function $f(\mathbf{k}) = \langle \tilde{n}(\mathbf{k}) \rangle$ entirely and self-consistently within the boson representation, still remains the most challenging part of the present theory.* But, to allow other important aspects of the theory to go forward, we have to interrupt this discussion and return to f only later. For present purposes we estimate the

---

[e]Consider a sum over N points, as in the following. Including an arbitrary factor $K(k)$, $\frac{1}{N} \sum_{\mathbf{k}} K(\mathbf{k})\tilde{n}(\mathbf{k}) = \frac{1}{N} \sum_{\mathbf{k}} K(\mathbf{k})\{\langle \tilde{n}(\mathbf{k}) \rangle + (\tilde{n}(\mathbf{k} - \langle \tilde{n}(\mathbf{k}) \rangle))\} = \frac{1}{N} \sum_{\mathbf{k}} K(\mathbf{k})f(\mathbf{k}) \pm \mathrm{O}\left(\frac{1}{\sqrt{N}}\right)$. The mean is O(1) and fluctuations about the mean, a random walk, adds up to $\pm \mathrm{O}(\sqrt{N})$ and vanishes as $1/\sqrt{N}$ after being divided by N. If the mean vanishes identically there is no simplification. This result roughly summarizes the Central Limit Theorem CLT.

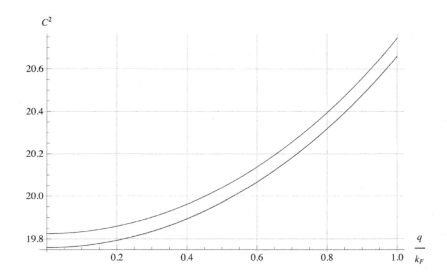

Fig. 1. *Normalization.* Plot of $C^2(q)$ versus q/k_F in Eq. (4), for an ideal 2D fermion gas at $kT = 0.05$ (lower curve) and 0.1. The two differ by a mere 0.2%. (We assume parabolic dispersion. Units are $k_F = 1$, $e_F = 1$).

various quantities by "back of the envelope" calculations using the *Fermi function* — the distribution function of an ideal fermion gas — $f_0(\mathbf{k}) = 1/(e^{\beta\varepsilon(\mathbf{k})} + 1)$, where $\beta = 1/k_B T$. Thus in the plot of $C^2(\mathbf{q})$ in $d = 2$ at two distinct temperatures T shown in Fig. 1, we assume no interactions. The calculations and the figure suggest that at sufficiently low temperatures ($kT < e_F/10$) the normalization parameter $C^2(\mathbf{q})$ is insensitive to the increased rounding of f with increasing temperature at the Fermi surface $\varepsilon = 0$. This may also reflect an insensitivity of $C^2(\mathbf{q})$ to internal forces H_2 when they are introduced (provided they are repulsive and weak, and so unable to erase the Fermi surface). Calculations in 3D have shown similar trends.

Actually, $a_0(\mathbf{q})$ is just the first in an infinite set ("field") of similarly constituted bosons, $a_j(\mathbf{q}) \equiv \frac{C(\mathbf{q})}{\sqrt{q \times \mathrm{Vol}}} \sum_k \theta(\mathbf{k} \cdot \mathbf{q}) \Phi_j(\mathbf{k}, \mathbf{q}) c^\dagger(\mathbf{k} - \mathbf{q}/2) c(\mathbf{k} + \mathbf{q}/2))$, whose "wavefunctions" Φ_j (typically real) are chosen from among a complete orthogonal set at each fixed $\mathbf{q}$. All obey $1 = \frac{C^2(\mathbf{q})}{|\mathbf{q}|(2\pi)^d} \int d^d k \theta(\mathbf{k} \cdot \mathbf{q}) \Phi_j^2(\mathbf{k}, \mathbf{q})(f(\mathbf{k} - \mathbf{q}/2) - f(\mathbf{k} + \mathbf{q}/2))$. This generalizes the previously introduced *normalization* requirement to $[a_j(\mathbf{q}), a_j^\dagger(\mathbf{q})] = 1$, for all $j = 0, 1, \ldots$ The existence of an infinite set of bosons *within* each sector labeled by $\mathbf{q}$ might appear to be an inconvenience; however, *it is the linchpin of the present theory.*

As is proved below, the functions $\Phi_j(\mathbf{k}, \mathbf{q})$ need to have a certain number of "radial" nodes[f] (roughly, in the direction of $\mathbf{q}$) or "transverse" nodes (roughly perpendicular to $\mathbf{q}$) in $\mathbf{k}$-space, to ensure the set of a_j's satisfy *the* Bose–Einstein algebra.

---

[f]The locus of points, lines or surfaces where Φ vanishes or changes sign.

This algebra is summarized by the following commutation relations:

$$[a_i(\mathbf{q}), a_j^\dagger(\mathbf{q}')] = \delta_{i,j}\delta_{\mathbf{q},\mathbf{q}'} \text{ and of course, } [a_i(\mathbf{q}), a_j(\mathbf{q}')] = [a_j^\dagger(\mathbf{q}'), a_i^\dagger(\mathbf{q})] = 0. \quad (5)$$

Within each sector all share a common $C(\mathbf{q})$. The a_0 operator is always, and uniquely, the *nodeless* one (the present choice $\Phi_0 = 1$ makes it obviously so!), a good and sufficient reason to single it out. Now we investigate the commutation relations (5) in somewhat more detail, to see what they can reveal.

3.1. *Two operators in two distinct sectors*

Equations (5) include a requirement that any two annihilation operators (and any two creation operators) commute. Let us prove this first by direct calculation involving two a_0 operators in two *distinct* sectors, q and q'. The generalization to arbitrary a_j's should then be obvious. Straightforwardly:

$$[a_0(\mathbf{q}), a_0(\mathbf{q}')] \equiv \frac{C(\mathbf{q})C(\mathbf{q}')}{\sqrt{q \times q' \times \mathrm{Vol}^2}} \sum_k \sum_{k'} \theta(\mathbf{k}\cdot\mathbf{q})\theta(\mathbf{k}'\cdot\mathbf{q}')$$
$$\times [c^\dagger(\mathbf{k}-\mathbf{q}/2)c(\mathbf{k}+\mathbf{q}/2), c^\dagger(\mathbf{k}'-\mathbf{q}'/2)c(\mathbf{k}'+\mathbf{q}'/2)]$$
$$= \frac{C(\mathbf{q})C(\mathbf{q}')}{\sqrt{q \times q' \times \mathrm{Vol}^2}} \sum_k \sum_{k'} \theta(\mathbf{k}\cdot\mathbf{q})\theta(\mathbf{k}'\cdot\mathbf{q}')$$
$$\times \{\delta_{\mathbf{k}+\mathbf{q}/2,\mathbf{k}'-\mathbf{q}'/2}c^\dagger(\mathbf{k}-\mathbf{q}/2)c(\mathbf{k}'+\mathbf{q}'/2)$$
$$- \delta_{\mathbf{k}'+\mathbf{q}'/2,\mathbf{k}-\mathbf{q}/2}c^\dagger(\mathbf{k}'-\mathbf{q}'/2)c(\mathbf{k}+\mathbf{q}/2)\}.$$

This looks impossibly messy. But after regrouping terms, redefining dummy indices of summation and performing some trivial algebra, one gains the simplified expression:

$$[a_0(\mathbf{q}), a_0(\mathbf{q}')]$$
$$= \frac{1}{\sqrt{\mathrm{Vol}}} \frac{C(\mathbf{q})C(\mathbf{q}')}{\sqrt{qq'\mathrm{Vol}}} \sum_k \left(\theta\left(\left(k - \frac{\mathbf{q}'}{2}\right)\cdot\mathbf{q}\right) \theta\left(\left(k + \frac{\mathbf{q}'}{2}\right)\cdot\mathbf{q}'\right) \right.$$
$$\left. - \theta\left(\left(k - \frac{\mathbf{q}}{2}\right)\cdot\mathbf{q}'\right) \theta\left(\left(k + \frac{\mathbf{q}'}{2}\right)\cdot\mathbf{q}\right) \right)$$
$$\times c^\dagger\left(k - \frac{q+q'}{2}\right) c\left(k + \frac{q+q'}{2}\right)$$
$$= \frac{1}{\sqrt{\mathrm{Vol}}}(A_i a_i(\mathbf{q}+\mathbf{q}') + A_j a_j^\dagger(-\mathbf{q}-\mathbf{q}')), \quad (6)$$

in which the operators a_i and $a_j^\dagger$ on the right-hand side of (6) are bosons defined in their respective sectors $\pm(\mathbf{q} + \mathbf{q}')$, assumed normalized but otherwise unspecified. The arguments are dictated by conservation of momentum. Coefficients A_i and A_j are both finite and real (but not necessarily positive) functions and are in units of $L^{d/2}$.

In the special case $\mathbf{q} + \mathbf{q}' = 0$ both these coefficients are identically zero,[g] hence $[a_0(\mathbf{q}), a_0(-\mathbf{q})] \equiv 0$ *identically*. In the limit of $\mathbf{q} \to \mathbf{q}'$, the summand in (6) *also* $\to 0$. Other than those special cases the right-hand side of (6) *still* vanishes — but only *in the thermodynamic limit*.

That is, given that both quantities A_i and A_j are nonsingular and that $\langle a_i^\dagger a_i \rangle$ and $\langle a_j^\dagger a_j \rangle$ also are finite, then the over-all factor $1/\sqrt{\mathrm{Vol}} \to 0$ causes the product (6) to vanish in the thermodynamic limit.

These results generalize, *mutatis mutandis*, to the full sets of a_j operators in any two distinct sectors. We can therefore assume that $[a_i(\mathbf{q}), a_j(\mathbf{q}')] = 0$ for *all* labels i, q, j, q'. However, it is important to distinguish *exact* commutation relations $[a_i(\mathbf{q}), a_i(\mathbf{q})] \equiv 0$ from the *weaker* but more general commutation relations $[a_i(\mathbf{q}), a_j(-\mathbf{q}')] \propto \frac{\text{operator}}{L^{d/2}} \to 0$ that are satisfied only in the thermodynamic limit (and only under the supposition that the "operator" in the numerator remains finite in this limit). Although such a distinction might be viewed as mathematical hair-splitting in the context of the many-body problem, if the theory is ever to be applied to *small* systems, e.g., atoms, nuclei, small molecules or quantum dots, it suggests caution.

3.2. *Two operators within one given sector*

Generalizing Eq. (6), one shows that two annihilation operators inhabiting a common sector must also commute, in the sense that, $[a_i(\mathbf{q}), a_j(\mathbf{q})] \propto \frac{\text{finite operator}}{L^{d/2}} \to 0$. Certainly, if $i = j$, the "finite operator" in the numerator vanishes *identically* — as mentioned previously. A second rule governs mixed commutators involving raising and lowering operators. When applied in the special case $i = j$ in a common sector q it reduces to the "normalization condition" introduced earlier, i.e., $[a_j(\mathbf{q}), a_j^\dagger(\mathbf{q})] = 1 = \frac{C^2(\mathbf{q})}{|\mathbf{q}|(2\pi)^d} \int d^d k\, \theta(\mathbf{k} \cdot \mathbf{q}) \Phi_j^2(\mathbf{k}, \mathbf{q})(f(\mathbf{k} - \mathbf{q}/2) - f(\mathbf{k} + \mathbf{q}/2))$ and serves to determine the magnitude — if not the functional form — of the Φ_j. More generally, *all* such commutators with distinct labels $i \neq j$ within a single fixed sector must vanish, and not just $[a_i(\mathbf{q}), a_j(\mathbf{q})]$. For example if $i \neq j$,

$$[a_i(\mathbf{q}), a_j^\dagger(\mathbf{q})] = \frac{C^2(\mathbf{q})}{|\mathbf{q}|(2\pi)^d} \int d^d k\, \theta(\mathbf{k} \cdot \mathbf{q}) \Phi_i(\mathbf{k}, \mathbf{q}) \Phi_j(\mathbf{k}, \mathbf{q})(f(\mathbf{k} - \mathbf{q}/2) - f(\mathbf{k} + \mathbf{q}/2))$$

$$= 0 \quad \text{if } i \neq j. \tag{7}$$

---

[g]Because $\theta(x) \cdot \theta(-x) \equiv 0$.

This vanishing should be interpreted as the "orthogonality" of two distinct functions Φ_i and Φ_j within a common sector.

An important consequence of (7): the integral on the right-hand side vanishes if either, but not both, Φ's are nodeless. Given any $j \neq 0$ and $\Phi_0 = 1$, one then deduces two identities:

$$[a_0(\mathbf{q}), a_j^\dagger(\mathbf{q})] = \frac{C^2(\mathbf{q})}{|\mathbf{q}|(2\pi)^d} \int d^d k \, \theta(\mathbf{k}\cdot\mathbf{q}) \Phi_j(\mathbf{k}, \mathbf{q})(f(\mathbf{k}-\mathbf{q}/2) - f(\mathbf{k}+\mathbf{q}/2)) = 0 \qquad (8A)$$

and

$$[a_0(\mathbf{q}), a_j(\mathbf{q})] = \frac{\text{finite operator}}{L^{d/2}} \to 0. \qquad (8B)$$

The second one, (8B), tells us nothing we did not already know. But for (8A) to vanish, it is necessary that the corresponding Φ_j have 1 or more nodes (i.e., changes of sign) as a function of $\mathbf{k}$, given that all other factors, $\theta(\mathbf{k}\cdot\mathbf{q})(f(\mathbf{k}-\mathbf{q}/2) - f(\mathbf{k}+\mathbf{q}/2))$, are positive semi-definite. This nodal structural requirement was stated earlier without proof. Next, as a matter of notation we shall distinguish *transverse* and *radial* nodes.

Depending on the particulars of H there *may* exist "good" quantum numbers that count the number of transverse nodes ($l = 0$ for s-waves, etc.) or radial nodes ($n = 0, 1, \ldots$) just as in atomic physics. The Φ_j's without a transverse node are "s-waves", those with one transverse node "p-waves", 2 transverse nodes "d-waves", etc. Generally the two-body interactions contain "exchange terms" that will mix them, but this mixing does not occur in the Hubbard model that we examine here, so l and n are good quantum numbers.

The vanishing of Eq. (7) proves that two distinct Φ_j's within a common sector will not have equal numbers of both radial and of transverse nodes. For — just as in ordinary quantum physics — whenever two wavefunctions occupying a common domain have the same number of transverse nodes they *must* have differing numbers of radial nodes — or *vice-versa* — in order to be orthogonal.

4. Decomposition of H into Noninteracting Sectors

Although it may seem obvious, the following statement, which *defines* the Abelian regime, is both seminal and crucial in the solution of the many-body problem. It is:

> Whenever the Hamiltonian can be decomposed into a sum over independent sectors, labeled by q, l, or whatever, such that each sector is diagonalizable independently of the others, the many-body problem is thereby "reduced to quadrature";[h] for then, the global wavefunctions are merely products of the eigenstates plucked from each sector.

---

[h] An aphorism probably dating to Euclid, meaning that a problem is essentially solved.

Let the following digression serve as an example of such factorization. Allow H_1 to be the surrogate for kinetic energy and H_2 for a momentum-conserving two-body interaction, both expressed in some hypothetically complete set of bosons b_j similar to the a_j introduced earlier. They are all labeled by q within one (arbitrarily chosen) hemisphere. Let $H_1 = \sum_{\mathbf{q}, q_z > 0} H_1(\mathbf{q})$, where $H_1(\mathbf{q}) = \sum_{i,j} \{ L_{i,j}(\mathbf{q}) b_i^\dagger(\mathbf{q}) b_j(\mathbf{q}) + L_{i,j}(-\mathbf{q}) b_i^\dagger(-\mathbf{q}) b_j(-\mathbf{q}) \}$ and similarly, $H_2 = \sum_{\mathbf{q}, q_z > 0} H_2(\mathbf{q})$ with $H_2(\mathbf{q}) = \sum_{i,j} \{ M_{i,j}(\mathbf{q}) b_i(\mathbf{q}) b_j(-\mathbf{q}) + \text{H.c.} \}$. Then,

$$H = \sum_{\mathbf{q}, q_z > 0} (H_1(\mathbf{q}) + H_2(\mathbf{q})) = \sum_{\mathbf{q}, q_z > 0} H(\mathbf{q}). \qquad (9)$$

That is, in each sector there are four compound operators:

$$H_{i,j}(\mathbf{q}) = (L_{i,j}(\mathbf{q}) b_i^\dagger(\mathbf{q}) b_j(\mathbf{q}) + L_{i,j}(-\mathbf{q}) b_i^\dagger(-\mathbf{q}) b_j(-\mathbf{q})) + (M_{i,j}(\mathbf{q}) b_i(\mathbf{q}) b_j(-\mathbf{q}) + \text{H.c.})$$

$$= \{ \quad 1 \quad + \quad 2 \quad + \quad 3 \quad + 4 \quad \}.$$

We wish to prove that $H(\mathbf{q})$ and $H(\mathbf{q}')$ commute, hence that H is *separable*. This implies that its eigenstates are all product states over the distinct sectors $\mathbf{q}$.

Given that $H(\mathbf{q})$ and $H(\mathbf{q}')$ are each quadratic in bosons, a proof that they commute requires showing that $[H_{i,j}(\mathbf{q}), H_{i',j'}(\mathbf{q}')] = [\{1 + 2 + 3 + 4\}, \{1' + 2' + 3' + 4'\}] = 0$, i.e.,

$$[((L_{i,j}(\mathbf{q}) b_i^\dagger(\mathbf{q}) b_j(\mathbf{q}) + L_{i,j}(-\mathbf{q}) b_i^\dagger(-\mathbf{q}) b_j(-\mathbf{q})) + (M_{i,j}(\mathbf{q}) b_i(\mathbf{q}) b_j(-\mathbf{q}) + \text{H.c.})),$$

$$((L_{i',j'}(\mathbf{q}') b_{i'}^\dagger(\mathbf{q}') b_{j'}(\mathbf{q}') + L_{i',j'}(-\mathbf{q}') b_{i'}^\dagger(-\mathbf{q}') b_{j'}(-\mathbf{q}'))$$

$$+ (M_{i',j'}(\mathbf{q}') b_{i'}(\mathbf{q}') b_{j'}(-\mathbf{q}') + \text{H.c.}))]$$

$$= 0. \qquad (10)$$

We shall prove a stronger result; not only does (10) vanish, but so does *every one* of the 16 individual commutators that comprise it. Take one of the 16 commutators, say $[1, 3']$, which is, $[AB, CD]$, where (omitting the irrelevant L or M coefficients) $A = b_i^\dagger(\mathbf{q})$, $B = b_j(\mathbf{q})$, $C = b_{i'}(\mathbf{q}')$ and $D = b_{j'}(-\mathbf{q}')$. It is evaluated with the help of an identity, $[AB, CD] \equiv A[B, CD] + [A, CD]B \equiv A\{[B, C]D + C[B, D]\} + \{[A, C]D + C[A, D]\}B$. A and B each are labeled $\mathbf{q}$, and C, D are labeled $\pm \mathbf{q}'$. We established in Eqs. (6) and (7) that whenever two such sectors are truly distinct *all* commutators involving the synthetic bosons (whether $[b_i(\mathbf{q}'), b_j(\mathbf{q})] = 0$ or $[b_i(\mathbf{q}'), b_j^\dagger(\mathbf{q})] = 0$) take the form $\frac{b \text{ and/or } b^\dagger}{L^{d/2}}$ and vanish in the thermodynamic limit. So $[AB, CD]$ decomposes into the sum of four operators, $A[B, C]D$. $AC[B, D]$, $[A, C]DB$ and $C[A, D]B$. *Each of these* is the product of three momentum-conserving operators, all presumably of O(1), situated in three possible sectors (q, q', and $\pm q \pm q'$), then divided by $L^{d/2}$, causing each of them to vanish in the limit.

This conclusion holds for each of the 16 terms in the expansion of Eq. (10), *QED*.

In the next section we *derive* precisely this sort of quadratic form $H(\mathbf{q})$ for the problem at hand. It shall be viewed as the Hamiltonian of a one-dimensional harmonic string that can be diagonalized by a variety of well known techniques.[11] Once the string's eigenstates are obtained explicitly, the complete set of eigenstates of the *total H* are known: they are product states, a factor taken from each of the individual sectors. The *extensive* properties — total energy, entropy, etc. — are sums of the respective quantities in the individual sectors.

Alas, there is a fly in the ointment: if *any* matrix element of H connects distinct sectors (such as "backward scattering" does in the one-dimensional Luttinger model), its very existence vitiates the simple *product state solutions* (even if other more complicated solutions can be found in specific instances). In an Appendix we assess such matrix elements to determine whether their importance rises to the level of their nuisance factor.

5. Density, Spin and Cooper-Pairing Channels

To simplify the various derivations we omitted the spin degrees of freedom in the preceding. However, the very definition of sectors depends on the form of the interaction. Originally quartic in the fermions the decomposition of H_2 into bilinear forms of bosons depends very much on the details. We examine fermions with spin (SU(2) particles), examples of which are the electrons or holes that live in metals or the neutrons and protons of nuclear physics. The simplest example of short-ranged two-body forces in such a system is provided by the Hubbard model,[i] used in the present paper to illustrate the separation into sectors and the methods of solution. In future publications we shall examine physically more relevant interactions, such as the Coulomb repulsion, requiring more complicated analysis.

Using density-fluctuation operators ϱ_σ, as defined just after Eq. (2), the interaction Hamiltonian of the Hubbard model is written,

$$H_2 = \frac{U}{\text{Vol}} \sum_{\mathbf{q}, q_z > 0} \{\rho_\uparrow(\mathbf{q})\rho_\downarrow(-\mathbf{q}) + \rho_\downarrow(-\mathbf{q})\rho_\uparrow(-\mathbf{q})\} \,. \tag{11}$$

Terms such as $\rho_\uparrow(\mathbf{q})\rho_\uparrow(-\mathbf{q})$ and $\rho_\downarrow(\mathbf{q})\rho_\downarrow(-\mathbf{q})$ are absent because $V(q) = U =$ constant in this model, hence all "direct" interactions among particles of the same species are precisely canceled by "exchange" terms and need not even be included

---

[i]Hubbard's model consists of fermions in an energy band of finite width with delta function interactions. The assumption is that the long-range interactions are screened or weakened by motion of the fermion fluid as a whole, hence that only the short-range forces need to be studied explicitly. But there is a *caveat* for free electrons (no upper energy cutoff): the delta function repulsion, which can be expressed as a hard-core repulsion in the limit that the radius of the hard-core r_0 shrinks to zero, has total scattering cross-section that vanishes as a power of $r_0 \to 0$ in 2D and 3D. Hence it cannot affect the many-body eigenstates in those cases — when it is treated correctly and completely. But with a momentum cutoff at a *finite* $q_0 \propto 1/r_0$ one gets around this objection while retaining the simplicity of a potential that, in reciprocal space, is constant wherever it is not zero.

ab initio.[j] The summand in (11) can be decomposed into bilinear forms of "particle density" and "spin density" interactions, each couched in its respective bosons. This is achieved by rewriting the curly brackets in the Eq. (11),

$$\{\rho_\uparrow(\mathbf{q})\rho_\downarrow(-\mathbf{q}) + \rho_\downarrow(\mathbf{q})\rho_\uparrow(-\mathbf{q})\} = \frac{1}{2}(\rho_\uparrow(\mathbf{q}) + \rho_\downarrow(\mathbf{q}))(\rho_\uparrow(-\mathbf{q}) + \rho_\downarrow(-\mathbf{q}))$$

$$- \frac{1}{2}(\rho_\uparrow(\mathbf{q}) - \rho_\downarrow(\mathbf{q}))(\rho_\uparrow(-\mathbf{q}) - \rho_\downarrow(-\mathbf{q})). \quad (12)$$

The first product of parentheses (with (+)'s) yields the density–density interactions that we denote $H_{2,\mathrm{dir}}$. We will start with that. The second parentheses (with (−)) yield just *one* of *three* components of the spin–spin exchange interactions — as we shall see further. This reformulation in a_0 comes about after comparing Eqs. (2) and (3) to identify density operators ϱ with the a_0 operators, i.e., $a_{0,\sigma}(\mathbf{q}) + a_{0,\sigma}^\dagger(-\mathbf{q}) = \frac{C(\mathbf{q})}{\sqrt{q\times\mathrm{Vol}}}\rho_\sigma(\mathbf{q})$ assuming $C(\mathbf{q})$ is invariant with respect to spin orientation σ. The total particle density is then,

$$\rho_\downarrow(\mathbf{q}) + \rho_\uparrow(\mathbf{q}) = \sqrt{\frac{q\times\mathrm{Vol}}{C^2(\mathbf{q})}}(a_{0,\downarrow}(\mathbf{q}) + a_{0,\uparrow}(\mathbf{q}) + a_{0,\downarrow}^\dagger(-\mathbf{q}) + a_{0,\uparrow}^\dagger(-\mathbf{q}))$$

$$= \sqrt{\frac{2\times q\times\mathrm{Vol}}{C^2(\mathbf{q})}}(a_0(\mathbf{q}) + a_0^\dagger(-\mathbf{q})). \quad (13)$$

In the second line of Eq. (13) we used the linear combination, $a_0(\mathbf{q}) \equiv \left(\frac{a_{0,\downarrow}(\mathbf{q})+a_{0,\uparrow}(\mathbf{q})}{\sqrt{2}}\right)$. The quadratic form $\frac{1}{2}(\rho_\uparrow(\mathbf{q}) + \rho_\downarrow(\mathbf{q}))(\rho_\uparrow(-\mathbf{q}) + \rho_\downarrow(-\mathbf{q}))$ yields the direct density–density (i.e., charge) interactions. Expressed in the spin-less a_0 boson operators, it is,

$$H_{2,\mathrm{dir}} = U \sum_{\mathbf{q},q_z>0} \frac{|q|}{C^2(q)} \{a_0^\dagger(\mathbf{q})a_0(\mathbf{q}) + a_0^\dagger(-\mathbf{q})a_0(-\mathbf{q}) + (a_0(\mathbf{q})a_0(-\mathbf{q}) + \mathrm{H.c.})\}. \quad (14)$$

Repeat the procedure with $\rho_\downarrow(\mathbf{q}) - \rho_\uparrow(\mathbf{q}) = \sqrt{\frac{2\times q\times\mathrm{Vol}}{C^2(\mathbf{q})}}(z_0(\mathbf{q}) + z_0^\dagger(-\mathbf{q}))$. These new operators are orthogonal to the a_0 and are given explicitly by:

$$z_0(\mathbf{q}) \equiv \frac{C(\mathbf{q})}{\sqrt{2\times q\times\mathrm{Vol}}} \sum_k \theta(\mathbf{k}\cdot\mathbf{q})(c_\uparrow^\dagger(\mathbf{k}-\mathbf{q}/2)c_\uparrow(\mathbf{k}+\mathbf{q}/2) - c_\downarrow^\dagger(\mathbf{k}-\mathbf{q}/2)c_\downarrow(\mathbf{k}+\mathbf{q}/2)).$$
$$(15)$$

---

[j]The exclusion principle prevents two parallel spin fermions from residing in the same cell. This is easy to see in coordinate space, using the identity $\tilde{n}_{j,\sigma}^2 = \tilde{n}_{j,\sigma}$ at each $\mathbf{R}_j$. The two-body interaction of Hubbard's model is $\frac{1}{2}U\sum_{\sigma,\sigma'}\tilde{n}_{j,\sigma}\tilde{n}_{j,\sigma'} = U\left(\tilde{n}_{j,\uparrow}\tilde{n}_{j,\downarrow} + \frac{1}{2}(\tilde{n}_{j,\uparrow} + \tilde{n}_{j,\downarrow})\right)$. Upon being summed over j the two terms linear in the occupation operator sum up to a constant of the motion that is absorbed into the definition of the chemical potential (Fermi level) and, effectively, disappears. After Fourier transformation the remaining terms, quadratic in the occupation numbers, sum up to H_2 in the form shown in the text.

If we assume both the one-body distribution function $f(\mathbf{k})$ and the normalization $C(\mathbf{q})$ to be isotropic and independent of spin orientation σ, then z_0 is normalized *ipso facto*. That is, $[z_0(\mathbf{q}), z_0^\dagger(\mathbf{q})] = 1$, by analogy with a_0. According to Eq. (12), the "longitudinal" spin contribution to H_2 is,

$$H_{2,z} = -U \sum_{\mathbf{q}, q_z > 0} \frac{|q|}{C^2(q)} \{z_0^\dagger(\mathbf{q})z_0(\mathbf{q}) + z_0^\dagger(-\mathbf{q})z_0(-\mathbf{q}) + (z_0(\mathbf{q})z_0(-\mathbf{q}) + \text{H.c.})\}. \quad (16)$$

However, this can only be *one part* of an "exchange" Hamiltonian, as symmetry requires the existence of additional bosons and their interactions, related to z_0 by $90°$ spatial rotations of the axes:

$$x_0(\mathbf{q}) \equiv \frac{C(\mathbf{q})}{\sqrt{2 \times q \times \text{Vol}}} \sum_k \theta(\mathbf{k} \cdot \mathbf{q})(c_\uparrow^\dagger(\mathbf{k} - \mathbf{q}/2)c_\downarrow(\mathbf{k} + \mathbf{q}/2) + c_\downarrow^\dagger(\mathbf{k} - \mathbf{q}/2)c_\uparrow(\mathbf{k} + \mathbf{q}/2))$$

$$(17)$$

and

$$y_0(\mathbf{q}) \equiv \frac{C(\mathbf{q})}{i\sqrt{2 \times q \times \text{Vol}}} \sum_k \theta(\mathbf{k} \cdot \mathbf{q})(c_\uparrow^\dagger(\mathbf{k} - \mathbf{q}/2)c_\downarrow(\mathbf{k} + \mathbf{q}/2) - c_\downarrow^\dagger(\mathbf{k} - \mathbf{q}/2)c_\uparrow(\mathbf{k} + \mathbf{q}/2)).$$

$$(18)$$

6. Abelian versus Non-Abelian

The algebra linking these various operators is of interest. All operators a_0, x_0, y_0 and z_0, were constructed *ab initio* as energy- and *momentum*-lowering operators and none of them is Hermitean. This set of operators commute among themselves but not with the set of their Hermitean conjugates. For example, $[z_0(\mathbf{q}), z_0^\dagger(\mathbf{q})] = 1$ as we have seen. Even more significantly, *mixed* commutators:

$$[z_0(\mathbf{q}), x_0^\dagger(\mathbf{q})] = iy_0(\mathbf{0})\frac{C(\mathbf{q})}{\sqrt{2 \times q \times \text{Vol}}} \quad (19)$$

and their cyclic permutations, also *might* fail to vanish if there is LRO. How could the right-hand side of (19) fail to vanish? Only if $y_0(0)$ assumes a macroscopic value $\geq \sqrt{\text{Vol}}$ instead of a finite value O(1), implying LRO along the y-axis.

The algebra that applies to such a set of noncommuting operators is *non-Abelian* and brings a new set of complications to bear, such as, causing the functions $C(\mathbf{q})$ and $f(k)$ to depend on spin or spatial orientation and mixing the normal modes from distinct sectors.

The same holds true for *anti*ferromagnetic or spiral LRO characterized by $\mathbf{Q}$'s. In such cases commutators from sectors separated by $\mathbf{Q}$, such as, $[z_0(\mathbf{q}), x_0^\dagger(\mathbf{q} + \mathbf{Q})]$, fail to vanish. Therefore we need to know whether any of the operators $x_0(\mathbf{Q})$, $y_0(\mathbf{Q})$, or $z_0(\mathbf{Q})$ can assume a macroscopic value $\geq \sqrt{\text{Vol}}$ and determine whether this occurrence is related to LRO.

Fortunately for this introductory essay, the *spontaneous* appearance of LRO is completely forbidden in $d = 1$ and 2 dimensions by virtue of the Mermin–Wagner theorem.[13] Of course, LRO could always be induced using external templates or ordering potentials. But in the absence of such external forces in $d = 1$ and 2 (also in many instances, in $d = 3$), the operators defined in distinct sectors *do* commute with one other. This is the Abelian case. Global solutions are then, quite simply, product states.

In 3D, we should expect spontaneously generated LRO under some circumstances, such as in the presence of strong repulsive interactions or arbitrary attractive interactions, at sufficiently low T. In such instances it is required to solve a specific non-Abelian algebra if we wish to find solutions to the entire model.

Relegating such complications to future publications, let us assume for the present that *there is no symmetry breaking of any kind and no macroscopic magnetization*. Then $x_0(\mathbf{q})$, $y_0(\mathbf{q})$ and $z_0(\mathbf{q})$ and $x_0^\dagger(\mathbf{q})x_0(\mathbf{q})$ etc, as well as $a_0(\mathbf{q})$ and $a_0^\dagger(\mathbf{q})a_0(\mathbf{q})$ all remain operators of magnitude O(1) at all q. Under those circumstances the three spin components, taken together with the density fluctuation operators a_0, comprise a set of orthonormal bosons that is resolutely Abelian.[k]

7.   The Complete Exchange Hamiltonian

When assembling all the interactive normal modes embedded in H_2, we find not just $H_{2,\mathrm{dir}}$ and $H_{2,z}$ as in the above construction, but additionally $H_{2,x}$ and $H_{2,y}$. These new contributions are predicted by symmetry — as was intimated in the preceding section — but they could also have been obtained directly by pairing fermions of spins "up" with spins "down" in the original quartic expression of H_2. These so-called "transverse" spin-dependent terms were inadvertently — and perhaps mistakenly — omitted in early formulations of Luttinger's model.[1] The following is a detailed derivation. Indicating pairing by underlining, in Eq. (11) one finds not just $+c_\uparrow^\dagger(\mathbf{k} + \mathbf{q}/2)c_\uparrow(\mathbf{k} - \mathbf{q}/2)\,c_\downarrow^\dagger(\mathbf{k}' - \mathbf{q}/2)c_\downarrow(\mathbf{k}' + \mathbf{q}/2)$ terms leading to Eq. (16) but *additionally*: $-c_\uparrow^\dagger(\mathbf{k} + \mathbf{q}/2)c_\downarrow(\mathbf{k} - \mathbf{q}/2)\,c_\downarrow^\dagger(\mathbf{k}' - \mathbf{q}/2)c_\uparrow(\mathbf{k}' + \mathbf{q}/2)$. This last is an "exchange" term, obtained by exchanging two fermion annihilation operators (hence the − sign) and redefining the dummy indices of summation. These elementary manipulations rely explicitly on $V(q)$ being constant $= U$ in the Hubbard model.

Finally after combining *all* the exchange terms that are required by rotational symmetry one obtains the correct, rotationally invariant, spin-dependent

---

[k]In this sense the spin bosons differ from ordinary spin-wave operators in the Heisenberg model. Additionally, the attentive reader may notice that spin raising or lowering (but energy-lowering) operators $\sigma_0^\pm(\mathbf{q}) = x_0(\mathbf{q}) \pm iy_0(\mathbf{q})$ are easily constructed from Eqs. (17) and (18) or easily derived from the original fermion representation. Such operators have proved useful in various theories of magnetism, *cf.* Ref. 14, although in the present context trivial but annoying notational hurdles need to be overcome (does one designate their energy-*raising* analogs $(\sigma_0^\pm)^\dagger$?), therefore they are neither convenient nor useful.

interaction. The *specific* expression valid in the present model is:

$$H_{2,\sigma} = -U \sum_{\mathbf{q}, q_z > 0} \frac{|q|}{C^2(q)} \{\boldsymbol{\sigma}_0^\dagger(\mathbf{q}) \cdot \boldsymbol{\sigma}_0(\mathbf{q}) + \boldsymbol{\sigma}_0^\dagger(-\mathbf{q}) \cdot \boldsymbol{\sigma}_0(-\mathbf{q}) + (\boldsymbol{\sigma}_0(\mathbf{q}) \cdot \boldsymbol{\sigma}_0(-\mathbf{q}) + \text{H.c.})\} .$$

(20)

The components of the vector spin operator $\boldsymbol{\sigma}_0(\mathbf{q}) = (x_0(\mathbf{q}), y_0(\mathbf{q}), z_0(\mathbf{q}))$ that appear here are those that were spelled out in Eqs. (15), (17) and (18). The vector-spin exchange Hamiltonian of Eq. (20), which respects the rotational symmetry of spin space, is what replaces Eq. (16).

Note that the negative coupling constant $-U$ seems to favor the creation of spontaneously nonvanishing magnitudes of $\boldsymbol{\sigma}_0(\mathbf{q})$ and not just at some discrete $q = 0$ or Q, but in *all* sectors. If this appears to challenge the original premise of *no* spontaneous magnetization or LRO, it is only because we have not yet considered H_1, the unperturbed Hamiltonian of the free fermions which *always* has its minimum energy for $\langle \boldsymbol{\sigma}_0(\mathbf{q}) \rangle \equiv 0$ at all $\mathbf{q}$.

Proof: The ground state of an even number of *free fermions* is a nondegenerate singlet. Only the introduction of two-body forces could conceivably engender a competition between two tendencies. But let us assume provisionally that even in 3D, the kinetic energy prevails and prevents LRO, just as was proved by Mermin and Wagner in dimensions $d \leq 2$ at all strengths of the interaction.

8. The Cooperons

But before proceeding with the magnetic degrees of freedom we pause to remark on *yet another* pairing allowed by Eq. (11), of type

$$+ \underbrace{c_\uparrow^\dagger(\mathbf{k} + \mathbf{q}/2) c_\downarrow^\dagger(-\mathbf{k} + \mathbf{q}/2)}_{} \underbrace{c_\downarrow(\mathbf{k}' + \mathbf{q}/2) c_\uparrow(-\mathbf{k}' + \mathbf{q}/2)}_{} .$$

First introduced in the BCS theory,[15] it is generally known as "Cooper pairing." Its inclusion with the three spin- pairings and the density-pairing adds up to the five degrees of freedom at the heart of the several SO(5) mean-field theories of high-T_c superconductivity that seemed so compelling at one time. Although a positive coupling constant $+U$ stifles formation of Cooper pairs (associated with "off-diagonal" long range order and superconductivity), nevertheless the ground-state might benefit from the zero-point contribution of what are *virtual* Cooper pairings or *virtual cooperons*.

Looking further into the cooperons one finds their designation to be not quite as straightforward as one would like them to be, for they apparently exist as *two* distinct species — each compatible with the above pairing. The lowering operators of the first, labeled (+), consist of the following linear combinations of elementary excitations:

$$\chi_{j,+}(\mathbf{q}) = \frac{D_+(\mathbf{q})}{\sqrt{\text{Vol}}} \sum_{\mathbf{k}} \theta(\varepsilon(\mathbf{k}+\mathbf{q}/2)+\varepsilon(-\mathbf{k}+\mathbf{q}/2))\Psi_{j,+}(\mathbf{q}|\mathbf{k})c_\downarrow(\mathbf{k}+\mathbf{q}/2)c_\uparrow(-\mathbf{k}+\mathbf{q}/2),$$

(21A)

that act only on states *outside* the joint Fermi sea of two particles ($\mathbf{k}$'s for which $\varepsilon(\mathbf{k} + \mathbf{q}/2) + \varepsilon(-\mathbf{k} + \mathbf{q}/2) > 0$, hence the subscripts "+"). These *lower* the total momentum by q and *also* lower the total energy by an amount which — while variable — is always positive.

The second set of operators act only *inside* the joint Fermi sea (note $c_\sigma^\dagger$). They too lower the total momentum by q while lowering the total energy by a variable (but always positive) amount. The subscript "−" indicates $\varepsilon(\mathbf{k}+\mathbf{q}/2)+\varepsilon(-\mathbf{k}+\mathbf{q}/2) < 0$, These operators are:

$$\chi_{j,-}(\mathbf{q}) = \frac{D_-(\mathbf{q})}{\sqrt{\text{Vol}}} \sum_{\mathbf{k}} \theta(-\varepsilon(\mathbf{k}-\mathbf{q}/2)-\varepsilon(-\mathbf{k}-\mathbf{q}/2))\Psi_{j,-}(\mathbf{q}|\mathbf{k})c_\uparrow^\dagger(-\mathbf{k}-\mathbf{q}/2)c_\downarrow^\dagger(\mathbf{k}-\mathbf{q}/2).$$

(21B)

Taken together the $\chi_\pm$'s define two complementary sets of cooperons. Their Hermitean conjugates are respectively *energy-* and *momentum-raising* operators,

$$\chi_{j,+}^\dagger(\mathbf{q}) = \frac{D_+(\mathbf{q})}{\sqrt{\text{Vol}}} \sum_{\mathbf{k}} \theta(\varepsilon(\mathbf{k}+\mathbf{q}/2)+\varepsilon(-\mathbf{k}+\mathbf{q}/2))\Psi_{j,+}^*(\mathbf{q}|\mathbf{k})c_\uparrow^\dagger(-\mathbf{k}+\mathbf{q}/2)c_\downarrow^\dagger(\mathbf{k}+\mathbf{q}/2)$$

(22A)

and

$$\chi_{j,-}^\dagger(\mathbf{q}) = \frac{D_-(\mathbf{q})}{\sqrt{\text{Vol}}} \sum_{\mathbf{k}} \theta(-\varepsilon(\mathbf{k}-\mathbf{q}/2) = \varepsilon(-\mathbf{k}-\mathbf{q}/2))\Psi_{j,-}^*(\mathbf{q}|\mathbf{k})c_\downarrow(\mathbf{k}-\mathbf{q}/2)c_\uparrow(-\mathbf{k}-\mathbf{q}/2).$$

(22B)

The choice of argument of the Heaviside functions $\pm(\varepsilon(\mathbf{k}+\mathbf{q}/2)+\varepsilon(-\mathbf{k}+\mathbf{q}/2))$ may seem arbitrary (why not just choose $k > k_F$ or $k < k_F$?) However, it does define the distinct sectors best. Also it leads to what is verifiably the correct normalization: $[\chi_\alpha, \chi_\beta^\dagger] = +\delta_{\alpha,\beta}$.[1] Just like the Φ's, the functions Ψ are, or can be chosen, real in most applications. Their magnitude is determined, as it was for the Φ_j's before, by the normalization. Their dependence on q and k depends on the functional form of the kinetic energy. However neither of the initial ($j = 0$) functions $\Psi_{0,\pm} = 1$ depends on k or q at all (except for the necessary cutoff to $\Psi_{0,+}$ at large k or $q = O(q_0)$).

Commutators of any two χ operators from distinct sectors vanish. Moreover *in the thermodynamic limit* the χ's commute with *all* other bosons — whether these deal with spin or with charge fluctuations, whether or not their q's are the same. By symmetry, $D_\pm(-\mathbf{q}) = D_\pm(+\mathbf{q})$, but because D_+ depends explicitly on the u-v

---

[1]Once the D's are calculated correctly. Unlike the C's these coefficients depend only weakly on q, which is why we do not include $\sqrt{q}$ in their definition.

cutoff q_0 while D_- has k_F as its sole cutoff, the two normalization constants are necessarily unequal.

We express the contribution to H_2 of the two species of cooperons in Hubbard's model as a positive definite, bilinear, form in the χ's,

$$H_{2,\chi} = +U \sum_{\mathbf{q}} \left(\frac{\chi_{0,-}(\mathbf{q})}{D_-(\mathbf{q})} + \frac{\chi_{0,+}^{\dagger}(-\mathbf{q})}{D_+(\mathbf{q})} \right) \cdot \left(\frac{\chi_{0,-}^{\dagger}(\mathbf{q})}{D_-(\mathbf{q})} + \frac{\chi_{0,+}(-\mathbf{q})}{D_+(\mathbf{q})} \right) . \qquad (23)$$

9. Construction of a String Theory

To calculate the restoring forces that prevent $\boldsymbol{\sigma}_0$ from growing arbitrarily large, and to complete the decomposition into nonoverlapping normal modes, one needs first to transform the kinetic energy — initially a bilinear form in fermions, $H_1 = \sum_k \sum_\sigma \varepsilon(\mathbf{k}) c_\sigma^\dagger(\mathbf{k}) c_\sigma(\mathbf{k})$ — into a bilinear form in *bosons*. It is then mapped onto strings. This procedure of "bosonization" is achieved solely with the aid of H_1, the kinetic energy. It can be understood as a mapping of fermionic states onto the elementary excitations and then mapping these onto low-lying boson excitations. The equations of motion connect $a_0(\mathbf{q})$ to the set $\{a_1(\mathbf{q}), a_2(\mathbf{q}), \ldots\}$. The equations are algebraically identical to those of the spin operators, hence the calculation needs only to be performed once. Equations of motion of the cooperons are also similar, although numerically the values of the coefficients differ due to the distinct geometry in those channels. Each set of normal modes lives in a sector having its own dedicated string.

We shall prove *by construction* that the "density" kinetic energy H_1 is isomorphic to a quadratic form constructed out of two sets of operators, $\{a_0(\mathbf{q}), a_1(\mathbf{q}), a_2(\mathbf{q}), \ldots\}$ and their Hermitean conjugates. The coefficients of this quadratic form are obtained numerically from an operator version of the *Lanczös procedure* — an algorithm that is often used in linear algebra and is a staple of every modern computer software library — wherein square matrices are reduced to tridiagonal form to facilitate the task of mapping them onto 1D or calculating their eigenvalues.

The construction of this set of operators starts with the $a_0(\mathbf{q})$ of Eq. (3) and its Hermitean conjugate. They are nodeless ($\Phi_0 = 1$) and are the only operators that appear in $H_{2,\mathrm{dir}}$, given that the interactions are not velocity-dependent. We use the $a_0(\mathbf{q})$ to create the first of a sequence of "equations of motion," *viz*:

$$[a_0(\mathbf{q}), H_1] = \frac{C(\mathbf{q})}{\sqrt{q \times \mathrm{Vol}}} \sum_k \theta(\mathbf{k} \cdot \mathbf{q})(\varepsilon(\mathbf{k} + \mathbf{q}/2) - \varepsilon(\mathbf{k} - \mathbf{q}/2)) c^\dagger(\mathbf{k} - \mathbf{q}/2) c(\mathbf{k} + \mathbf{q}/2) .$$

$$(24)$$

If we were to study Luttinger's original 1D model by this method, the parenthetical coefficient $(\varepsilon(\mathbf{k} + \mathbf{q}/2) - \varepsilon(\mathbf{k} - \mathbf{q}/2)) \propto \pm q$ in (24) would be independent of k. Then the entire right-hand side of (24) would be proportional to a_0 and there would be no need for any subsequent a_1, a_2, etc. With this immediate closure the length of the string reduces to a point, $j = 0$. Among the many ways of elucidating Luttinger's model that have been proposed over the past 50 years, this method appears to be the most succinct.

But as soon as we generalize to 2D or 3D (or even in 1D in the presence of "dispersion"[2-4]) the right-hand side of Eq. (24) needs to be decomposed into *two* parts: the first being proportional to the initial $a_0(\mathbf{q})$ and the second to some new operator which is arbitrarily *denoted* $a_1(\mathbf{q})$, of which nothing is known — except it is required to be normalized and orthogonal to $a_0(\mathbf{q})$. (Recall that by "orthogonal" it is meant that $a_1(\mathbf{q})$ and $a_1^\dagger(\mathbf{q})$ both commute with $a_0(\mathbf{q})$ and $a_0^\dagger(\mathbf{q})$).

That said, it is possible to recast (24) and the conjugate equation in a generic, more perspicuous, form, i.e.,

$$[a_0(\mathbf{q}), H_1] = A_0^0(\mathbf{q})a_0(\mathbf{q}) + A_0^1(\mathbf{q})a_1(\mathbf{q})\,, \tag{25A}$$

$$[H_1, a_0^\dagger(-\mathbf{q})] = A_0^0(\mathbf{q})a_0^\dagger(-\mathbf{q}) + A_0^1(\mathbf{q})a_1^\dagger(-\mathbf{q})\,. \tag{25B}$$

The A's are coefficients, the subscripts of which refer to the rank of the equation and the superscripts to the position within it. Thus, A_0^0 refers to the leading coefficient in the leading equation. All A's can be assumed real and symmetric, $A_i^j(\mathbf{q}) = A_j^i(\mathbf{q}) = (A_i^j(\mathbf{q}))^*$, mainly because H_1 itself is both real and Hermitean.

10. Constructing the Set of Coefficients A

The following illustrates the procedure by which the A coefficients in Eq. (25) and its iterations are calculated. First, $A_0^0(\mathbf{q})$ is isolated with the aid of the commutator of (25) with $a_0^\dagger(\mathbf{q})$,

$$[[a_0(\mathbf{q}), H_1], a_0^\dagger(\mathbf{q})]$$

$$= [A_0^0(\mathbf{q})a_0(\mathbf{q}) + A_0^1(\mathbf{q})a_1(\mathbf{q}), a_0^\dagger(\mathbf{q})] = A_0^0(\mathbf{q})$$

$$= \frac{C^2(\mathbf{q})}{|q|\text{Vol}} \sum_k \theta(\mathbf{k} \cdot \mathbf{q})(\varepsilon(\mathbf{k} + \mathbf{q}/2) - \varepsilon(\mathbf{k} - \mathbf{q}/2))(f(\mathbf{k} - \mathbf{q}/2) - f(\mathbf{k} + \mathbf{q}/2))\,. \tag{26}$$

The second line of (26) makes explicit use of orthogonality $[a_1(\mathbf{q}), a_0^\dagger(\mathbf{q})] = 0$ and of normalization $[a_0(\mathbf{q}), a_0^\dagger(\mathbf{q})] = 1$, while the third line shows the results obtained by using the fermions' anticommutator relations and subjecting the integrand to the Central Limit Theorem (CLT), which allows replacing operators $c^\dagger(\mathbf{k})c(\mathbf{k}) = \tilde{n}(\mathbf{k})$ by their quantum and thermodynamic averages $\langle \tilde{n}(\mathbf{k}) \rangle_{TA} \to f(\mathbf{k})$.

In using the CLT we have to distinguish the sum in Eq. (26) from that in, say, Eq. (3). We claim that for the purposes of applying CLT the right-hand side of

Eq. (26) is, in the thermodynamic limit, an integral whereas the right-hand side of Eq. (3) is not. How to tell? There are two ways; in Eq. (3) the sum is divided, not by Vol, but by $\sqrt{\text{Vol}}$. Additionally, in (3) the averages are zero but the fluctuations are not — exactly the opposite of (26).

Once $A_0^0(\mathbf{q})$ is known it becomes a simple matter to obtain $A_0^1(\mathbf{q})$ and to set up recursion equations for all successive A's, as we shall show. But, before proceeding to this next step, let us show how this procedure can be conceptually and algebraically simplified by the introduction of a more elegant and symbolic notation that is exhibited next.

11. Moments

Let us define a nonnegative "probability function," $P_{\mathbf{q}}(\mathbf{k}) = \frac{C^2(\mathbf{q})}{|q|\text{Vol}}\theta(\mathbf{k} \cdot \mathbf{q})(f(\mathbf{k} - \mathbf{q}/2) - f(\mathbf{k} + \mathbf{q}/2))$, at each density sector q. The average of an arbitrary function $G(\mathbf{k})$ subjected to this probability is:

$$\langle G(\mathbf{k})\rangle ==\equiv \frac{C^2(\mathbf{q})}{|q|\text{Vol}} \sum_k \theta(\mathbf{k} \cdot \mathbf{q})(G(\mathbf{k}))(f(\mathbf{k} - \mathbf{q}/2) - f(\mathbf{k} + \mathbf{q}/2)). \qquad (27)$$

Obviously this average is, generally, also a function of q.

Suppose we test it by using the constant "function" $G(\mathbf{k}) = 1$; then Eq. (27) yields an identity $\langle 1\rangle = 1$, a neat restatement of the "normalization condition."

We also make free use of a streamlined notation, $\Delta\varepsilon \equiv \varepsilon(\mathbf{k} + \mathbf{q}/2) - \varepsilon(\mathbf{k} - \mathbf{q}/2)$, when there is no ambiguity. For example, Eq. (26) for $A_0^0(\mathbf{q})$ is now,

$$A_0^0(\mathbf{q}) = \langle \varepsilon(\mathbf{k} + \mathbf{q}/2) - \varepsilon(\mathbf{k} - \mathbf{q}/2)\rangle \equiv \langle \Delta\varepsilon\rangle. \qquad (28)$$

This quantity is plotted in Fig. 2 as function of q and seen to be proportional to $|q|$ in the long-wavelength limit. For the purpose of calculating $A_0^1(\mathbf{q})$ we rearrange Eq. (25): $A_0^1(\mathbf{q})a_1(\mathbf{q}) = [a_0(\mathbf{q}), H_1] - A_0^0(\mathbf{q})a_0(\mathbf{q})$, such that only known quantities appear on the right-hand side.

$$A_0^1(\mathbf{q})a_1(\mathbf{q})$$

$$= \frac{C(\mathbf{q})}{\sqrt{|q|\text{Vol}}} \sum_k \theta(\mathbf{k} \cdot \mathbf{q})(\varepsilon(\mathbf{k} + \mathbf{q}/2) - \varepsilon(\mathbf{k} - \mathbf{q}/2) - A_0^0(\mathbf{q}))c^\dagger(\mathbf{k} - \mathbf{q}/2)c(\mathbf{k} + \mathbf{q}/2)$$

$$= \frac{C(\mathbf{q})}{\sqrt{|q|\text{Vol}}} \sum_k \theta(\mathbf{k} \cdot \mathbf{q})(\Delta\varepsilon - \langle\Delta\varepsilon\rangle)c^\dagger(\mathbf{k} - \mathbf{q}/2)c(\mathbf{k} + \mathbf{q}/2). \qquad (29)$$

Then, without knowing anything more about a_1 *except that it has been presumed to be normalized*, we can evaluate the commutator of (29) with its Hermitean conjugate to obtain $|A_0^1(\mathbf{q})|^2$: that is, $[A_0^1(\mathbf{q})a_1(\mathbf{q}), A_0^1(\mathbf{q})^*a_1^\dagger(\mathbf{q})] = |A_0^1(\mathbf{q})|^2$. Because $A_0^1(\mathbf{q})$ is real, it is:

$$A_0^1(\mathbf{q}) = \sqrt{\langle\Delta\varepsilon^2\rangle - \langle\Delta\varepsilon\rangle^2} = \sqrt{\langle(\Delta\varepsilon - \langle\Delta\varepsilon\rangle)^2\rangle}. \qquad (30)$$

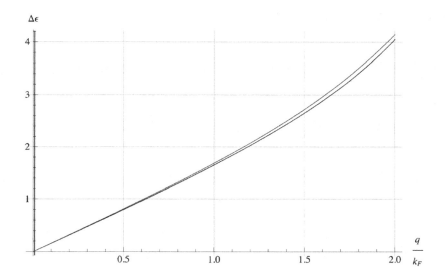

Fig. 2. *Averaged Dispersion.* Plot of $A_0^0 = \langle \Delta \varepsilon \rangle$ as calculated in Eqs. (26) or (28), as function of q/k_F, in 2D at $kT = 0.1$ (lower curve) and 0.2 (upper) in the same units as in Fig. 1.

We see this, the first nondiagonal, nontrivial, A coefficient, to be the *variance* in the distribution of the elementary excitations $\Delta \varepsilon$. Were we to repeat this in the original 1D Luttinger model in which Eq. (30) vanishes because there was no dispersion, we would find that the boson chain terminated rigorously at the first equation.

One subtle point: the decoupling of $j = 0$ from $j \geq 1$ in the original dispersionless 1D Luttinger model at all q differs qualitatively from what happens in higher dimensions. Although $A_0^1(\mathbf{q})$ vanishes as $q \to 0$ in the long wavelength limit, as seen in Fig. 3 for $d = 2$, even at $q = 0$ we cannot yet state that the chains terminate at $j = 1$, given that the other A's also vanish in that limit — including $A_0^0(\mathbf{q})$ shown in Fig. 2. It has not yet been determined whether the higher coefficients vanish *even faster* with decreasing q. The devil is in the details.

Once $A_0^1(\mathbf{q}) \neq 0$, division of (29) by (30) allows *explicit* construction of the new operator:

$$a_1(\mathbf{q}) = \frac{A_0^1(\mathbf{q}) a_1(\mathbf{q})}{A_0^1(\mathbf{q})}$$

$$= \frac{\dfrac{C(\mathbf{q})}{\sqrt{|q|\mathrm{Vol}}} \sum_k \theta(\mathbf{k} \cdot \mathbf{q})(\Delta \varepsilon - \langle \Delta \varepsilon \rangle) c^\dagger(\mathbf{k} - \mathbf{q}/2) c(\mathbf{k} + \mathbf{q}/2)}{\sqrt{\langle (\Delta \varepsilon - \langle \Delta \varepsilon \rangle)^2 \rangle}} . \tag{31A}$$

It is explicitly normalized *by construction* and the corresponding Φ_1 (which can be read off (31A)) exhibits a single "radial" node at $\Delta \varepsilon = \langle \Delta \varepsilon \rangle$:

$$\Phi_1(\mathbf{k}, \mathbf{q}) = \frac{(\Delta \varepsilon - \langle \Delta \varepsilon \rangle)}{\sqrt{\langle (\Delta \varepsilon - \langle \Delta \varepsilon \rangle)^2 \rangle}} . \tag{31B}$$

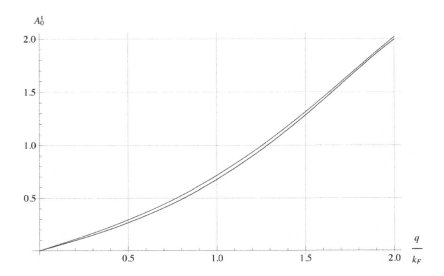

Fig. 3.   *The Variance.* A second coefficient, A_0^1 versus q/k_F, at temperatures $kT = 0.1$ (lower curve) and 0.2 (upper) in 2D, using the same units as in Figs. 1 and 2.

We could continue in this vein, but at this juncture it is helpful once again to interrupt the narrative to introduce additional useful concepts and notation before proceeding.

12.   Node Counting

The preceding suggests an interesting and most productive way of classifying the Φ_j's. In the numerator of Eqs. (31) some of the elementary excitations $\Delta\varepsilon = \varepsilon(\mathbf{k} + \mathbf{q}/2) - \varepsilon(\mathbf{k} - \mathbf{q}/2)$ have energy lower than average and some are higher. This shows that Φ_1 possesses at least *one "radial"* node, i.e., a change of sign in the direction of increasing $\Delta\varepsilon$ (relative to $\Phi_0 = 1$ which has none).

This radial node is the locus of the *curve* (in 2D) or *surface* (in 3D) defined by solutions of $\Phi_1 = 0$, where the equation $\Delta\varepsilon - \langle\Delta\varepsilon\rangle = \varepsilon(\mathbf{k}+\mathbf{q}/2) - \varepsilon(\mathbf{k}-\mathbf{q}/2) - A_0^0 = 0$ is satisfied. Clearly a_2 has to have *two* nonoverlapping radial nodes to be orthogonal to both a_0 and a_1; one extrapolates that for all $j \geq 2$, the function Φ_j has j radial nodes in k-space. So, in 2D, Φ_j vanishes on j distinct *curves* in k-space, in 3D on j distinct *surfaces*. This may be deduced from Φ_j being constructed as a polynomial of degree j in having j real roots.

If radial nodes are necessary, then why not transverse? (Recall, these are defined as changes in sign in Φ_j along trajectories of *constant* $\Delta\varepsilon = e(\mathbf{k}+\mathbf{q}/2) - e(\mathbf{k}-\mathbf{q}/2)$). Now, given that $\Phi_0 = 1$ has neither transverse nor radial nodes, and that the equations of motion that allow Φ_1 to acquire one *radial* node, Φ_2 two *radial* nodes, etc, never after the number of transverse nodes, should we regard the latter as superfluous?

Generally *no*, because "completeness" compels us to use *all* possible normal modes in the calculation of certain nontrivial correlation functions and dynamical quantities. The Hubbard model is an exception, in that the number of transverse nodes is unaffected by H_2, so $l = 0$ *is* a constant of the motion. For more structured interactions, boson operators a with 1 or more transverse nodes *do* appear in the equations of motion, as they are connected to $j = 0$ bosons by "exchange" terms. This provides an additional reason, if such were needed, for dwelling on the simpler Hubbard model which, by avoiding such complications, is best suited to the present expository text.

13. The Second Equation of Motion and Beyond

Next, we pursue the narrative of Sec. 11 to calculate successive commutator bracket "equations of motion," $[a_1(\mathbf{q}), H_1]$ and beyond, in the density sectors of the Hubbard model (actually, the same equations also apply to the *spin* components), using the "averaging" notation of Sec. 11.

In words: The first and successive iterations produce three operators. The first is already known: $A_1^0 a_0$, rewritten $A_0^1 a_0$. The second is a multiple of $a_1(\mathbf{q})$, $A_1^1 a_1(\mathbf{q})$, which defines A_1^1. Finally, there may be a remainder. If so, it has to be orthogonal to both a_0 and a_1 and is denoted $A_1^2 a_2(\mathbf{q})$. The operator $a_2(\mathbf{q})$ is a new boson, presumed normalized and orthogonal to a_0 and a_2, which remains to be identified. A_1^1 and A_1^2 are two new coefficients, functions of q. (We again emphasize that in the Hubbard model, and only in this model, do the A coefficients calculated for the x, y and z bosons satisfy the same equations as do the a's, hence obviating separate calculations).

Algebraically (i.e., symbolically): We can express exactly the same facts much more succinctly, by writing the equation:

$$[a_1(\mathbf{q}), H_1] = A_0^1(\mathbf{q})a_0(\mathbf{q}) + A_1^1(\mathbf{q})a_1(\mathbf{q}) + A_1^2(\mathbf{q})a_2(\mathbf{q}) \, .$$

$A_0^1(\mathbf{q})a_1(\mathbf{q})$ is known from Eqs. (29) and (31), $a_0(\mathbf{q})$ and $a_1(\mathbf{q})$ have both been previously defined (cf. Eq. (3)) or derived (cf. Eqs. (31)). We project $A_1^1(\mathbf{q})$ out using (29) and the nested commutator $A_1^1(\mathbf{q}) = \frac{1}{(A_0^1)^2}[[A_0^1 a_1(\mathbf{q}), H_1], A_0^1 a_1^\dagger(\mathbf{q})]$. Explicitly,

$$A_1^1(\mathbf{q}) = \frac{1}{(A_0^1(\mathbf{q}))^2} \langle (\varepsilon(\mathbf{k} + \mathbf{q}/2) - \varepsilon(\mathbf{k} - \mathbf{q}/2) - A_0^0(\mathbf{q}))^2 (\varepsilon(\mathbf{k} + \mathbf{q}/2) - \varepsilon(\mathbf{k} - \mathbf{q}/2)) \rangle$$

$$= \frac{\langle (\Delta\varepsilon - \langle\Delta\varepsilon\rangle)^2 \cdot (\Delta\varepsilon) \rangle}{\langle (\Delta\varepsilon - \langle\Delta\varepsilon\rangle)^2 \rangle} = \frac{\langle \Delta\varepsilon^3 - 2\langle\Delta\varepsilon^2\rangle\langle\Delta\varepsilon\rangle + \langle\Delta\varepsilon\rangle^3 \rangle}{\langle \Delta\varepsilon^2 - \langle\Delta\varepsilon\rangle^2 \rangle} \, . \tag{32}$$

This is a real quantity not unlike a virial coefficient in statistical mechanics.

The square of $A_1^2(\mathbf{q})$ can be calculated in two steps *before* even knowing anything about a_2 except that it is normalized. First, rearrange the terms in the equation of motion $[a_1(\mathbf{q}), H_1] = A_0^1(\mathbf{q})a_0(\mathbf{q}) + A_1^1(\mathbf{q})a_1(\mathbf{q}) + A_1^2(\mathbf{q})a_2(\mathbf{q})$ to solve for $A_1^2(\mathbf{q})a_2(\mathbf{q})$,

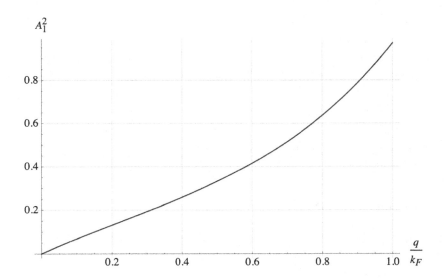

Fig. 4. *The Next Coefficient.* Coefficient $A_1^2(\mathbf{q})$ as a function of q/k_F in 2D, at $kT = 0.1$, in same units as the preceding figures.

then multiply and divide the right-hand side by $A_0^1(\mathbf{q})$ so that we can use our prior knowledge of $A_0^1(\mathbf{q})a_1(\mathbf{q})$ obtained in Eq. (29). That is,

$$A_1^2(\mathbf{q})a_2(\mathbf{q}) = \frac{1}{A_0^1(\mathbf{q})}([A_0^1(\mathbf{q})a_1(\mathbf{q}), H_1] - (A_0^1(\mathbf{q}))^2 a_0(\mathbf{q}) - A_1^1(\mathbf{q})A_0^1(\mathbf{q})a_1(\mathbf{q}))$$

$$= \frac{1}{A_0^1(\mathbf{q})}\frac{C(\mathbf{q})}{\sqrt{q \times \text{Vol}}} \sum_k \theta(\mathbf{k} \cdot \mathbf{q})$$

$$\times \{(\varepsilon(\mathbf{k} + \mathbf{q}/2) - \varepsilon(\mathbf{k} - \mathbf{q}/2) - A_0^0(\mathbf{q}))(\varepsilon(\mathbf{k} + \mathbf{q}/2)$$

$$-\varepsilon(\mathbf{k} - \mathbf{q}/2) - A_1^1(\mathbf{q})) - (A_0^1(\mathbf{q}))^2\}c^\dagger(\mathbf{k} - \mathbf{q}/2)c(\mathbf{k} + \mathbf{q}/2). \quad (33A)$$

Then

$$[A_1^2(\mathbf{q})a_2(\mathbf{q}), A_1^2(\mathbf{q})a_2^\dagger(\mathbf{q})] = |A_1^2(\mathbf{q})|^2 = \frac{1}{(A_0^1(\mathbf{q}))^2}\langle((\Delta\varepsilon - A_0^0)(\Delta\varepsilon - A_1^1) - (A_0^1)^2)^2\rangle,$$

the positive root of which yields,

$$A_1^2(\mathbf{q}) = \frac{1}{A_0^1(\mathbf{q})}\sqrt{\langle((\Delta\varepsilon - \langle\Delta\varepsilon\rangle)(\Delta\varepsilon - A_1^1(\mathbf{q})) - (A_0^1(\mathbf{q}))^2)^2\rangle}. \quad (33B)$$

This coefficient has been expressed in terms of two previously calculated $A_n^m(\mathbf{q})$ coefficients, each of which has been expressed in the $\langle function(\Delta\varepsilon)\rangle$ notation. We plot our calculation of $A_1^2(\mathbf{q})$ in Fig. 4 above.

Equations (32) and (33) also allow us to extract the normalized, two-radial-node boson operator a_2 as a sum over elementary excitations, in terms of the known quantities $A_0^0(\mathbf{q})$, $A_0^1(\mathbf{q})$ $(= A_1^0(\mathbf{q}))$, $A_1^1(\mathbf{q})$ and $A_1^2(\mathbf{q})$ $(= A_2^1(\mathbf{q}))$. With the aid of these coefficients we can also extract the wavefunction Φ_2 from Eqs. (33). Leaving derivation and verification as exercises for the reader, we find:

$$a_2(\mathbf{q}) = \frac{1}{A_1^2(\mathbf{q})A_0^1(\mathbf{q})}([A_0^1(\mathbf{q})a_1(\mathbf{q}), H_1] - (A_0^1(\mathbf{q}))^2 a_0(\mathbf{q}) - A_1^1(\mathbf{q})A_0^1(\mathbf{q})a_1(\mathbf{q})).$$

Similarly, at the nth turn we obtain successive $A_n^m = A_m^n$ coefficients with $m = n-1$, n, and $n+1$. Direct calculation shows all these coefficients to be real.

The generalized equation of motion are, for arbitrary n:

$$[a_n(\mathbf{q}), H_1] = A_{n-1}^n(\mathbf{q})a_{n-1}(\mathbf{q}) + A_n^n(\mathbf{q})a_n(\mathbf{q}) + A_n^{n+1}a_{n+1}(\mathbf{q}), \qquad (34A)$$

with $A_0^{-1} = A_{-1}^0 \equiv 0$ serving as the initial (boundary) condition. The Hermitean conjugate equations are:

$$[H_1, a_n^\dagger(-\mathbf{q})] = A_{n-1}^n(\mathbf{q})a_{n-1}^\dagger(-\mathbf{q}) + A_n^n(\mathbf{q})a_n^\dagger(-\mathbf{q}) + A_n^{n+1}(\mathbf{q})a_{n+1}^\dagger(-\mathbf{q}). \qquad (34B)$$

14. The Harmonic String

Consider a quadratic form of *boson* operators constructed with the aid of coefficients calculated above,

$$\hat{A}(\mathbf{q}) = \sum_{n=0}^{\infty} A_n^n(\mathbf{q})(a_n^\dagger(\mathbf{q})a_n(\mathbf{q}) + a_n^\dagger(-\mathbf{q})a_n(-\mathbf{q}))$$

$$+ \sum_{n=0}^{\infty}(A_n^{n+1}(\mathbf{q})(a_n^\dagger(\mathbf{q})a_{n+1}(\mathbf{q}) + a_n^\dagger(-\mathbf{q})a_{n+1}(-\mathbf{q})) + \text{H.c.}) \qquad (35)$$

This quadratic form, $\hat{A}(\mathbf{q})$, is the Hamiltonian of an "*harmonic string.*" The equations of motion of $a_0(\mathbf{q})$, $a_1(\mathbf{q}), \ldots, a_n(\mathbf{q})$ with respect to this Hamiltonian are:

$$[a_n(\mathbf{q}), \hat{A}(\mathbf{q})] = A_{n-1}^n(\mathbf{q})a_{n-1}(\mathbf{q}) + A_n^n(\mathbf{q})a_n(\mathbf{q})) + A_n^{n+1}a_{n+1}(\mathbf{q}). \qquad (36)$$

As far as the bosons are concerned $\hat{A}(\mathbf{q})$ is interchangeable with H_1, given that the right-hand side of (36) is identical with that of Eq. (34A). Similarly, equations of motion of the conjugate operators $a_n^\dagger$ with the same $\hat{A}(\mathbf{q})$ are identical with those of (34B).

We established previously that the equations of motion do not change the initial number of transverse nodes — as opposed to radial nodes. This number is therefore a "constant of the motion". The transversally nodeless bosons here are "s-waves." (Bosons with $l = 1$ transverse node and $n = 0, 1, 2, \ldots$ radial nodes are p-waves, those with $l = 2$ transverse nodes are d-waves, etc. Each has its own set of A

coefficients. Numerically, these will differ from each other and from those of the s-waves, so for a complete identification we *should* write them $A_n^m(l, \mathbf{q})$, as they depend on l as well as on q. However, in the present context only $l = 0$ appears and this additional notation will not be necessary.

Before proceeding, let us take one additional detour to examine the remaining channels: those of Cooper pairs.

15. Cooperon Strings

The equations of motion of cooperons are similar to the preceding but their coefficients, that we shall denote B, differ from the A's numerically and, in their dependence on q, geometrically.

Start with the $(+)$, particle-like cooperons of Eq. (21):

$$\chi_{j,+}(\mathbf{q}) = \frac{D_+(\mathbf{q})}{\sqrt{\mathrm{Vol}}} \sum_{\mathbf{k}} \theta(\varepsilon(\mathbf{k}+\mathbf{q}/2) + \varepsilon(-\mathbf{k}+\mathbf{q}/2)) \Psi_{j,+}(\mathbf{q}|\mathbf{k}) c_\downarrow(\mathbf{k}+\mathbf{q}/2) c_\uparrow(-\mathbf{k}+\mathbf{q}/2) .$$

Upon being written out, the first equation, that of the nodeless mode $\chi_{0,+}(\mathbf{q})$, yields a multiple of $\chi_{0,+}(\mathbf{q})$ and a new operator, $\chi_{1,+}(\mathbf{q})$.

$$[\chi_{0,+}(\mathbf{q}), H_1] = B_0^0(\mathbf{q})\chi_{0,+}(\mathbf{q}) + B_0^1(\mathbf{q})\chi_{1,+}(\mathbf{q}) .$$

In the fermion operators,

$$[\chi_{0,+}(\mathbf{q}), H_1] = \frac{D_+(\mathbf{q})}{\sqrt{\mathrm{Vol}}} \sum_{\mathbf{k}} \theta(\varepsilon(\mathbf{k} + \mathbf{q}/2) + \varepsilon(-\mathbf{k} + \mathbf{q}/2))$$

$$\times (\varepsilon(-\mathbf{k} + \mathbf{q}/2) + \varepsilon(\mathbf{k} + \mathbf{q}/2)) c_\downarrow(\mathbf{k} + \mathbf{q}/2) c_\uparrow(-\mathbf{k} + \mathbf{q}/2) , \qquad (37)$$

The first coefficient in the string, $B_0^0(\mathbf{q})$, is obtained from $[[\chi_{0,+}(\mathbf{q}), H_1], \chi_{0,+}^\dagger(\mathbf{q})]$, and is:

$$B_0^0(\mathbf{q}) = \frac{(D_+)^2}{\mathrm{Vol}} \sum_{\mathbf{k}} \theta(\varepsilon(\mathbf{k} + \mathbf{q}/2) + \varepsilon(-\mathbf{k} + \mathbf{q}/2))$$

$$\times (\varepsilon(-\mathbf{k} + \mathbf{q}/2) + \varepsilon(\mathbf{k} + \mathbf{q}/2))(1 - f(\mathbf{k} + \mathbf{q}/2) - f(-\mathbf{k} + \mathbf{q}/2)) . \quad (38A)$$

Let us define a cooperon probability function by analogy with Sec. 11. Thus,

$$P_{\mathbf{q},+}^C(\mathbf{k}) = \frac{(D_+)^2}{\mathrm{Vol}} \theta(\varepsilon(\mathbf{k}+\mathbf{q}/2) + \varepsilon(-\mathbf{k}+\mathbf{q}/2)) \times (1 - f(\mathbf{k}+\mathbf{q}/2) - f(-\mathbf{k}+\mathbf{q}/2)) . \quad (39A)$$

This probability satisfies two important requirements: it is everywhere nonnegative and its sum over k is 1 (equivalent to the normalization of $\chi_{0,+}(\mathbf{q})$). Making use of this P and writing symbolically, $\varepsilon + \varepsilon' \equiv \varepsilon(\mathbf{k} + \mathbf{q}/2) + \varepsilon(-\mathbf{k} + \mathbf{q}/2)$, we follow the earlier procedure to obtain the next coefficient, which is $B_0^0(\mathbf{q}) = \langle \varepsilon + \varepsilon' \rangle$. Also,

$$(B_0^1(\mathbf{q}))^2 = [B_0^1(\mathbf{q})\chi_{1,+}(\mathbf{q}), (B_0^1(\mathbf{q})\chi_{1,+}(\mathbf{q}))^\dagger] = \langle(\varepsilon + \varepsilon' - \langle\varepsilon + \varepsilon'\rangle)^2\rangle, \text{ i.e.,}$$

$$B_0^1(\mathbf{q}) = \sqrt{\langle(\varepsilon + \varepsilon' - \langle\varepsilon + \varepsilon'\rangle)^2\rangle}. \tag{40}$$

One major difference with the density or spin-modes concerns the cutoff; at large values of the k-space variable $\varepsilon + \varepsilon' \equiv \varepsilon(\mathbf{k} + \mathbf{q}/2) + \varepsilon(-\mathbf{k} + \mathbf{q}/2)$, the probability density (39) tends to a constant, hence P cannot be normalized without a cutoff at $k = q_0$. Such a cutoff comes naturally in the Hubbard model as q_0 is inversely proportional to a lattice parameter.

There is, however, no such conundrum concerning the hole-like branch of the cooperons, as k_F provides an intrinsic cutoff. There the corresponding probability function is:

$$P_{\mathbf{q},-}^C(\mathbf{k}) = \frac{(D_-)^2}{\text{Vol}}\theta(-\varepsilon(\mathbf{k} + \mathbf{q}/2) - \varepsilon(-\mathbf{k} + \mathbf{q}/2)) \times (f(\mathbf{k} + \mathbf{q}/2) + f(-\mathbf{k} + \mathbf{q}/2) - 1). \tag{39B}$$

Now in the $(-)$ branch the ε's are < 0, but their $B_0^0(\mathbf{q})$ also acquires an extra minus sign,

$$B_0^0(\mathbf{q}) = -\frac{(D_-)^2}{\text{Vol}}\sum_{\mathbf{k}}\theta(-\varepsilon(\mathbf{k} + \mathbf{q}/2) - \varepsilon(-\mathbf{k} + \mathbf{q}/2))$$

$$\times(\varepsilon(-\mathbf{k} + \mathbf{q}/2) + \varepsilon(\mathbf{k} + \mathbf{q}/2))(f(\mathbf{k} + \mathbf{q}/2) + f(-\mathbf{k} + \mathbf{q}/2) - 1) \tag{38B}$$

hence the coefficient $B_0^0(\mathbf{q}) = \langle-(\varepsilon + \varepsilon')\rangle$ remains resolutely positive and the formula for $B_0^1(\mathbf{q})$ remains formally identical to (40) except that averages must be taken with respect to probabilities $P_{\mathbf{q},-}^C(\mathbf{k})$. The plots in Fig. 5 show $B_0^1(\mathbf{q})$ for the $(-)$ species has an approximately semicircular dependence on q.

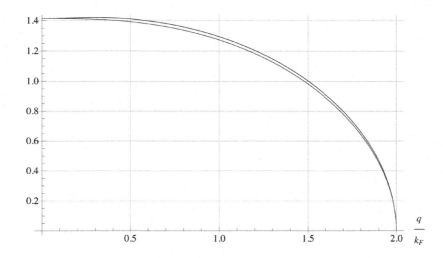

Fig. 5. *Cooperons.* Coefficient $B_0^1(\mathbf{q})$ versus q/k_F for the hole-like cooperons in 2D, at two temperatures, $kT = 0.1$ (upper curve) and 0.2, in the same units as previously.

Further iterations proceed in the same manner as they did in the density and spin sectors.

16. Stitching of Pairs of Strings by H_2

We return to the equations of motion and replace H_1 by $H = H_1 + H_2$. This modifies them in two ways.

The original strings devolved from initial operators (a_0, x_0, etc) culled from the interaction H_2 — a felicitous choice, in that H_2 can then affect coefficients only at the initial sites of each string. As an example, consider modifications caused by introducing $H_{2,\mathrm{dir}}$ of Eq. (14) in Sec. 3, into the equations of motion. The interaction $H_{2,\mathrm{dir}}$ takes the form:

$$H_{2,\mathrm{dir}} = U \sum_{\mathbf{q}, q_z > 0} \frac{|q|}{C^2(q)} \{ a_0^\dagger(\mathbf{q}) a_0(\mathbf{q}) + a_0^\dagger(-\mathbf{q}) a_0(-\mathbf{q}) + (a_0(\mathbf{q}) a_0(-\mathbf{q}) + a_0^\dagger(-\mathbf{q}) a_0^\dagger(\mathbf{q})) \}.$$

The diagonal operators $a_0^\dagger(\mathbf{q}) a_0(\mathbf{q}) + a_0^\dagger(-\mathbf{q}) a_0(-\mathbf{q})$ just add $U\frac{|q|}{C^2(q)}$ to $A_0^0(\mathbf{q})$ (displayed in Eq. (28)) without modifying any of the other A's.

Nondiagonal operators $a_0(\mathbf{q}) a_0(-\mathbf{q}) + a_0^\dagger(-\mathbf{q}) a_0^\dagger(\mathbf{q})$ are also new, but they are nontrivial in that they connect *pairs of strings* that were *formerly disjoint*, the one for annihilation operators at q and the other for creation operators at $-q$. These pairs of strings merge at $j = 0$. Momentum conservation dictates that the string that commences with $a_0(\mathbf{q})$ can connect only to the string that commences with $a_0^\dagger(-\mathbf{q})$. Thus it is that strings get "stitched" in pairs at $j = 0$. This is now shown in detail.

First, two equations of motion replace Eq. (25). The first is:

$$[a_0(\mathbf{q}), H_1] = \left(A_0^0(\mathbf{q}) + U \frac{|q|}{C^2(q)} \right) a_0(\mathbf{q}) + A_0^1(\mathbf{q}) a_1(\mathbf{q}) + \left(\frac{U|q|}{C^2(q)} \right) a_0^\dagger(-\mathbf{q}) \quad \text{(41A)}$$

and the other, under the assumption that the A's are real, is:

$$[H, a_0^\dagger(-\mathbf{q})] = \left(A_0^0(\mathbf{q}) + U \frac{|q|}{C^2(q)} \right) a_0^\dagger(-\mathbf{q}) + A_0^1(\mathbf{q}) a_1^\dagger(-\mathbf{q}) + \left(\frac{U|q|}{C^2(q)} \right) a_0(\mathbf{q}). \quad \text{(41B)}$$

Subsequent iterations connecting sites 1 to 2 etc. are unaffected by H_2; they remain independent of U and are, symbolically and numerically, the same as in (25). The strings of annihilation operators at $\mathbf{q}$ connect to strings of creation operators at $-\mathbf{q}$ at the origin, $j = 0$, but nowhere else.

The preceding suggests renumbering the strings of creation operators emanating from (41B), changing their subscripts j to $-j$, i.e., renaming $a_j^\dagger(-\mathbf{q}) \to a_{-j}^\dagger(-\mathbf{q})$ for all $j \geq 1$. In this way, at each q we map two semi-infinite strings onto *a single, infinite,* string, numbered from $-\infty < j < +\infty$. All the effects of the two-body interactions $H_{2,\mathrm{dir}}$ are made to reside on the single, central, site, $j = 0$.

We define the numerical coefficients at negative indices as $A_{-i}^{-j}(-\mathbf{q}) = A_i^j(\mathbf{q}) = (A_i^j(\mathbf{q}))^*$. Then, replacing $\hat{A}(\mathbf{q})$, the complete pseudo-Hamiltonian at $\mathbf{q}$ for the *density–density* interactions is:

$$\hat{H}_{\text{dens}}(\mathbf{q}) = \sum_{n=0}^{\infty} \left(A_n^n(\mathbf{q}) + \delta_{n,0} \frac{U|q|}{C^2(q)} \right) a_n^\dagger(\mathbf{q}) a_n(\mathbf{q})$$

$$+ \sum_{n=0}^{\infty} (A_n^{n+1}(\mathbf{q}) a_n^\dagger(-\mathbf{q}) a_{n+1}(\mathbf{q}) + \text{H.c.})$$

$$+ \sum_{n=0}^{\infty} \left(A_n^n(\mathbf{q}) + \delta_{n,0} \frac{U|q|}{C^2(q)} \right) a_{-n}^\dagger(-\mathbf{q}) a_{-n}(-\mathbf{q})$$

$$+ \sum_{n=0}^{\infty} (A_n^{n+1}(\mathbf{q}) a_{-n}^\dagger(-\mathbf{q}) a_{-n-1}(-\mathbf{q}) + \text{H.c.})$$

$$+ \frac{U|q|}{C^2(q)} (a_0(\mathbf{q}) a_0(-\mathbf{q}) + a_0^\dagger(-\mathbf{q}) a_0^\dagger(\mathbf{q})). \tag{42}$$

For *spin–spin* interactions we replace U by $-U$, but retaining the same set of A's:

$$\hat{H}_\sigma(\mathbf{q}) = \sum_{n=0}^{\infty} \left(A_n^n(\mathbf{q}) - \delta_{n,0} \frac{U|q|}{C^2(q)} \right) \sigma_n^\dagger(\mathbf{q}) \cdot \sigma_n(\mathbf{q})$$

$$+ \sum_{n=0}^{\infty} (A_n^{n+1}(\mathbf{q}) \sigma_n^\dagger(\mathbf{q}) \cdot \sigma_{n+1}(\mathbf{q}) + \text{H.c.})$$

$$+ \sum_{n=0}^{\infty} \left(A_n^n(\mathbf{q}) - \delta_{n,0} \frac{U|q|}{C^2(q)} \right) \sigma_{-n}^\dagger(-\mathbf{q}) \cdot \sigma_{-n}(-\mathbf{q})$$

$$+ \sum_{n=0}^{\infty} (A_n^{n+1}(\mathbf{q}) \sigma_{-n}^\dagger(-\mathbf{q}) \cdot \sigma_{-n-1}(-\mathbf{q}) + \text{H.c.})$$

$$- \frac{U|q|}{C^2(q)} (\sigma_0(\mathbf{q}) \cdot \sigma_0(-\mathbf{q}) + \sigma_0^\dagger(-\mathbf{q}) \cdot \sigma_0^\dagger(\mathbf{q})). \tag{43}$$

Cooperon strings present a distinct challenge discussed further, in Sec. 20.

17. The Solutions: Generalities

We have the equations but how do we solve them? At finite q it may not be practical to diagonalize expressions such as (42) or (43); or it may be unwieldy, as this requires knowing all the coefficients A_n^m as functions of q, most notably in the asymptotic regions $n \to \pm\infty$. So let us take the time to examine other possibilities. The most obvious is to express the raising/lowering operators of the quadratic forms (42) or (43) as linear combinations of the bare operators. For example, for one of the spin modes, define what would be an exact lowering operator of a longitudinal magnon:

$$\xi(\mathbf{q}) = \sum_{j=0}^{\infty} \{ F_j z_j(\mathbf{q}) + G_j z_{-j}^\dagger(-\mathbf{q}) \}. \tag{44}$$

For this to be true, ξ has to satisfy an eigenvalue equation: $[\xi(\mathbf{q}), \hat{H}_z(\mathbf{q})] = \lambda \xi(\mathbf{q})$, in which $\lambda > 0$. This also, automatically, makes it a lowering operator of $\hat{H}_\sigma(\mathbf{q}) = \hat{H}_x + \hat{H}_y + \hat{H}_z$ (given that in the absence of LRO it commutes with $\hat{H}_x + \hat{H}_y$), and therefore, of the full H.

Assuming we were able to solve this commutator equation for all its $N \to \infty$ roots, $\lambda_0, \lambda_1, \ldots$, we could use the results to transform $\hat{H}_z(\mathbf{q})$ into a diagonal form:

$$\hat{H}_z(\mathbf{q}) \Rightarrow \sum_{n=0}^{\infty} \lambda_n \{\xi_n^\dagger(\mathbf{q})\xi_n(\mathbf{q}) + \xi_n^\dagger(-\mathbf{q})\xi_n(-\mathbf{q}) + K_n\}. \tag{45}$$

in which $\lambda_n K_n$ are individual contributions to the zero-point energy, the λ_n being the eigenvalues (always chosen as positive roots in the event of ambiguity).

But how do we even calculate these λ_n? It is frequently possible to solve equations of motion using the two point recursions, as roots of continued fractions. However this procedure requires knowledge of all the A_n^m's.

If, at small q, we discarded A_0^1 and all successive coefficients, the surviving strings at $j = 0$ have zero length and are solved trivially, as in the original Luttinger model — except that, unlike the original, this renders service in two or three dimensions. To the extent that discarded coefficients vanish faster at small q than those that are retained, this truncation could allow the calculation of asymptotic quantities such as long-range correlations, etc. The eigenstates we seek come in two varieties: (1) "bound-state" solutions that one can characterize as "magnons", "plasmons" or whatever, whose energies $\lambda_j(q)$ lie outside the continuum of unperturbed energies $\{\Delta\varepsilon\}$ ("outside" can mean either above or below the continuum, but always above 0, as previously remarked), or (2) as "scattering-state" solutions, the energies of which interlace the continuum. These are comprised of an "incoming wave" (a given linear combination of elementary excitations) accompanied by a cloud of interactions.

For the bound states the best method is also the one most easily implemented. It regards $\xi(\mathbf{q})$ as the sum of two arbitrary functions, which for spin excitations are of the type,

$$\xi(\mathbf{q}) = \xi_1(\mathbf{q}) + \xi_2^\dagger(-\mathbf{q})$$

$$= \frac{C(\mathbf{q})}{\sqrt{2 \times q \times \mathrm{Vol}}} \sum_k \{\theta(\mathbf{k} \cdot \mathbf{q})\Phi_1(\mathbf{k})(c_\uparrow^\dagger(\mathbf{k} - \mathbf{q}/2)c_\uparrow(\mathbf{k} + \mathbf{q}/2)$$

$$- c_\downarrow^\dagger(\mathbf{k} - \mathbf{q}/2)c_\downarrow(\mathbf{k} + \mathbf{q}/2)) + \theta(-\mathbf{k} \cdot \mathbf{q})\Phi_2(\mathbf{k})(c_\uparrow^\dagger(\mathbf{k} - \mathbf{q}/2)c_\uparrow(\mathbf{k} + \mathbf{q}/2)$$

$$- c_\downarrow^\dagger(\mathbf{k} - \mathbf{q}/2)c_\downarrow(\mathbf{k} + \mathbf{q}/2))\}. \tag{46}$$

The solution of $[[\xi(\mathbf{q}), \hat{H}_z(\mathbf{q})] - \lambda(\mathbf{q})\xi(\mathbf{q}), c_\sigma^\dagger(\mathbf{k} - \mathbf{q}/2)c_\sigma(\mathbf{k} + \mathbf{q}/2)] = 0$ at each (k, σ) yields the amplitudes Φ_1 and Φ_2 as functions of k and q. The corresponding eigenvalues $\lambda(q)$ can be computed and, provided they lie outside — either higher or lower than — the continuum, they are real and > 0. (Within the continuum scattering theory has to be used, requiring an incoming wave and a cloud of states scattered by H_2).

18. The Solutions: Do Magnons Exist in a Hubbard Metal?

By "exist" is meant, are the magnons stationary states (with $\lambda > 0$)? H consists of two parts: H_1 and H_2. Let us start with just H_1 in (A), following which we will add H_2 into the mix in (B). The first part of the calculation is trivial, $[\xi(\mathbf{q}), \hat{H}_{1,z}(\mathbf{q})] - \lambda(\mathbf{q})\xi(\mathbf{q})$. Making use of (43), this is:

$$\frac{C(\mathbf{q})}{\sqrt{2 \times q \times \text{Vol}}} \sum_k \{\theta(\mathbf{k} \cdot \mathbf{q})\Phi_1(\mathbf{k})((2\mathbf{k} \cdot \mathbf{q}) - \lambda)(c_\uparrow^\dagger(\mathbf{k} - \mathbf{q}/2)c_\uparrow(\mathbf{k} + \mathbf{q}/2)$$

$$- c_\downarrow^\dagger(\mathbf{k} - \mathbf{q}/2)c_\downarrow(\mathbf{k} + \mathbf{q}/2)) + \theta(-\mathbf{k} \cdot \mathbf{q})\Phi_2(\mathbf{k})$$

$$\times ((2\mathbf{k}\cdot\mathbf{q}) - \lambda)(c_\uparrow^\dagger(\mathbf{k} - \mathbf{q}/2)c_\uparrow(\mathbf{k} + \mathbf{q}/2) - c_\downarrow^\dagger(\mathbf{k} - \mathbf{q}/2)c_\downarrow(\mathbf{k} + \mathbf{q}/2))\}. \quad (47\text{A})$$

The second part, $[\xi(\mathbf{q}), \hat{H}_{2,z}(\mathbf{q})]$, invokes Eqs. (16) or (46). It is,

$$-U\frac{|q|}{C^2(q)}[\{\xi_1(\mathbf{q}) + \xi_2^\dagger(-\mathbf{q})\}, \{z_0^\dagger(\mathbf{q})z_0(\mathbf{q}) + z_0^\dagger(-\mathbf{q})z_0(-\mathbf{q}) + (z_0(\mathbf{q})z_0(-\mathbf{q}) + \text{H.c.})\}]. \quad (47\text{B})$$

Calculation of the nonvanishing contributions to (47B) involve such quantities as:

$$[\xi_1(\mathbf{q}), z_0^\dagger(\mathbf{q})] = \frac{C^2(\mathbf{q})}{q \times \text{Vol}} \sum_k \theta(\mathbf{k} \cdot \mathbf{q})\Phi_1(\mathbf{k})\{f(\mathbf{k} - \mathbf{q}/2) - f(\mathbf{k} + \mathbf{q}/2)\} = \langle\Phi_1(\mathbf{k})\rangle_+$$

and $[\xi_2^\dagger(-\mathbf{q}), z_0(-\mathbf{q})] = -\langle\Phi_2(\mathbf{k})\rangle_-$ where $\langle\cdots\rangle_\pm$ indicates that the averaging is over the outside or the inner of the two nonintersecting ensembles, $\mathbf{k} \cdot \mathbf{q} > 0$ for Φ_1 and $\mathbf{k} \cdot \mathbf{q} < 0$ for Φ_2. (The new notation proves its worth in intricate calculations such as this!) Thus, (47B) works out to be,

$$-U\frac{|q|}{C^2(q)}(\langle\Phi_1\rangle - \langle\Phi_2\rangle)(z_0^\dagger(-\mathbf{q}) + z_0(\mathbf{q})). \quad (47\text{C})$$

In Eq. (47A) the coefficient of the following operator:

$$\frac{C(\mathbf{q})}{\sqrt{2 \times q \times \text{Vol}}}\theta(\mathbf{k} \cdot \mathbf{q})(c_\uparrow^\dagger(\mathbf{k} - \mathbf{q}/2)c_\uparrow(\mathbf{k} + \mathbf{q}/2) - c_\downarrow^\dagger(\mathbf{k} - \mathbf{q}/2)c_\downarrow(\mathbf{k} + \mathbf{q}/2))$$

is $\Phi_1(\mathbf{k}\cdot\mathbf{q})((2\mathbf{k}\cdot\mathbf{q}) - \lambda)$. In Eqs. (47B) or (47C) the coefficient of the same operator is:

$$-U\frac{|q|}{C^2(q)}(\langle\Phi_1\rangle_+ - \langle\Phi_2\rangle_-).$$

For the sum to vanish as required, the following must hold:

$$\Phi_1(\mathbf{k}) = U \frac{|q|}{C^2(q)} \frac{(\langle\Phi_1\rangle_+ - \langle\Phi_2\rangle_-)}{(2\mathbf{k}\cdot\mathbf{q} - \lambda)} \tag{48A}$$

and similarly,

$$\Phi_2(\mathbf{k}) = U \frac{|q|}{C^2(q)} \frac{(\langle\Phi_1\rangle_+ - \langle\Phi_2\rangle_-)}{(-|2\mathbf{k}\cdot\mathbf{q}| - \lambda)}. \tag{48B}$$

Averaging each Φ_j over its appropriate $(\pm)$ ensemble, we obtain two equations in two unknowns:

$$\langle\Phi_1\rangle_+ = (\langle\Phi_1\rangle_+ - \langle\Phi_2\rangle_-) \frac{U}{\text{Vol}} \sum_{\mathbf{k}} \theta(\mathbf{k}\cdot\mathbf{q}) \frac{f(\mathbf{k} - \mathbf{q}/2) - f(\mathbf{k} + \mathbf{q}/2)}{(2\mathbf{k}\cdot\mathbf{q} - \lambda)} \tag{49A}$$

and

$$\langle\Phi_2\rangle_- = (\langle\Phi_1\rangle_+ - \langle\Phi_2\rangle_-) \frac{U}{\text{Vol}} \sum_{\mathbf{k}} \theta(-\mathbf{k}\cdot\mathbf{q}) \frac{f(\mathbf{k} + \mathbf{q}/2) - f(\mathbf{k} - \mathbf{q}/2)}{(-|2\mathbf{k}\cdot\mathbf{q}| - \lambda)}. \tag{49B}$$

By an obvious symmetry the integral on the right-hand side of (49B) can also be rewritten in the form of (49A):

$$\frac{U}{\text{Vol}} \sum_{\mathbf{k}} \theta(\mathbf{k}\cdot\mathbf{q}) \frac{f(\mathbf{k} - \mathbf{q}/2) - f(\mathbf{k} + \mathbf{q}/2)}{(-2\mathbf{k}\cdot\mathbf{q} - \lambda)}.$$

Thus, for (49A) and (49B) to have a simultaneous solution, either $(\langle\Phi_1\rangle - \langle\Phi_2\rangle)$ vanishes, or:

$$1 = \frac{4U}{\text{Vol}} \mathbf{q} \cdot \sum_{\mathbf{k}} \mathbf{k} \left(\theta(\mathbf{k}\cdot\mathbf{q}) \frac{f(\mathbf{k} - \mathbf{q}/2) - f(\mathbf{k} + \mathbf{q}/2)}{4(\mathbf{k}\cdot\mathbf{q})^2 - \lambda(q)^2} \right). \tag{50}$$

Equation (50) takes the form, $\frac{1}{U} = M(q, \lambda(q))$. In 2D, in the thermodynamic limit and denoting the integral in (50) by M, we find it is,

$$M(q, \lambda(q)) = \frac{1}{q\pi^2} \int dk_y \int_0^\infty dk_x k_x \frac{f(\mathbf{k} - \mathbf{q}/2) - f(\mathbf{k} + \mathbf{q}/2)}{4k_x^2 - (\lambda(q)/q)^2}.$$

The x-axis has been chosen along $\mathbf{q}$.

Even at $T = 0$ there are no bound states at small q. Plotting M as function of its arguments (q and λ) shows us that at small $q < 2$, either M is complex or, if real, that it is *negative* over the entire range.

However, we also find that at large q and U there *do* exist solutions of Eq. (50), provided U is sufficiently large ($U > 18$, approximately) and $q > 2$. The solutions detach from the continuum and can be obtained from Fig. 6 as follows.

Drawing a horizontal line at a given U one can read off the range of allowed q's for that value of U: magnons exist for those values of q in a range $q_{\text{min}} < q < q_{\text{max}}$. These limits are shown as the two separate curves in the figure. The magnon branch

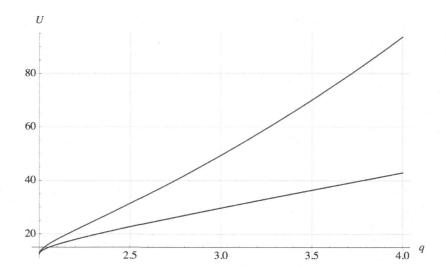

Fig. 6. *Magnon Spectrum.* Showing curves of q_{min} and q_{max} at a fixed U in 2D. *At fixed U*, the q of the *magnon* eigenstates ranges from a minimum value given by the upper curve (at which the eigenvalue λ is a maximum and lies just below continuum) to a maximum q given by the lower curve (at which the magnon eigenvalue $\lambda = 0$). *For example*, at fixed $U = 40$, $q_{min} = 2.75$ (at which $\lambda = 5.5e_F$) and $q_{max} = 3.75$ (where $\lambda = 0$). It is notable that the magnons cease to exist for values of $U < 18$ approximately.

that lies between these two delimiting curves is described qualitatively as follows: (1) its spectrum has negative dispersion and (2) it emanates from the continuum of elementary excitations at finite λ at q_{min}, and then drops to $\lambda = 0$ at q_{max}.

We conclude that, absent of any external agency or symmetry breaking of *some* sort, the Hubbard model does not — and cannot — spontaneously sustain a stable spectrum of *long wavelength* magnons. However, this does not preclude the existence of a continuum spectrum of *magnetic scattering states*, the energies of which interlace the continuum but which, unlike their short-wavelength brethren, cannot be distinguished as individual particles or as stationary states due to their short lifetimes. Because the Hamiltonian is quadratic in the bosons, such scattering states can be described as exactly as they are in one-body quantum mechanics.

One starts by selecting an "incoming" operator, say,

$$\Theta_{\mathbf{q}}(\mathbf{k}) \equiv \frac{\theta(\mathbf{k} \cdot \mathbf{q})}{\sqrt{2}} (c_\uparrow^\dagger(\mathbf{k} - \mathbf{q}/2)c_\uparrow(\mathbf{k} + \mathbf{q}/2) - c_\downarrow^\dagger(\mathbf{k} - \mathbf{q}/2)c_\downarrow(\mathbf{k} + \mathbf{q}/2)) \qquad (51)$$

accompanied by a continuum of "outgoing" scattering states, and combine them into an *exact* expression that includes all scattering processes,

$$\Omega_{\mathbf{q}}(\mathbf{k}) = \Theta_{\mathbf{q}}(\mathbf{k}) + \frac{1}{\text{Vol}} \sum_{\mathbf{k}' \neq \mathbf{k}} L_{\mathbf{q}}(\mathbf{k}, \mathbf{k}')\Theta_{\mathbf{q}}(\mathbf{k}') + \frac{1}{\text{Vol}} \sum_{k''} M_{\mathbf{q}}(\mathbf{k}, \mathbf{k}'')\Theta_{-\mathbf{q}}^\dagger(\mathbf{k}''). \qquad (52)$$

The energy eigenvalue of $\Omega_{\mathbf{q}}(\mathbf{k})$ is, to $O(1/\text{Vol})$, the same as that of (51), which is $2\mathbf{k} \cdot \mathbf{q}$. So the equations of motion principally serve to calculate the scattering amplitudes $L_{k,k'}$ and $M_{k,k''}$ and the elastic scattering cross-section, $\frac{1}{\tau_{\mathbf{q}}(\mathbf{k})}$. This last is in turn related to integrals that take the form, $\frac{1}{\text{Vol}}\text{Im}\{\sum_{k'} T_q(\mathbf{k}, \mathbf{k}')\}$ and can be calculated in closed form. Here the actual calculations would take us too far afield.

19.   The Solutions: Zero Sound

The equations of motion of the particle density modes a_j are, after replacement of U by $-U$ in all the equations, identical to those of the magnons. So, whereas long wavelength magnons could not be sustained in the latter, here we find that long wavelength density-waves *are* coherent, stationary states in this model (and conversely at short wavelengths). In a charge-neutral gas with short-ranged two-body repulsions this type of normal mode is called *"zero-sound."* In Fig. 7 we plot $s = s(0)$, the speed of zero-sound as a function of U at $q = 0$, being the numerical solution of $\frac{-1}{U} = M$ at $q = 0$ in 2D.

The figure shows that s is a monotonically increasing function of the coupling constant but is never less than 2, the Fermi velocity in the present units ($k_F = 1$, $e_F = 1$, $\nu_F = 2$). In addition, where the energy $\lambda(q)$ of zero-sound waves coincides with the upper edge of the continuum their eigenstates become scattering states, to be treated similarly to Eqs. (51) and (52).

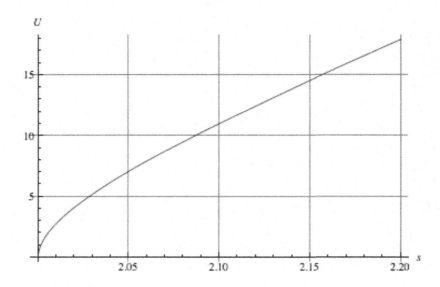

Fig. 7. *Speed of Sound.* The parameter U (in arbitrary units) versus s, the speed of zero-sound at $q = 0$ in the Hubbard model, at $T = 0$, in 2D. (The speed of a particle at the Fermi surface is $\nu_F = 2$, seen to be the limiting value of our solution for s at $U = 0$).

20. The Solutions: Are There Any Cooperons?

By analogy with the preceding paragraphs, we seek to solve the equations of motion of stable cooperon collective modes — if they exist — not by diagonalizing their strings but directly by solving for the envelope wavefunctions of the corresponding collective boson operators. We set the annihilation operator $\chi(\mathbf{q}) \equiv (\chi_+(\mathbf{q}) + \chi_+^\dagger(-\mathbf{q}) + \chi_-(\mathbf{q}) + \chi_-^\dagger(-\mathbf{q}))$ as some arbitrary linear combination of outer and inner cooperon operators. That is,

$$\xi(\mathbf{q}) = \frac{D_+(\mathbf{q})}{\sqrt{\text{Vol}}} \sum_{\mathbf{k}} \theta\left(2\mathbf{k}^2 + \frac{\mathbf{q}^2}{2} - 2\right)$$

$$\times (\Psi_1(\mathbf{k})c_\downarrow(\mathbf{k}+\mathbf{q}/2)c_\uparrow(-\mathbf{k}+\mathbf{q}/2) + \Psi_2(\mathbf{k})c_\uparrow^\dagger(-\mathbf{k}-\mathbf{q}/2)c_\downarrow^\dagger(\mathbf{k}-\mathbf{q}/2))$$

$$+\frac{D_-(\mathbf{q})}{\sqrt{\text{Vol}}} \sum_{\mathbf{k}} \theta\left(2 - 2\mathbf{k}^2 - \frac{\mathbf{q}^2}{2}\right)$$

$$\times (\Psi_3(\mathbf{k})c_\downarrow(\mathbf{k}+\mathbf{q}/2)c_\uparrow(-\mathbf{k}+\mathbf{q}/2) + \Psi_4(\mathbf{k})c_\uparrow^\dagger(-\mathbf{k}-\mathbf{q}/2)c_\downarrow^\dagger(\mathbf{k}-\mathbf{q}/2)). \quad (53)$$

Then, to find its eigenvalue $\lambda(q) > 0$, this field operator has to satisfy:

$$[\chi(\mathbf{q}), H] - \lambda(\mathbf{q})\chi(\mathbf{q}) = 0. \quad (54)$$

The part of the commutator that involves the interaction is, explicitly:

$$[\chi(\mathbf{q}), H_{2,\chi}] = +U\left[\chi(\mathbf{q}), \left(\frac{\chi_{o,-}(-\mathbf{q})}{D_-(\mathbf{q})} + \frac{\chi_{o,+}^\dagger(\mathbf{q})}{D_+(\mathbf{q})}\right) \cdot \left(\frac{\chi_{o,-}^\dagger(-\mathbf{q})}{D_-(\mathbf{q})} + \frac{\chi_{o,+}(\mathbf{q})}{D_+(\mathbf{q})}\right)\right]$$

$$+U\left[\chi(\mathbf{q}), \left(\frac{\chi_{o,-}(\mathbf{q})}{D_-(\mathbf{q})} + \frac{\chi_{o,+}^\dagger(-\mathbf{q})}{D_+(\mathbf{q})}\right) \cdot \left(\frac{\chi_{o,-}^\dagger(\mathbf{q})}{D_-(\mathbf{q})} + \frac{\chi_{o,+}(-\mathbf{q})}{D_+(\mathbf{q})}\right)\right]. \quad (55)$$

The rest involves the kinetic energy, that is, $[\chi(\mathbf{q}), H_1]$,

$$[\xi(\mathbf{q}), H_1] = \frac{D_+(\mathbf{q})}{\sqrt{\text{Vol}}} \sum_{\mathbf{k}} \theta(\varepsilon(\mathbf{k}+\mathbf{q}/2) + \varepsilon(-\mathbf{k}+\mathbf{q}/2))(\varepsilon(\mathbf{k}+\mathbf{q}/2) + \varepsilon(-\mathbf{k}+\mathbf{q}/2))$$

$$\times (\Psi_1(\mathbf{k})c_\downarrow(\mathbf{k}+\mathbf{q}/2)c_\uparrow(-\mathbf{k}+\mathbf{q}/2) - \Psi_2(\mathbf{k})c_\uparrow^\dagger(-\mathbf{k}-\mathbf{q}/2)c_\downarrow^\dagger(\mathbf{k}-\mathbf{q}/2))$$

$$+\frac{D_-(\mathbf{q})}{\sqrt{\text{Vol}}} \sum_{\mathbf{k}} \theta(-\varepsilon(\mathbf{k}+\mathbf{q}/2) - \varepsilon(-\mathbf{k}+\mathbf{q}/2))(\varepsilon(\mathbf{k}+\mathbf{q}/2) + \varepsilon(-\mathbf{k}+\mathbf{q}/2))$$

$$\times (\Psi_3(\mathbf{k})c_\downarrow(\mathbf{k}+\mathbf{q}/2)c_\uparrow(-\mathbf{k}+\mathbf{q}/2) - \Psi_4(\mathbf{k})c_\uparrow^\dagger(-\mathbf{k}-\mathbf{q}/2)c_\downarrow^\dagger(\mathbf{k}-\mathbf{q}/2)) \quad (56)$$

and finally, the term $-\lambda(\mathbf{q})\chi(\mathbf{q})$, that we won't bother writing out.

After some algebra, Eqs. (55) and (56) lead to the solutions of Eq. (54). They are,

$$\text{Either} \quad \left(\frac{\langle \Psi_1 \rangle_+}{D_+(\mathbf{q})} - \frac{\langle \Psi_3 \rangle_-}{D_-(\mathbf{q})} \right) = 0 \quad \text{or} \quad \frac{1}{U} = \left(\frac{S_-(-\lambda(q))}{D_-} - \frac{S_+(\lambda(q))}{D_+} \right), \quad (57A)$$

and/or,

$$\left(\frac{\langle \Psi_2 \rangle_+}{D_+(\mathbf{q})} - \frac{\langle \Psi_4 \rangle_-}{D_-(\mathbf{q})} \right) = 0 \quad \text{or} \quad \frac{1}{U} = \left(\frac{S_+(-\lambda(q))}{D_+} - \frac{S_-(\lambda(q))}{D_-} \right), \quad (57B)$$

where

$$S_-(\lambda(q)) = \frac{1}{\text{Vol}} \sum_{\mathbf{k}} \frac{\theta \left(2 - 2k^2 - \frac{q^2}{2} \right) (f(\mathbf{k}+\mathbf{q}/2) + f(\mathbf{k}-\mathbf{q}/2) - 1)}{2 - 2k^2 - \frac{q^2}{2} - \lambda(q)}$$

and

$$S_+(\lambda(q)) = \frac{1}{\text{Vol}} \sum_{\mathbf{k}} \frac{\theta \left(2k^2 + \frac{q^2}{2} - 2 \right) (1 - f(\mathbf{k}+\mathbf{q}/2) - f(\mathbf{k}-\mathbf{q}/2))}{2k^2 + \frac{q^2}{2} - 2 - \lambda(q)}$$

assuming, as always, $\varepsilon(k) = k^2 - 1$, and using the P's of Sec. 15 for evaluation of the averages. Here, as was previously the case for long wavelength magnons, denominators change sign when $\lambda > 0$ and cause the integrals $S_\pm(\lambda)$ to be complex although both integrals $S_\pm(-\lambda)$ are real. Because $1/U$ is real, none of the equations (57) has a solution and thus cooperons do not have stationary states in this model.

21. Collective Modes in the Presence of Coulomb or Other Forces

For arbitrary potentials the *direct interaction* terms — that is, those terms that are not related either to exchange or pairing phenomena — *always* involve just the density operators ϱ. These are proportional to what we denoted the particle-density operators a_0. Replacing Eq. (14) we find the more general direct interaction,

$$H_{2,\text{dir}} = \sum_{\mathbf{q}, q_z > 0} \frac{V(\mathbf{q})|q|}{C^2(q)} \{ a_0^\dagger(\mathbf{q}) a_0(\mathbf{q}) + a_0^\dagger(-\mathbf{q}) a_0(-\mathbf{q}) + (a_0(\mathbf{q}) a_0(-\mathbf{q}) + \text{H.c.}) \}.$$

The associated density string is therefore the same as before, *except* that the site at $j = 0$ hosts a coupling constant $V(q)$ at each q instead of a constant U. That would be a minor change indeed, if that were all that happened.

But, in fact, the *exchange* and *pairing* correlations are profoundly affected by any changes we bring to the structure of the two-body potentials. For *arbitrary* potentials the exchange and pairing terms can no longer be understood solely using strings with just nearest-neighbor connections. Even if we were interested in just the exchange terms, these could only be found as solutions to some integral equations with a kernel related to $V(q)$. (The reader may wish to mull on this and we also intend to return to this topic in a separate publication, but at this juncture let us just appreciate that it is solely the Hubbard model, with its structureless interactions in reciprocal space, that allows the many-body problem to be solved explicitly in *all* its channels).

22. Green Functions and the Free Energy

For what purposes can we use a solution? From elementary thermodynamics we know that the free energy F is an important quantity that reveals many of the collective properties of a system: the total internal energy $E = \frac{\partial(\beta F)}{\partial \beta}$ (where $\beta = 1/kT$) and specific heat $c_v(T) = \frac{1}{\text{Vol}} \frac{dE}{dT}$, the system's entropy $S(T) = -\frac{\partial F}{\partial T}$ and so on. It is generally advantageous to separate F into two parts: (1) the ideal-gas contribution F_0 of a noninteracting system having the same number of particles in the same volume and (2) ΔF, the additional contributions attributed to the interactions among the particles.

ΔF has six contributions here: one attributable to density fluctuations (zero sound), three to spin fluctuations (magnons) and one in each of two distinct cooperon channels. In the absence of LRO that might tie some of them together, all these channels are independent of one another.

With curly brackets $\{\ldots\}$ indicating the nature of the bosons that are involved in creating each expression, these individual contributions are: $\Delta F_1 = \Delta F(U, T, \{a(q)\})$, $\Delta F_2 = \Delta F(U, T, \{\sigma_x(q)\})$, $\Delta F_3 = \Delta F(U, T, \{\sigma_y(q)\})$, $\Delta F_4 = \Delta F(U, T, \{\sigma_z(q)\})$, $\Delta F_5 = \Delta F(U, T, \{\chi_-(q)\})$ and $\Delta F_6 = \Delta F(U, T, \{\chi_+(q)\})$.

Their values are not all distinct. As we remarked earlier, $\Delta F_2 = \Delta F_3 = \Delta F_4$ and $\Delta F_2(U, T) = \Delta F_1(-U, T)$. Thus $F = F_0 + \Delta F_1(U, T) + 3\Delta F_1(-U, T) + \Delta F_5(U, T) + \Delta F_6(U, T)$, where

$$F_0 = -2kT \sum_{\mathbf{k}} \log(1 + e^{-\beta \varepsilon(\mathbf{k})}). \tag{58}$$

For the remainder let us use Feynman's theorem and integrate on the coupling constant U. For example, using $\langle \text{op} \rangle_u$ to stand for "op" averaged over states appropriate to a coupling constant u, we can express the density operators' contribution to the *free energy*, as,

$$\Delta F_1(U, T) = \sum_{\mathbf{q}, q_z > 0} |q| \int_0^U \frac{du}{C^2(q)} \langle \{ a_0^\dagger(\mathbf{q}) a_0(\mathbf{q}) + a_0^\dagger(-\mathbf{q}) a_0(-\mathbf{q})$$

$$+ (a_0(\mathbf{q}) a_0(-\mathbf{q}) + a_0^\dagger(-\mathbf{q}) a_0^\dagger(\mathbf{q})) \} \rangle_u \tag{59}$$

to be distinguished from the corresponding interaction *energy* which is,

$$\Delta E_1(U,T) = U \sum_{\mathbf{q}, q_z > 0} \frac{|q|}{C^2(q)} \langle \{ a_0^\dagger(\mathbf{q}) a_0(\mathbf{q}) + a_0^\dagger(-\mathbf{q}) a_0(-\mathbf{q})$$

$$+ (a_0(\mathbf{q}) a_0(-\mathbf{q}) + a_0^\dagger(-\mathbf{q}) a_0^\dagger(\mathbf{q})) \} \rangle_U \, .$$

The expectation values $\langle \dots \rangle$ in either case are obtained with the aid of the two-time Green functions $\langle\!\langle \dots | \dots \rangle\!\rangle (\omega)$ of many-body physics. Harmonic forces are easily and satisfactorily disposed of, by this method.[20] In the present example of interacting bosons,

$$\langle a_0^\dagger(\mathbf{q}) a_0(\mathbf{q}) \rangle_u = \frac{1}{2\pi i} \int d\omega \frac{1}{e^{\beta\omega} - 1} \{ \langle\!\langle a_0(\mathbf{q}) | a_0^\dagger(\mathbf{q}) \rangle\!\rangle_u (\omega - i\delta)$$

$$- \langle\!\langle a_0(-\mathbf{q}) | a_0^\dagger(\mathbf{q}) \rangle\!\rangle_u (\omega + i\delta) \} \, , \tag{60}$$

where $\langle\!\langle a_0(\mathbf{q}) | a_0^\dagger(\mathbf{q}) \rangle\!\rangle_u$ stands for the appropriate "retarded" Green function, evaluated at coupling constant u and frequency $\omega - i\delta$, where $\delta > 0$ is infinitesimal. It can be calculated *exactly* from the equations of motion when the pseudo-Hamiltonian is quadratic in the harmonic operators. So it would seem we have an exact and complete solution of the many-body problem specified at the beginning of this chapter.

There is just one fly in the ointment: the value of $C^2(q)$ in the denominator of (59), like some other aspects of the model, requires an accurate knowledge of $f(k)$ at nonzero U and T. To the extent that $C^2(q)$ and $f(\mathbf{k})$ depend on U they affect integrals such as (59). Therefore we turn to the evaluation of this one-particle correlation function next.

23. The One-Body Distribution Function $f(k)$

To obtain the one-body correlation function $f(\mathbf{k})$, it is advantageous to compute the *functional derivative* of the free energy, i.e., $\langle \tilde{n}(\mathbf{k}) \rangle = f(\mathbf{k}) = \frac{1}{2} \frac{\delta F}{\delta \varepsilon(\mathbf{k})}$ (using the factor $1/2$ to take spin degeneracy into account and treating the individual $\varepsilon(k)$ as an independent coupling constant). If this calculation is not practical one could take advantage of circular (or spherical) symmetry to evaluate a simpler quantity,

$$\frac{1}{\mathrm{Vol}} \frac{1}{2\rho(\varepsilon)} \frac{\partial F}{\partial \varepsilon} = \frac{1}{2\rho(\varepsilon)} \frac{1}{\mathrm{Vol}} \sum_{\mathbf{k}} \sum_{\sigma = \pm} \delta(\varepsilon - \varepsilon(\mathbf{k})) f(\mathbf{k}) = f(\varepsilon) \tag{61}$$

where $\varrho(\varepsilon)$ is the (known) density of states.

$F = F_0 + \Delta F$ and $f = f_0 + \Delta f$, with $f_0 = \frac{1}{e^{\beta\varepsilon}+1}$ being the ordinary Fermi function. The ensuing calculation might appear to be a formidable task but, actually, in the case of the two-dimensional Hubbard model the relevant calculations were already performed — some two decades ago — by Hua Chen[21] while studying the normal phase of high-T_c superconductors within the random-phase approximation

(RPA). Calculations based on the present string theory are similar if not identical; therefore in what follows we just quote the original results.[21]

Defining the drop in the distribution function at the Fermi level as Z_F ($0 < Z_F < 1$) one finds Z_F to be 1 at $T = 0$ when $U = 0$, then to decrease smoothly as U is increased until at $U = U_c$ it vanishes. This critical point was calculated at $T = 0$ to be approximately $U_c = 18$ in units $e_F = 1$. Although it was obtained by entirely different means, this value of U_c is in good accord with the results summarized in Fig. 6.

For $U < U_c$, in the usual metallic Landau Fermi liquid phase,[20] the inverse lifetime $1/\tau$ of a particle near the Fermi surface is $1/\tau \propto \varepsilon^2$, i.e., $1/\tau_F \propto T^2$.

But, as U enters a different Fermi liquid phase at $U \geq U_c$, $f(\varepsilon)$ becomes continuous. Its dependence on ε near the Fermi surface is somewhat less singular, approximately[m] $f = \frac{1}{2}(1 \pm A|\varepsilon|^\delta)$, with inverse lifetime $1/\tau \propto |\varepsilon|$, i.e., $1/\tau_F \propto T$ (at all values of $\delta \geq 0$).

This so-called "marginal Fermi gas" behavior has in fact been identified in the normal phase of optimally doped two-dimensional high-T_c superconductors.[22]

The one-body distribution function $f(\mathbf{k})$ satisfies an integral equation in which a kernel J receives contribution from all six channels discussed previously. However, we find that only the spin channels contribute substantially at the Fermi surface; numerically they are — almost entirely — responsible for changes in the discontinuity (Z_F) at the Fermi level from 1 (at $U = 0$) to 0 (at $U = U_c$). The relevant integral equation is:

$$f(\mathbf{k}) = f_0(\mathbf{k}) + \frac{3}{2\pi \text{Vol}} \text{Re} \left\{ \sum_{k'} \int_0^\infty dv J(iv, |\mathbf{k} - \mathbf{k}'|) \frac{f(\mathbf{k}) - f(\mathbf{k}')}{(iv + e_k - e_{k'})^2} \right\}. \quad (62)$$

The essential contribution to J in Eq. (62) is

$$J(\omega, |\mathbf{q}|) = \frac{U^2 \Pi(\omega, |\mathbf{q}|)}{1 + U\Pi(\omega, |\mathbf{q}|)},$$

in which the Lindhard function or *"polarization part,"* is

$$\Pi(\omega, |\mathbf{q}|) = \frac{1}{\text{Vol}} \sum_k \frac{f(\mathbf{k} + \mathbf{q}/2) - f(\mathbf{k} - \mathbf{q}/2)}{\omega + e_{\mathbf{k}+\mathbf{q}/2} - e_{\mathbf{k}-\mathbf{q}/2}}.$$

By this means, the one-body distribution function, essential to the normalization of the boson normal modes, can be obtained from functional differentiation of the free energy. An alternative method obtains f from the one-particle Green function and *its* equations of motion. The results should be comparable to the above.

Finally, one might think this distribution function could be obtained directly by the method of *fermionization*: the name given to a procedure whereby fermions are

---

[m]The exponent δ increases linearly as $\propto (U - U_c)$ with increasing U in the range $U > U_c$.

expressed in the language of the boson operators. This last method works well in the context of the original Luttinger model[23, 24] but it is not yet clear how to implement it in $d > 1$ dimensions. A discussion follows.

24.   Fermionization in the Original Luttinger Model

To evaluate f what we need is $\langle c_\sigma^\dagger(\mathbf{k})c_\sigma(\mathbf{k})\rangle$ where $c_\sigma(\mathbf{k}) = \frac{1}{\sqrt{L^d}}\sum_j e^{-i\mathbf{k}\cdot\mathbf{r}_j}\psi_\sigma(\mathbf{r}_j)$, in which an operator $\psi_\sigma(\mathbf{r}_j)$ annihilates a fermion of spin σ at the jth cell (site) and averages, i.e., correlation functions $\langle\psi^\dagger(\mathbf{r})\psi(\mathbf{r}')\rangle$, are taken in the interacting regime (finite U) and temperature T. If the Hamiltonian is to be expressed entirely in terms of strings of operators defined previously and symbolized as a's, σ's and χ's we shall have to express $\psi_\sigma(\mathbf{r})$ in all these bosons. A minimum requirement is that commutators such as, $[\psi_\sigma(\mathbf{r})$, *any* relevant *boson*$]$ are preserved for all bosons that appear in H (which identifies them as being *relevant*).

In the original Luttinger model only the $a_0(q)$ were relevant, so it was a relatively easy task to construct a fermion operator using these operators. The argument goes as follows: consider right-hand goers at first, together with a representative commutator that involves the right-hand-going particle-density operator in the original model. (We can assume right-going fermions commute with left-going densities). Using only fermion operators,

$$\left[\psi(x), \sum_k c^\dagger(k-q/2)c(k+q/2)\right] = \left[\frac{1}{\sqrt{L}}\sum_{k'} c(k')e^{ik'x}, \sum_k c^\dagger(k-1/2)c(k+q/2)\right]$$

$$= \frac{e^{-qx}}{\sqrt{L}}\sum_k c(k)e^{ikx} = e^{-iqx}\psi(x). \tag{63}$$

One sees that $\psi(x)$ is a lowering operator of each $\varrho(q)$ with eigenvalue e^{-iqx}.

Now, when written in terms of fermions, a *right-hand-going density* boson in 1D is $a_0(q) = \sqrt{\frac{2\pi}{Lq}}\sum_k c^\dagger(k-q/2)c(k+q/2)$, assuming a dispersionless $\varepsilon = ck$ for all k. The inverse problem consists of expressing $\psi(x)$ in bosons (let's call this $\tilde\psi(x)$)), in such a way that satisfies $[\tilde\psi(x), q_0(q)] = e^{-iqx}\sqrt{\frac{2\pi}{Lq}}\tilde\psi(x)$. While a solution may not *appear* obvious, it really is!

Recall from studies of elementary harmonic oscillators that, in commutator brackets, a_0 acts like a derivative with respect to $a_0^\dagger$. For example, $[a_0(q), a_0^\dagger(q)] = \left(\frac{\partial}{\partial a_0^\dagger(q)}a_0^\dagger(q)\right) = 1$. Therefore Eq. (63) resembles nothing so much as a homogeneous first-order differential equation. Its solution has to be an exponentiated expression which, after appropriate normalization, is[23, 24]

$$\tilde\psi(x) = \frac{1}{\sqrt{L}}e^{ik_F x}e^{-\sqrt{\frac{2\pi}{L}}\sum_{q>0}\frac{e^{-qx}}{\sqrt{q}}a_0^\dagger(q)}e^{+\sqrt{\frac{2\pi}{L}}\sum_{q>0}\frac{e^{-qx}}{\sqrt{q}}a_0(q)}. \tag{64}$$

Remarkably this representation also satisfies *anticommutation* relation $\{\tilde{\psi}(x), \tilde{\psi}^{\dagger}(y)\} = \delta(x - y)$. Similar solutions apply to left-hand goers. It is dispersion that changes the algebra quite radically as there is no equivalent to the simple Eq. (63) for velocity-dependent modes $a_1(q)$, $a_2(q)$, ... But even without it we have not found the appropriate generalization valid at all in $d = 2$ or 3 dimensions except for the models of Refs. 5–7, that factor into 1D models by construction. For if we wrote $\Psi(\mathbf{r}) = \tilde{\psi}(x)\tilde{\psi}(y)\tilde{\psi}(z)$ in 3D to satisfy the anticommutation relations, this highly anisotropic expression, which explicitly violates rotational symmetry, *commutes* with all density operators at $\mathbf{q}$'s that are not aligned with the x, y or z axes. Short of a breakthrough, fermionization in $d > 1$ appears to present difficulties that, for now, appear insurmountable.

25. Conclusion

As we have just witnessed, the "string" version of the many-fermion problem is, in practice, not much different from the 1950's era RPA of Bohm and Pines, albeit with conceptual differences.

1. Here, decoupling into independent sectors is not approximate, but it is rigorous.
2. Nor is anything "random", with the exception of $(\tilde{n}(\mathbf{k}) - \langle \tilde{n}(\mathbf{k}) \rangle)$ which — when it appears under an integral sign — is subject to the CLT and therefore has to vanish.[e]
3. If it were not for the dependence of the one-body distribution function on the coupling constant, the decoupling of the dynamics into a set of individual "boson strings" (or modified one-dimensional Luttinger models!) would be the end of the story.
4. Asymptotic properties that can be calculated at small $|q|$ could be obtained numerically and with the least amount of effort in $d > 1$, if the effective *string lengths* decrease sufficiently fast in the limit as $|q| \to 0$. *We have not determined this to be so.* But if it were, it might allow application of the original dispersionless Luttinger model to the calculations of some *asymptotic* correlations in dimensions $d > 1$.

Obviously much other work remains to be done. For example, one would wish to make firm connections with conventional many-body perturbation theory. Also it would be good to express localized annihilation or creation *quasiparticle* operators, $\Psi_{\sigma}(\mathbf{r})$ and $\Psi_{\sigma}^{\dagger}(\mathbf{r})$ as functions of *all* normal modes, generalizing to $d > 1$ the fermionization that was illustrated briefly in Sec. 24, however difficult this may seem at the present time.

But already at this stage we can see that a model of interacting fermions that initially could only be conceived in 1D, is capable of illuminating aspects of the many-fermion problem in 2D and 3D. Stated somewhat differently: on this, the jubilee of Luttinger's discovery (or should we say, invention?) it is amazing indeed to find this concept is still teeming with possibilities.

Appendix A

The formulation of this higher-dimensional extension of the canonical one-dimensional fermion gas interacting *via* short-range interactions owes its success to the absence or neglect of certain terms in the interaction Hamiltonian, an operator that is quartic in the field operators. (This is also true of the RPA, despite its many useful applications in the 70 years of *its* history). We are, of course, referring to those terms that, in the one-dimensional models, are called "backward scattering" (they must also be treated separately) and to those that involve *umklapp*.[n] Leaving aside the latter because they are inconsequential in a low-density Fermi gas, let us look at the former. Among offending terms we find (omitting spin labels),

$$V(\mathbf{q})c^\dagger(\mathbf{k}-\mathbf{q})c^\dagger(\mathbf{k}'+\mathbf{q})c(\mathbf{k}')c(\mathbf{k}) \quad \text{where} \quad \mathbf{k}\cdot\mathbf{q} > 0 \quad \text{and} \quad |k| < |q|, \text{ while } \mathbf{k}'\cdot\mathbf{q} > 0. \tag{A.1}$$

i.e., operators that are not easily expressed in the boson picture.

The question is, do they invalidate the theory as it is worked out in the text? A quick answer is: not at small q because this also implies small k, and for that there is an insignificant amount of phase space. Thus the theory, as it was derived in the text, remains valid *to some leading order* in the cutoff q_0, regardless of these bad actors.

A deeper answer is: to the extent that such terms appear that change the fermion occupation number in any given sector, they do signal a failure — already present in the original model — in expressing a fluid of fermions purely in terms of normal modes. But one should draw an analogy to Grüneisen's parameter γ, a small number (order of 1/10) that measures the extent of nonlinearity in the elastic spectrum of ordinary solids. These nonlinearities are what permits thermal expansion — impossible in an elastic solid — and give a finite lifetime to normal modes, thereby limiting their thermal conductivity to a finite value.

It is similar here. Logic dictates that a picture of the normal modes be drawn first *before* one investigates their bad behavior and that, of course, is what we have attempted to do in the present text.

Acknowledgements

The author thanks Drs. Peter Levy and Vieri Mastropietro and a perspicuous (but anonymous) referee, for many helpful remarks that helped shape what was originally

[n] *Umklapp* concerns the lack of momentum convervation in a solid (in amounts equal to a vector of the reciprocal lattice.) Backward scattering is discussed in Ref. 25.

an overambitious manuscript into a more focused text. We are also very grateful to World Scientific Publishers for their helpful cooperation. All figures in this paper were computed and drawn using Mathematica software on an iMac personal computer.

References

1. J. M. Luttinger, *J. Math. Phys.* **4**, 1154 (1963) (as first solved by D. C. Mattis and E. H. Lieb, *J. Math. Phys.* **6**, 304 (1965)).
2. D. C. Mattis, *Int. J. Mod. Phys. B* **26**, 124407 (2012).
3. M. Pustilnik, M. Khodas, A. Kamenev and L. I. Glazman, *Phys. Rev. Lett.* **96**, 196405 (2006).
4. A. Imambekov and L. I. Glazman, *Science* **323**, 228 (2009).
5. D. C. Mattis, *Phys. Rev. B* **36**, 745 (1987), *inter alia*.
6. J. De Woul and E. Langmann, *Commun. Math. Phys.* **314**(1), 1 (2012).
7. J. De Woul and E. Langmann, *Int. J. Mod. Phys. B* **26**(22), 1244005 (2012).
8. M. Apostol, *Rev. Roum. Phys.*, Tome 22, No. 10, p. 1045 (1977), P. Kopietz, *Bosonization of Interacting Fermions in Arbitrary Dimensions* (1997), arXiv:cond-mat/0605402v1.
9. A. Houghton, H.-J. Kwon and J. B. Marston, *Multidimensional Bosonization*, arXiv:cond-mat/9810388v2, published later in *Adv. Phys.* **49**, 141 (2000).
10. D. C. Mattis, *J. Math. Phys.* **53**, 095212 (2012).
11. D. C. Mattis, *The Many Body Problem,an Encyclopedia of Exactly Solved Models in One Dimension* (World Scientific Publ. Co., Singapore, 2009).
12. J. Hubbard, *Proc. Roy. Soc. A* **243**, 336 (1957).
13. N. Mermin and H. Wagner, *Phys. Rev. Lett.* **17**, 1133 (1966).
14. D. C. Mattis, *Theory of Magnetism Made Simple* (World Scientific, Singapore, 2006).
15. J. Bardeen, L. N. Cooper and J. R. Schrieffer (BCS), *Phys. Rev.* **108**, 1175 (1957).
16. C. N. Yang and S.-C. Zhang, *Mod. Phys. Lett. B* **4**, 759 (1990).
17. S.-C. Zhang, *Science* **275**, 1089 (1997).
18. L.-A. Wu, M. Guidry, Y. Sun and C.-L. Wu, *Phys. Rev. B* **67**, 014515 (2003).
19. J. Bardeen, *Handbuch der Physik* **15**, 274 (1950).
20. G. D. Mahan, *Many-Particle Physics*, 2nd Edn. (Plenum Press, New York, 1991).
21. H. Chen and D. C. Mattis, *Mod. Phys. Lett. B* **7**, 723 (1993).
22. W. E. Pickett *et al.*, *Science* **255**, 46 (1992).
23. K. D. Schotte and U. Schotte, *Phys. Rev.* **182**, 479 (1969).
24. D. C. Mattis, *J. Math. Phys.* **115**, 609 (1974).
25. A. Luther and V. J. Emery, *Phys. Rev. Lett.* **33**, 589 (1974).